Chemistry and Biochemistry of Winemaking, Wine Stabilization and Aging

Edited by Fernanda Cosme, Fernando M. Nunes and Luís Filipe-Ribeiro

Published in London, United Kingdom

IntechOpen

Supporting open minds since 2005

Chemistry and Biochemistry of Winemaking, Wine Stabilization and Aging
http://dx.doi.org/10.5772/intechopen.87471
Edited by Fernanda Cosme, Fernando M. Nunes and Luís Filipe-Ribeiro

Contributors
Vanda Oliveira, Helena Pereira, Liliana Godoy, Andrea Acuña-Fontecilla, Daniela Catrileo, Antonio Morata, Iris Loira, Buenaventura Guamis, Javier Raso, Juan Manuel Del Fresno, Carlos Escott, María Antonia Bañuelos, Ignacio Alvarez, Wendu Tesfaye, Carmen González, Jose Antonio Suárez Lepe, Jesus Simal-Gandara, Anita Gonzalez Pereira, Maria Fraga, Paula Garcia-Oliveira, Cecilia Jimenez-Lopez, Catarina Lourenço-Lopes, Miguel Ángel Ángel Prieto Lage, María Carpena Rodriguez, Bin Tian, Roland Harrison, Pierre-Louis Teissedre, Ruth Hornedo-Ortega, Kleopatra Chira, Reyes Gonzalez, Michael Jourdes, Alessandra Rinaldi, Luigi Moio, Maurizio Petrozziello, Tiziana Nardi, Andriani Asproudi, Federica Bonello, Maria Carla Cravero, Fernanda Cosme, Luís Filipe-Ribeiro, Fernando M. Nunes, Juliana Milheiro, Lillian Barros, Isabel C.F.R. Ferreira, Jesus Simal-Gandara

First published in London, United Kingdom, 2021 by IntechOpen
IntechOpen is the global imprint of INTECHOPEN LIMITED, registered in England and Wales, registration number: 11086078, 5 Princes Gate Court, London, SW7 2QJ, United Kingdom
Printed in Croatia

British Library Cataloguing-in-Publication Data
A catalogue record for this book is available from the British Library

Additional hard and PDF copies can be obtained from orders@intechopen.com

Chemistry and Biochemistry of Winemaking, Wine Stabilization and Aging
Edited by Fernanda Cosme, Fernando M. Nunes and Luís Filipe-Ribeiro
p. cm.
Print ISBN 978-1-83962-575-6
Online ISBN 978-1-83962-576-3
eBook (PDF) ISBN 978-1-83962-577-0

For EU product safety concerns:
IN TECH d.o.o., Prolaz Marije Krucifikse Kozulić 3, 51000 Rijeka, Croatia, info@intechopen.com or visit our website at intechopen.com.

Meet the editors

Fernanda Cosme is an Assistant Professor at the University of Trás-os-Montes and Alto Douro in the area of food science and technology/oenology, mainly in wine stabilisation. Her research lines include grapes and wines with respect to grape composition, wine technology, wine stability, quality, and ageing. Recently, she edited two books on oenology as a scientific editor. She is the author of 4 patents and 37 articles in international journals cited in the Science Citation Index. She has also published 19 book chapters, 7 articles in open access journals with peer-review and she participates regularly in international and national congresses as a speaker. She has participated in several funded research projects and supervised PhD and MSc student theses in food science and oenology. (orcid.org/0000-0001-5924-1050).

Fernando M. Nunes is an Assistant Professor with Habilitation at the University of Trás-os-Montes and Alto Douro in the area of food chemistry and analytical chemistry. He has experience in the analysis of phenolic compounds and aromas namely in wine, carbohydrate analysis, Maillard reactions, and melanoidins. He is the author of 11 patents and 131 articles in international journals cited in the Science Citation Index in the areas of food chemistry, carbohydrate chemistry, polysaccharide chemistry, phenolic compounds, and biological activities of these compounds. He has been the principal investigator of several research projects with industrial partners. (orcid.org/0000-0001-5540-318X; Scopus ID 7102392847)

Luís Filipe-Ribeiro has a degree in oenology, a Master in viticulture and oenology, and a Doctorate in chemical and biological sciences. He was an oenology Professor in the University of Trás-os-Montes, and Alto Douro since 1997-98 where he participated in many subjects (vinification, wine stabilisation, analytical control, sensorial analysis, winery building and equipment). He is a researcher in the Chemistry Research Centre of Vila Real where he participates in many projects, research works, and student orientations. He is the Technical Director of SAI Enology and is responsible for the Laboratory and R&D. He is the author of several scientific (SCI) and technical publications, patents and he participates regularly in international and national congresses as a speaker. He was a consultant in viticulture and oenology from 1999-2007, having been the responsible winemaker in several wineries. (orcid.org/0000-0002-6744-4082)

Contents

Preface

In recent years, there has been great progress in the understanding of wine chemical and biochemical reactions that occur during winemaking, stabilisation, and wine ageing processes. This book intends to provide to the reader a deep and comprehensive overview of the latest developments in winemaking, stabilisation, and ageing chemistry and biochemistry focused on the most important developments in this critically important oenological area. Thus, the chapters in this book cover different topics of recent advances in the chemical and biochemical knowledge related to winemaking, stabilisation, and the wine ageing process.

The impact of phenolic compounds and their transformation products in wine sensory properties is currently a hot topic, therefore Chapter 1 presents the most relevant information about the composition, variations, and transformations of phenolic compounds in the winemaking process (alcoholic and malolactic fermentation) and wine ageing with or without wood contact, that result in complex transformations (copigmentation, cycloaddition, polymerisation, and oxidation) of polyphenols that can influence their sensory properties. Emerging non-thermal technologies open new possibilities in winemaking technology, facilitating the control of indigenous microorganisms and the use of new biotechnologies such as the fermentation with non-*Saccharomyces* yeasts or the use of yeast-bacteria co-inoculations. Chapter 2 reviews the main non-thermal technologies (HHP, UHPH, PEF, eBeam, and US) with application in oenology and their impact in the extraction of phenolic, aroma, and flavour compounds from the grape skins, the elimination of indigenous microorganisms that can result in SO_2 reduction during winemaking.

Nowadays, the quality of wines has improved due to the deep knowledge of the factors involved in wine quality, with many of them related to the grape composition. Despite the advances in our understanding of how the grape berry is developed and which chemical components are important for wine aroma and flavour, there is still much to explore. The study of the compounds responsible for aroma and flavour, as well as their correlation with the wine quality, is ongoing. The advances that are being carried out in different fields such as the determination of aromas and their evolution will allow us to continue expanding this knowledge and these advances are described in Chapter 3. Non-*Saccharomyces* yeasts, considered until recently as undesirable or spoilage yeasts, can improve the chemical composition and aroma profile of the wine. The activity of these yeasts is considered essential for the final wine aroma profile. Given the relevance of its contribution to the quality and typicity of wines, more studies with genetic approaches that explain the metabolic diversity of these yeasts are required. Chapter 4 summarises relevant aspects and pathways involved in the synthesis of aroma compounds by non-*Saccharomyces* yeasts as well as studies at the genetic and transcriptional level associated with their formation. In modern winemaking, oak products alternative to barrels are useful and flexible tools for wineries, enabling them to meet the needs of an increasingly wide and varied market. Although there are observed differences in the wine compositional and sensory properties between wines aged in barrels and those aged with alternative products, Chapter 5 deals with the main oenological uses of oak wood chips, the chemical transformations that underlie this practice, and the effect of their utilization on wine quality.

Port wine is a traditional and worldwide renowned fortified wine produced in the Douro Demarcated Region (DDR) Northeast of Portugal. The quality and uniqueness of the different Port wine styles are highly dependent on the ageing process, namely time, temperature, and oxygen levels. The chemical changes occurring during the Port wine ageing process result in significant changes in colour and aroma according to the different ageing conditions employed for the different Port wine styles. In Chapter 6, a broad and generic overview of the known and potential chemical changes occurring during ageing of Port wine is described, as well as the known chromatic and aroma characteristics of the different Port wine styles and the specific ageing processes, reductive or oxidative way. However, more studies are still needed to deepen our knowledge in order to understand and control the reactions involved in the Port wine ageing process that contribute to its uniqueness.

Astringency is a tactile sensation evoked in the mouth by polyphenols, mainly tannins. Tannins can provoke negative sensations as shrinking, drawing, or puckering of the epithelium, due to the interactions between salivary proteins and tannins. Nevertheless, in some red wines, astringency can be felt as pleasant mouth-feelings of richness, fullness, mouth-coating, and velvet in the mouth. In Chapter 7, an overview of the research concerned with molecular and sensory mechanisms of astringency is updated, as well as the several methods that have been developed to measure the intensity of the sensation, namely the saliva precipitation index (SPI), based on the binding and precipitation of human saliva with grape and wine tannins.

In Chapter 8, the authors describe the main physicochemical wine instabilities and defects that can appear during wine production. Although the best approach is to prevent their appearance in wines, some of the defects are intrinsic to the grape composition and/or wine production process therefore they must be removed or minimised before bottling. The current research trend is focused on the development of fining agents, additives, or technological solutions with improved specificity that will allow the removal of the defect without changing the other wine characteristics, and at the same time to explore low cost, natural, or renewable materials that will allow a lower environmental impact of the stabilisation process. The preventive measures and treatments, either by using fining agents, additives, or other technological solutions, to avoid them or reduce their impact on wine quality are presented in this chapter. Wine protein instability is a frequent problem in white wine production mainly related to the presence of grape pathogenesis-related (PR) proteins. The development of PR proteins in grapes during ripening and the extraction of PR proteins from grapes into juice can largely affect the concentration of PR proteins in the final wine, which consequently influences wine protein stability. Bentonite fining is the most common method to remove proteins in white wine before bottling, but it can cause the loss of wine volume and the removal of beneficial aromas. Thus, a number of alternative methods have been proposed for their potential to replace bentonite fining and these methods are reviewed in Chapter 9.

Finally, the natural cork stoppers are world known as closures for high-quality wines. Cork is a cellular material with chemical inertia and a set of physical and mechanical properties that provide outstanding performance for in-bottle wine ageing, by combining minute oxygen transfer with sealing, durability, and chemical stability, for example, inertness toward the liquid content and along with storage, preventing sensory deterioration. The cork industry has steadily improved production processes and adopted innovative technologies, quality control measures, and certification, regarding the prevention of potential wine taints and off-flavours. Chapter 10 reviews cork stoppers, their properties, and quality and role for wine bottle ageing, tackling their importance for wine aroma and off-flavours while presenting the latest advances in cork research.

Summing up the wide range of issues presented in the book covering different research hot topics from winemaking, stabilization and ageing of wines, this book is addressed to quite a broad audience, including both academic researchers and practitioners in halls of wine industries interested in gaining current knowledge. Also, it is heartily recommended to graduate and PhD students in oenology and food science. This book is the result of many collaborating parties and we gratefully acknowledge all the authors who contributed to this book and the IntechOpen team for this opportunity.

Fernanda Cosme, Fernando M. Nunes and Luís Filipe-Ribeiro
Chemistry Research Centre - Vila Real, (CQ-VR),
Food and Wine Chemistry Laboratory,
University of Trás-os-Montes and Alto Douro,
Vila Real, Portugal

Chapter 1

Phenolic Compounds of Grapes and Wines: Key Compounds and Implications in Sensory Perception

Ruth Hornedo-Ortega, María Reyes González-Centeno, Kleopatra Chira, Michaël Jourdes and Pierre-Louis Teissedre

Abstract

Phenolic compounds are a wide family of thousands of natural bioactives well-known for their overwhelming demonstrated health benefits. Particularly in wines, polyphenols and quality are closely interconnected. Indeed, these compounds possess a critical role due to their contribution to organoleptic wine quality as color, astringency, and bitterness. The profile or the composition of certain polyphenols has been even proposed as an analytical tool for authenticity certification. In this sense, although important progress has been achieved, the understanding of the relationship between the quality of a particular wine and its phenolic composition remains one of the major challenges in enology research. But why? If there is an adjective to define wine, it is "complex." This final complexity of a wine begins with the enormous polyphenolic variability that may be present in grapes influenced by ripening, genetic, or environmental factors, among others. Winemaking process (alcoholic and malolactic fermentation) and wine aging with or without wood contact produce endless reactions giving rise to complex transformations (copigmentation, cycloaddition, polymerization, and oxidation) of polyphenols. This chapter gathers the most relevant information about the composition, variations, and transformations of phenolic compounds from grape to wine including their influence on sensory properties.

Keywords: phenolic compounds, grapes, evolution, wine, oak wood, sensory perception

1. Introduction

Grape and wine phenols represent a large family of compounds with a great diversity of chemical structures and degrees of complexity. The term "polyphenols" or "phenolics" is used to define a group of plant secondary metabolites that presents one or more than one hydroxyl (–OH) groups attached to one or more benzene rings [1].

Polyphenols are synthetized by phenylpropanoid pathway, being the amino acid phenylalanine (a shikimate pathway product) their common precursor. They can be divided into flavonoid (anthocyanins, flavan-3-ols, flavonols, flavanones, flavanonols, flavones, and chalcones) and non-flavonoid (hydroxybenzoic and hydroxycinnamic acids and stilbenes) families [2].

These compounds are critically important for wine quality, due to their contribution to their sensory properties: color, taste, mouthfeel, flavor, astringency, and bitterness [3, 4]. For this reason, the understanding of the relationship between wine quality and its phenolic composition is considered, nowadays, one of the major challenges in enology research.

Furthermore, fermentation, maturation, and/or aging of wine may be performed in contact with oak wood. Spontaneous clarification, slow and continuous oxygen diffusion through the oak wood pores (for barrels and casks), and extraction of many volatile and nonvolatile (mainly ellagitannins) compounds are observed. As a result, wine undergoes a modulation of its quality and complexity with regard to aroma, structure, astringency, bitterness, persistence, and color stability [5].

The objective of this book chapter is to examine the key phenolic compounds in grapes and in oak wood used for the maturation of wine. Likewise, the evolution of these compounds during winemaking and wine aging and their impact in the sensory properties of wine will be discussed.

2. Key grape phenolic compounds

The most important grape polyphenols comprise anthocyanins, flavan-3-ols, proanthocyanidins and flavonols (flavonoid family), and phenolic acids and stilbenes (non-flavonoid family). Each family can be present in their free or conjugated forms, differing by their hydroxylation level and by the substitution of the hydroxy groups (methylation, glycosylation, acylation) and even forming adducts between them (e.g., phenolic acids with anthocyanins; condensed tannins). This fact explains the great chemical diversity of polyphenols in grapes [6].

It is noteworthy that polyphenolic composition in grapes is highly affected by different factors such as viticulture practices, environmental conditions (soil, climate), and pathogen attacks [7]. Although, one of the most important factors is undoubtedly the varietal or genetic differences [8] as well as the winemaking process.

2.1 Flavonoid grape polyphenols: Anthocyanins, flavan-3-ols, procyanidins, and flavonols

Flavonoids are basically formed by a structure of 15 carbons (C6-C3-C6) divided in 2 aromatic rings, A and B, which are joined by a 3-carbon chain that is part of a heterocyclic C ring (**Figure 1**). Depending on the oxidation state of C ring, this family can be subdivided in anthocyanins, flavan-3-ols, or flavonols.

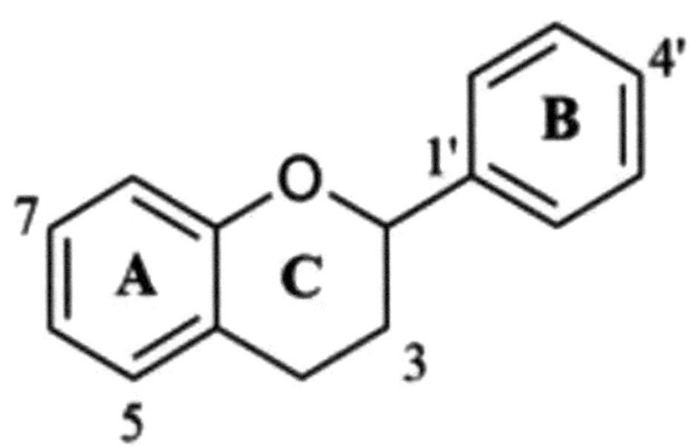

Figure 1.
General chemical structure of flavonoid family.

2.1.1 Anthocyanins

Structurally, anthocyanins are mainly present in nature in the form of heterosides. The aglycone form of anthocyanins, also called anthocyanidin, is based on the flavylium or 2-phenylbenzopyrilium ion having hydroxyl and methoxyl groups in different positions.

Anthocyanins are the most important natural pigments in wine grapes. These compounds are predominately accumulated in skins of red grapes during the ripening, and they are also present in the flesh of "teinturier" varieties [9]. In addition, it has been recently demonstrated that certain white grape cultivars (Sauvignon

Compound	g/kg dm			Grape variety	References
	Pomace	Seeds	Skins		
Total polyphenol content[a]	19–40.5	36.6–88.7	20.2–52.3	Grenache, Syrah, Carignan noir, Mourvèdre, Counoise	[13]
Anthocyanins					
Total	11.47–29.82	11–47–29.82	ND	Cabernet Mitos, Lemberger, Spätburgunder, Schwarzriesling, Trollinger	[12, 14]
Delphidin-3-*O*-glucoside	0.44–1.11	0.44–1.11	ND		
Cyanidin-3-*O*-glucoside	1.51–3.81	1.51–3.81	ND		
Petunidin-3-*O*-glucoside	0.53–1.34	0.53–1.34	ND		
Peonidin-3-*O*-glucoside	0.99–2.49	0.99–2.49	ND		
Malvidin-3-*O*-glucoside	4.12–10.19	4.12–10.19	ND		
Total acetylated	3.88–10.88	3.88–10.88	ND		
Flavan-3-ols/proanthocyanidins					
Catechin	0–0.3	2.14–2.42	0–0.3	Grenache, Syrah, Carignan noir, Mourvèdre, Counoise Cabernet Sauvignon, Chardonnay, Sauvignon blanc, Moscatel de Grano Menudo, Gewurztraminer, Riesling, Viognier, Merlot, Cencibel	[13, 15–19]
Epicatechin	0–0.2	0.88–1.60	0–0.13		
Epigallocatechin	0–0.05	0.05	ND		
Epigallocatechin-3-*O*-gallate	0–0.007	0.06–0.07	ND		
Epicatechin-3-*O*-gallate	0.003	0.25–0.31	0.04		
Procyanidin B1	0.11–0.60	0.14–0.17	0.18–0.6		
Procyanidin B2	0.01–0.84	0.04–0.18	0.01–0.84		
Total tannins	39.1–105.8	62.3–167.8	44.9–73.0		
Flavonols					
Total	0.03–0.63	0.48–0.63	0.02–0.05	Merlot Weisser Riesling	[14]
Phenolic acids					
Total	0.03–8.31	0.10–0.11	0.17–8.23	Cabernet Sauvignon, Merlot, Cabernet Mitos	[12, 14, 19]
Gallic acid	0.03–0.11	0.03	0.01–0.11		
Coutaric acid	0–1.23	0.03–1.23	—		
Caftaric acid	0–6.97	0.11–6.97	—		

[a]*Gallic acid equivalents (GAE) per g dry weight.*

Table 1.
Main phenolic compounds in different grape parts (expressed in g/kg of dm).

Blanc, Riesling, and Chardonnay) can contain measurable traces of anthocyanins [10]. Several factors can influence the anthocyanin biosynthesis in grapes such as origin and type of the grape vine, degree of maturity, and weather conditions like temperature, water availability, or the light exposure and intensity [11].

Regarding total anthocyanins, their quantities vary between 11.47 and 29.83 g/kg of dry matter (dm) in red grape skins [12] (**Table 1**). The principal individual anthocyanins in *Vitis vinifera* cultivars are the 3-*O*-monoglucosides (glucose linked through glucosidic bonds at the C3 positions of C ring) of delphinidin, cyanidin, petunidin, peonidin, and malvidin (**Figure 2**). Among these, malvidin-3-*O*-glucoside is generally the most abundant with values of 4.12–10.19 g/kg dm [14]. More recently, He and co-workers demonstrated for the first time the presence of pelargonidin-3-*O*-glucoside at trace levels on berry skins of Cabernet Sauvignon and Pinot noir cultivars [20]. Moreover, the monoglucoside forms can be acylated at the C6″ position of the glucose moiety with both aromatic (*p*-coumaric, caffeic, ferulic, and sinapic acid) or aliphatic acids (acetic, malic, malonic, oxalic, and succinic acid). The most common acylated anthocyanins in *V. vinifera* grape includes 3-*O*-(6″-*p*-coumaroyl)-glucosides, 3-*O*-(6″-acetyl)-glucosides, and 3-*O*-(6-caffeoyl)-glucosides of delphinidin, petunidin, peonidin, and malvidin [8, 12, 14, 21]. To go further, even anthocyanin dimers (malvidin-3-*O*-glucoside dimer and malvidin-3-*O*-glucoside-peonidin-3-*O*-glucoside) have been identified in grape skins [8, 22]. The presence of these acetylated forms is important for the color stabilization and intensity of wines [23]. The color intensity increases with the number of substituted groups on the B ring (di-oxygenated forms are redder, while tri-oxygenated are more purple) and with the replacement of hydroxyl by methoxyl groups (i.e., malvidin has the darkest color). Moreover, methoxylated anthocyanins (malvidin and peonidin) are more stable than hydroxylated ones to environmental and viticultural factors [24]. Additionally, anthocyanins can be found as 3,5-*O*-diglucosides or acylated 3,5-*O*-diglucosides, which are considered as marker compounds of non-*V. vinifera* species or hybrid red grapes [25].

In general, anthocyanin concentration is maximized under nonirrigated conditions in all cultivars, but anthocyanin profile and relative distribution of individual anthocyanins among irrigation treatments are influenced principally by the cultivar. In fact, Cabernet Sauvignon, Merlot, Syrah, and Tempranillo are characterized by a major proportion of malvidin forms, while in Nebbiolo (Italian cultivar) peonidin-3-*O*-glucoside is the most prevalent anthocyanin [11]. Other varieties, for example, Pinot noir, red Chardonnay, and pink Sultana (white red-colored mutants), are not able to synthetize acetylated anthocyanins [26]. In consequence, the anthocyanins profile in grapes can be used as an authentication tool of varietal wines [27].

2.1.2 Flavan-3-ols and proanthocyanidins (oligomeric proanthocyanidins or condensed tannins)

Flavan-3-ols are monomeric flavonoids formed by a benzopyran unit (rings A and C) with an aromatic cycle (ring B) linked to the carbon C-2 of the pyranic cycle (ring C). The presence of two chiral centers on the molecule (C2 and C3) gives rise to four possible configurations for a single monomer. These monomeric structures may be joined together forming dimers, oligomers (3–10 units of flavan-3-ols), and polymers (more than 10 units of flavan-3-ols). All these more complex structures are so-called condensed tannins. If they are formed by (+)-catechin and (−)-epicatechin and their gallic esters, they are named procyanidins, while when they are

DOI: http://dx.doi.org/10.5772/intechopen.93127

Anthocyanins

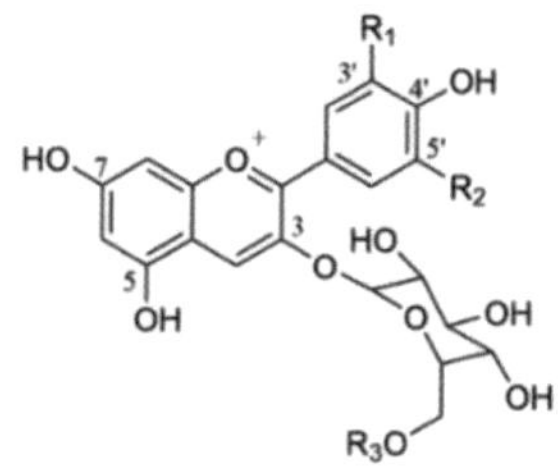

	R1	R2
Cyanidin-3-*O*-glucoside	OH	H
Delphinidin-3-*O*-glucoside	OH	OH
Peonidine-3-*O*-glucoside	OCH3	H
Petunidin-3-*O*-glucoside	OCH3	OH
Malvidin-3-*O*-glucoside	OCH3	OCH3

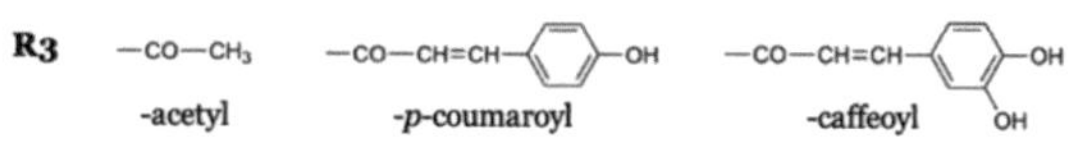

Flavan-3-ols and proanthocyanidins

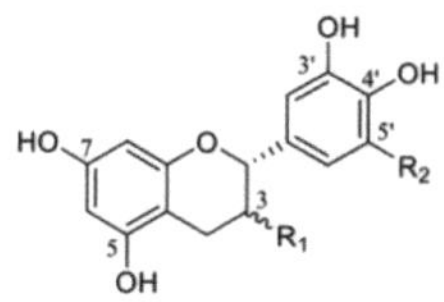

	R1	R2
Catechin	β-OH	H
Epicatechin	α-OH	H
Epigallocatechin	α-OH	OH
Epigallocatechin gallate	α-O-gallyl	OH
Epicatechin gallate	α-O-gallyl	H

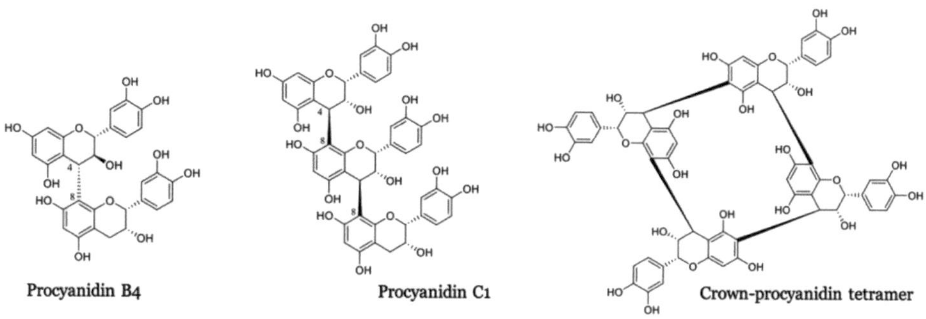

Flavonols

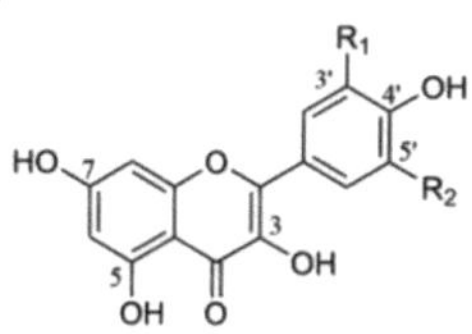

	R1	R2		R1	R2
Quercetin	OH	H	Isorhamnetin	OCH3	H
Myricetin	OH	OH	Laricitrin	OCH3	OH
Kaempferol	H	H	Syringetin	OCH3	OCH3

Phenolic acids

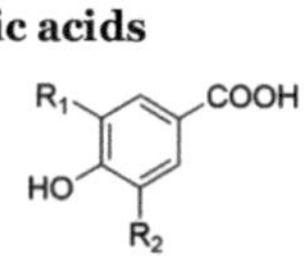

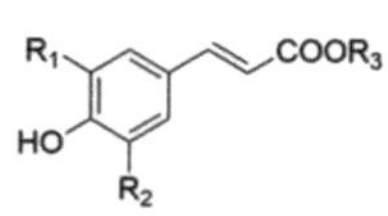

Hydroxybenzoic acids	R1	R2
p-hydroxybenzoic acid	H	H
protocatechuic acid	OH	H
vanillic acid	OCH3	H
gallic acid	OH	OH
syringic acid	OCH3	OCH3

Hydroxycinnamic acids	R1	R2	R3
p-coumaric acid	H	H	H
caffeic acid	OH	H	H
ferulic acid	OCH3	H	H
sinapic acid	OCH3	OCH3	H
p-coumaroyltartaric acid (coutaric acid)	H	H	$C_4H_5O_5$
caffeoyltartaric acid (caftaric acid)	OH	H	$C_4H_5O_5$
feruloytartaric acid (fertaric acid)	OCH3	H	$C_4H_5O_5$

Stilbenes

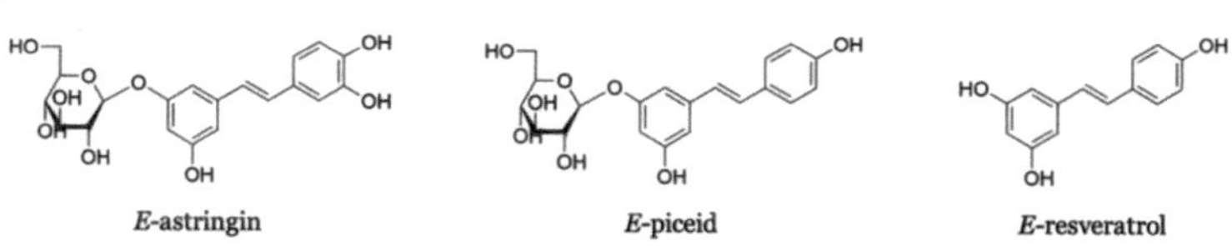

Figure 2.
Chemical structures of major grape phenolic compounds.

constituted by (+)-gallocatechin and (−)-epigallocatechin and their galloylated derivatives, the term used is prodelphinidins [28].

They are located in all grape clusters solid parts (skins, seeds, stalks) and are responsible for the stabilization of wines' both color and sensory characteristics due to their astringent and bitter properties [29]. Five monomeric flavan-3-ols are commonly present in grapes (**Figure 2**): (+)-catechin and its stereoisomer (−)-epicatechin as the predominant ones in seeds (2.14–2.44 g/kg dm and 0.88–1.60 g/kg dm, respectively) (**Table 1**) (+)-gallocatechin, (−)-epigallocatechin, and (+)-catechin-3-*O*-gallate [13, 15–19].

As explained above, condensed tannins are oligomers of flavan-3-ol monomer units. These units can be linked by C-4 → C-6 or C-4 → C-8 bonds, so-called B-type proanthocyanidins. A-type condensed tannins are characterized by the presence of a second interflavonoid bond by C–O oxidative coupling (C-2 → O-7 on the basic flavan-3-ol unit) [28]. B-type proanthocyanidins, and in particular, dimers as B1, B2, B3, and B4 or trimer C1 are mainly located in grape skins (0.01–0.86 g/kg dm) and, in a lower extent, in seeds (0.04–0.18 g/kg dm) [16–18]. On the contrary, complex procyanidins ($n > 3$) are more abundant in seeds (58–163 g/kg dm) than in skins (45–71 g/kg dm) (**Table 1**).

Tannins' structure is characterized by the nature of its constitutive extension and terminal units, its mean degree of polymerization (mDP; average number of units in the polymer), and its degree of galloylation (%G; percentage of subunits bearing gallic acid esters). In the case of skins, the percentage of (−)-epigallocatechin (EGC) or also called the percentage of prodelphinidins (%P) is also used for characterization purposes. Condensed tannins with different mDPs may have different organoleptic properties. Generally, astringency increases with tannin concentration, molecular size, and %G [29]. Polymerized procyanidins are increasingly reactive with proteins and, therefore, have a more important astringent character [30]. Proanthocyanidins' molecular size could also affect bitterness since monomers are considered to be more bitter than oligomers and polymers. Therefore, the estimation of both mDP and %G of procyanidins could be a useful parameter to evaluate the type of procyanidins present in a sample.

The quantity of flavan-3-ols and proanthocyanidins varies during ripening being higher at flowering and lower as the grapes maturate [31]. Both flavan-3-ols and proanthocyanidins are the major polyphenolic compounds in *V. vinifera* grapes. The greatest content is observed in seeds (62–168 g/kg dm) followed by skins (45–73 g/kg dm) (**Table 1**). This amount can be also variable depending on the grape variety and vintage [13].

Very recently, a new condensed tannin called "crown" proanthocyanidin tetramer has been isolated for the first time in grape skins of Cabernet Sauvignon cultivar. This tetramer is totally absent in seeds differentiating it from the rest of proanthocyanidins. This name "crown" is associated to an unusual macrocyclic carbon skeleton that has never been characterized before in the plant kingdom [32].

2.1.3 Flavonols

Flavonols constitute a group of flavonoids, which have the peculiarity to present a double bound between C2 and C3 and a hydroxyl group in C3. They vary in color from white to yellow and possess an important role in the color stabilization of young red wines, through copigmentation interaction with anthocyanins [3], and in the sensory perception of astringency and bitterness [33].

Conventionally, flavonols are present in berry skins of both white and colored grapes, and their total flavonoid content varies notably depending on cultivars and ripening stage [34]. In relation with cultivars, more quantities of flavonols have been reported, for example, in *V. vinifera* French varieties (Syrah, Petit Verdot, Cabernet Sauvignon, and Merlot), than with Spanish ones (Tempranillo, Garnacha, and Garnacha Tintorera) [35]. The total amount of flavonols in grapes varies from 1 to 80 mg/kg of fresh berry, being the red cultivars regularly richer than the white ones [35, 36].

Flavonols are found in grape berry skins in 3-*O*-glycoside forms. The main flavonols reported in red grapes are the dihydroxylated quercetin-3-*O*-glucoside and 3-*O*-glucuronide and the trihydroxylated myricetin 3-*O*-glucoside. In addition, other compounds such as kaempferol and the methylated isorhamnetin, laricitrin, and syringetin 3-*O*-glucosides have also been identified [35]. Furthermore, kaempferol and laricitin-3-*O*-galactosides, kampferol-3-*O*-glucuronide, and even quercetin and siringetin-3-*O*-(6″-acetyl)-glucoside have been identified in Cabernet Sauvignon grapes in lower quantities [37]. Interestingly and more recently, laricitrin-3-*O*-galactoside and syringetin-3-*O*-galactoside were reported in red grapes for the first time. With regard to white grapes cultivars, myricetin, laricitrin, and syringetin have not been identified [36].

2.2 Non-flavonoid grape polyphenols: phenolic acids and stilbenes

2.2.1 Phenolic acids

Phenolic acids can be classified in two main groups: hydroxybenzoic acids (C6-C1) and hydroxycinnamic acids (C6-C3). This family is found in skins, pulp, and seeds of grapes, being generally most numerous in skins (0.2–8.2 g/kg dm) (**Table 1**). The quantities of total hydroxycinnamic or hydroxybenzoic acids in grape skins vary depending on cultivar and origin. For example, hydroxycinnamic acids are more predominant in *V. vinifera* East Asian or North American grapes than in European grapes in which these phenolic acids are only present in trace levels. However, the hydroxybenzoic acid amounts are similar between cultivars [38].

Individually, the most important hydroxybenzoic acids in grapes are gallic, vanillic, and syringic acids. Predominantly present in grape seeds, gallic acid is considered the most important phenolic acid (100–230 mg/kg dm), being the precursor of all hydrolyzable tannins [39]. In lower quantities, protocatechuic acid and *p*-hydroxybenzoic acids are also present [39, 40] (**Figure 2**).

Regarding hydroxycinnamic acids, they are principally located in skins, being *p*-coumaric, caffeic, ferulic, and sinapic acids the most significant. It should be reminded that *p*-coumaric and caffeic acids can be found esterified by the glucose of the anthocyanin monoglucosides forming their acylated derivates. In grapes (mainly white) and also in wines, hydroxycinnamic acids are mainly esterified with tartaric acid forming caftaric, *p*-coutaric, or fertaric acids (from caffeic, *p*-coumaric, and ferulic acids, respectively) [18] (**Figure 2**).

Phenolic acids, and overall, hydroxycinnamic acids can act as copigments. Indeed, they are implicated on the formation of new more stable pigments (pyranoanthocyanins) in wine and, in consequence, are considered as color stabilizer agents of young red wines, through copigmentation with anthocyanins [3]. Moreover, they are also associated with the sensory perception of astringency and bitterness [41].

2.2.2 Stilbenes

Stilbenes (1,2-diphenylethylene) are formed from two phenyl rings linked together by an ethylene bridge generating a C6—C2—C6 structure. The aromatic rings are generally substituted by different functions such as hydroxyl, methyl, methoxyl, prenyl, or geranyl groups. Moreover, monomeric units (resveratrol) can also be coupled, leading to the formation of more complex stilbenes. Their composition and content are extremely variable depending on different biotic (attack of plant pathogens) and abiotic factors including grape variety and ripening stage [42].

In grape berries, stilbenes are mainly concentrated in skins [43]; only trace amounts are reported in seeds [44]. In addition, red varieties seem to present higher stilbene content than white ones [45]. The major stilbenes in grapes are the glucosides piceid (mean 1.3 mg/kg fresh weight (fw)), resveratrol (mean 1.1 mg/kg fw), and astringin (mean 0.5 mg/kg fw) [46] (**Figure 2**). Furthermore, other minor monomers were identified such as pterostilbene and isorhapontigenin [47]. Finally, different dimers as pallidol, ε-viniferin and δ-viniferin, trimers such as miyabenol C and α-viniferin, and tetramers as hopeaphenol and isohopeaphenol have also been reported [46] (**Figure 2**).

3. Key phenolic compounds in oak wood

Oak wood is the preferred material for the manufacture of barrels, casks, or whatever derived wood product (chips, blocks, winewoods, tankstaves, etc.) used during fermentation and/or aging of wines. Resistance, flexibility, easy handling, and low permeability make oak wood particularly suitable for wine maturation and storage, in relation to mechanical, physical, and chemical properties provided by other woods [48].

Regardless of the species, oak heartwood is basically composed by cellulose, hemicelluloses, and lignin, representing approximately 90% of dry wood and acting as key structural polymers of wood matrix. The remaining 10% of dry wood corresponds to an extractable fraction, mainly consisting of phenolic compounds but also presenting low molecular weight compounds and volatile compounds. Lignans, coumarins, phenolic acids, and phenolic aldehydes may be found in the oak wood phenolic fraction, but hydrolyzable tannins are the major constituents [49].

Depending on the release of gallic or ellagic acid under acidic conditions, hydrolyzable tannins may be classified, respectively, as either gallotannins or ellagitannins [50]. Gallotannins are the simplest hydrolyzable tannins with a structure consisting of polygalloyl esters of glucose. The oxidative coupling of their galloyl groups converts gallotannins to the related ellagitannins [51].

Ellagitannins are the major nonvolatile extractives from oak heartwood (**Figure 3**). These oak phenolics have a specific structure, consisting of an glucose open-chain esterified at positions 4 and 6 by a hexahydroxydiphenoyl unit (HHDP) and a nonahydroxyterphenoyl unit (NHTP) esterified at positions 2, 3, and 5 with a *C*-glycosidic bond between the carbon of the glucose and position 2 of trihydroxyphenoyl unit [52]. Among ellagitannins, the monomers castalagin and vescalagin

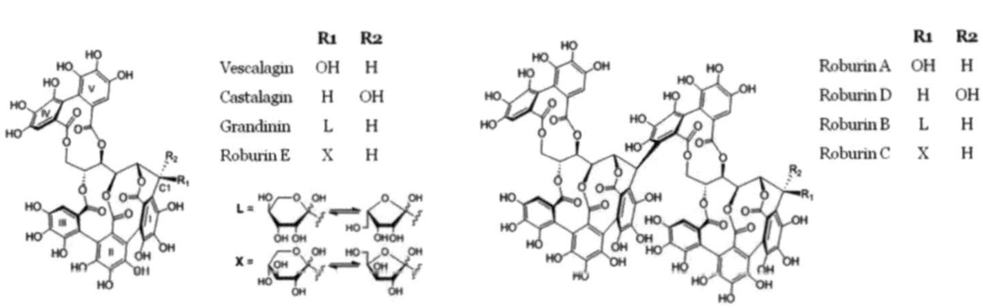

Figure 3.
Chemical structure of main ellagitannins present in oak wood.

Treatment	Species	Monomers		Pentosylated monomers		Dimers		Pentosylated dimers		Total ellagitannins	References
		Castalagin	Vescalagin	Roburin E	Grandinin	Roburin A	Roburin D	Roburin B	Roburin C		
Untreated wood	*Quercus robur*	5.82–27.50	1.76–23.90	0.69–13.70	1.00–10.40	0.39–6.64	0.95–5.34	0.44–4.67	0.47–6.29	13.47–91.02	[54, 56–59]
	Quercus petraea	1.63–11.76	1.05–8.01	0.78–4.44	0.76–5.47	0.13–2.05	0.17–3.27	0.29–2.41	0.06–2.25	5.87–34.41	
	Quercus alba	1.19–5.29	0.70–2.83	0.29–1.38	0.38–2.17	0.31–0.70	0.08–0.67	0.13–0.83	0.04–0.54	3.48–13.14	
Seasoned wood	*Quercus robur*	1.43–13.00	0.85–10.60	0.77–9.20	0.36–6.15	0.09–1.91	0.18–2.52	0.09–0.48	0.14–1.72	3.93–43.97	[53, 55, 56, 60–62]
	Quercus petraea	0.90–15.91	0.39–11.76	0.39–7.98	0.12–5.80	0.05–2.73	0.02–2.58	0.04–1.84	0.01–1.97	1.98–39.77	
	Quercus alba	0.41–4.45	0.15–6.44	0.15–2.30	0.07–1.81	0.03–0.40	0.02–0.84	0.02–0.79	0.01–0.20	0.89–8.93	
Toasted wood	*Quercus robur*	3.60–5.44	0.89–1.15	1.80–2.39	0.28–0.66	0.41–0.47	0.45–0.50	0.17–0.20	0.12–0.19	7.72–11.00	[53, 60, 62, 63]
	Quercus petraea	1.75–3.79	0.36–1.46	0.09–2.23	0.13–0.59	0.20–0.50	0.16–0.44	0.10–0.22	0.03–0.17	3.53–8.96	
	Quercus alba	0.21–0.44	0.08–5.28	0.04	0.02	ND	0.04	ND	0.01	0.55–5.72	

All values are displayed in mg/g of wood. ND, not detected.

Table 2.
Ellagitannin concentration in untreated, seasoned, and toasted oak wood from the main Quercus *species used in cooperage.*

largely predominate in oak wood, representing from 40 to 65% by weight of total ellagitannins [53–58] (**Table 2**). Six additional ellagitannins have been identified in oak wood: the lyxose/xylose-bearing monomers grandinin and roburin E, the dimers roburins A and D, and the lyxose/xylose-bearing dimers roburins B and C [64].

Since ellagitannins are very soluble in wines and spirits, with a high reactivity, their levels in oak-aged beverages are much lower than what could be expected. When comparing both main monomers, vescalagin presents a more polar configuration that confers it a lower stability in hydro-alcoholic solutions [65]. From a sensory point of view, their level and profile may affect the astringency and bitterness of wine [66].

The level of ellagitannins in oak heartwood depends on the botanical species, the geographical origin, the single-tree variability, the sampling position in the tree, the grain, and the processing of wood in cooperage, notably the type and length of both seasoning and toasting periods.

3.1 Influence of botanical species

Among the more of 150 oak species classified in the genus *Quercus*, the most frequently used in cooperage for winemaking are *Quercus robur* (pedunculate oak) and *Quercus petraea* (sessile oak), both growing in Europe, and *Quercus alba*, commonly known as American white oak, growing in the United States [5]. American oaks differ from European species not only because of their mechanical properties (higher density and resistance and lower porosity and permeability) [60] but also for the chemical composition of their phenolic fraction. Ellagitannin concentration is generally lower in *Q. alba* than European species, which in turn show a greater ellagitannin content in pedunculate oak than in sessile oak (**Table 2**) [53, 55–58].

3.2 Influence of geographical origin

Until recently, French and American oak forests have been the quintessential source of wood for cooperage. Meanwhile, over the last few years, a huge number of studies on pedunculate and sessile oaks from different European origins (Hungary, Poland, Russia, Romania, Slovenia, Spain, Ukraine, and Moldova, among others) confirm their prospective use for maturation of quality wines [60]. Oaks from these new European origins appear to present ellagitannin concentrations halfway between French and American oaks [5, 58].

3.3 Influence of single-tree variability

The width of oak wood rings ("grain") is of great importance for the choice of oak wood for barrel and cask making, since it influences the wood chemical composition and affects the contribution of oak aging to wine quality. The higher the grain size, the larger the amount of ellagitannins released and the faster that release [67]. Furthermore, the grain size also exerts an effect on the oxygen transfer ratio (OTR): the smaller the grain size, the greater the OTR, and the faster the wine maturation [68].

3.4 Influence of cooperage operations

Fresh wood cannot be directly used in winemaking, due to the great percentage of humidity (up to 70%), an excess of phenolic compounds, and a shortage of aromatic constituents. Oak wood conditioning in cooperage includes two stages that will determine the enological quality of wood. Both seasoning and toasting affect the structure and final chemical composition of the wood that is going to be in contact with wine.

Seasoning allows, not only reduction of humidity in wood, but also fiber contraction and wood maturation. During this process, an important decline of ellagitannin content is observed due to different physical, chemical, and biochemical mechanisms involved: stave leaching by rain, hydrolytic oxidative degradation, polymerization, and fungal enzymatic activity [56]. These phenomena occur particularly in the surface of wood and, in a lesser degree, but uniformly, in the inner wood [53]. Among the different oak wood seasoning methods, natural seasoning in open air seems to be more effective than mixed and artificial methods in reducing the excess of ellagitannins [56].

Toasting also induces an important modification of wood chemical composition, including an additional decline of the ellagitannin content. During toasting, castalagin is mainly oxidized in dehydrocastalagin, whereas its diastereomer vescalagin is reduced in deoxyvescalagin [69]. Similar deoxy- and dehydro-derivates have been observed for roburins A and D, respectively.

Ellagitannin degradation, and in turn their sensory impact on wines, may be modulated by changing the toasting thermal profile (temperature and length) [5]. In this sense, an ever-widening range of toasting levels is available at cooperages for all oak wood products. The higher the toasting level, the greater the ellagitannin decomposition, via dimerization and hydrolysis reactions, as well as formation of copolymers with cell-wall components [64].

Recently, new compounds that showed [M-H]-ion peak at *m/z* 1055.0631 (compound A) and at *m/z* 1011.0756 (compound B) have been identified as a result of thermodegradation of ellagitannins. The A compound corresponds to castacrenin E which is the oxidation product of castacrenin D, a vescalagin with an additional aromatic ring (gallic acid) to the C-1 through a C–C bond. The levels of these compounds, found under experimental conditions and further searched in commercial oak wood, are dependent on both oak wood size and toasting method [70].

But, the extraction of oak phenolics into wine depends not only on the pool of potential extractible compounds originally present in wood, determined by the abovementioned factors, but also on the wine matrix, the aging length, and the exposed wood area to wine volume ratio.

4. From grapes to wine aging: phenolic compound evolution

During the grape ripening phase, the physiological and biochemical changes determine grape quality. The first period of grapes growth consists mostly of cell division and expansion, followed by a rapid growth phase during which the berry is formed and the seed embryos are produced. In this period, several compounds are accumulated in the berries, especially the tartaric and malic acids, conferring the acidity of the future wine. During the first growth period, several polyphenolic compounds increased like hydroxycinnamic acids in grapes' pulp and skin and tannins and catechins in the skin and seed. The most important changes in grapes composition happen during the second growth phase (the ripening stage). Grapes switch from small, hard, and acidic berries to larger, softer, sweeter, less acidic, flavored, and colored ones. The majority of the solutes accumulated during the first growing phase remain at harvest. During the second period, the malic acid is metabolized and used as an energy source, its proportion decreasing toward the tartaric acid concentration, which remains almost unchanged. In general, the chemical composition of the final product is much more complex than in the raw material, due to the formation of new compounds [71].

4.1 Phenolic compound changes during winemaking

Winemaking techniques involve the extraction of juice from ripe grapes, fermentation with yeast, and changes in polyphenolic composition that occur due to the participation of these compounds in various reactions such as copigmentation, cycloaddition, polymerization, and oxidation. These reactions begin after grape crushing, followed by fermentation and aging, contributing to the sensory properties of wines, mainly color and mouthfeel sensation. The total extractable phenolic content in grapes is encountered in seeds (60–70%), in the skin (28–35%), and in the pulp (about 10% or less). In the seeds, the phenolic content may range between 5 and 8%, by weight [72].

The understanding of the relationship between the quality of a particular wine and its phenolic composition remains one of the major challenges in enological research. For example, the anthocyanin fingerprints of varietal wines are proposed as an analytical tool for authenticity certification [27]. Patterns of some classes of flavonoids are under strict genetic control, and their distribution varies considerably among different grape cultivars [73, 74].

Several factors impact the wine phenolic composition, including the "terroir," the grape variety and its degree of maturation before harvesting, or the winemaking process with its specific conditions of fermentation or aging [75]. Certain technological procedures, such as addition of sulfur dioxide (SO_2) and/or ascorbic acid prior to crushing the grapes, maceration, yeast strain utilization and alcoholic fermentation, oxidation, or adsorption, can also influence the levels of phenolics during the winemaking process [76]. The addition of SO_2 and pectolytic enzymes before fermentation caused an increase in color intensity, color stability, total phenolics, anthocyanins, catechin, and epicatechin in a red Italian wine [77].

In white grape musts, the predominant phenolic compounds are hydroxycinnamic tartaric acid esters as catechins and proanthocyanidins which are found mainly in their skins. The must fermentation of red wines is realized in the presence of both grape skins and seeds. During this process, phenolic compounds such as anthocyanins are subjected to various reactions, such as enzymatic oxidation, nucleophilic substitution, degradation, and cycloaddition of the carbonyl compounds leading to the formation of vitisins (A and B). These pyranoanthocyanins in red wine are mainly orange pigments. Moreover, the red wine color evolution and stabilization are mainly induced by the formation of polymerized pigments. The acidic hydrolysis of proanthocyanidins leads to the formation of flavan-3-ol unit or tannin oligomers with a carbocation in C4 position which can be attacks by positions C6 and C8 of another proanthocyanidins or an anthocyanin. This reaction will induce the modification of the condensed tannin polymerization degree or the formation of the polymerized pigment. These newly formed purple pigments induce the color modification of the young red wine. Moreover, polymerization through acetaldehyde between two condensed tannins or between condensed tannins and anthocyanins also occurs during the winemaking process and aging. The formation of these ethyl bridge compounds will also produce modification in the organoleptic properties of the wine and the color stabilization since the ethyl-bridged anthocyanin-tannin compounds also exhibit purple color [72]. During the red wine maturation in bottles, all these newly formed purple polymerized pigments will undergo slight oxidative reaction to slowly form some more orange pigments which together with pyranoanthocyanins are forming the color of old wine.

4.2 Using oak wood during winemaking

The winemaking process involves the alcoholic fermentation of must, often followed by malolactic fermentation (MLF). When MLF is completed, the wine is subjected to different clarification and stabilization treatments and/or is stored in oak barrels for aging for a variable period of time. MLF and aging in oak barrels are two enological processes which modify the composition and sensory characteristics of the wines [5, 78–80]. When oak wood derivatives like chips are added after fermentation, wines seem to have a greater aging potential compared to the wines fermented with chips due to their higher ellagitannin content and enhanced condensation reactions. On the other hand, color stabilization and tannin polymerization occur faster when chips are added during fermentation, resulting in shorter aging periods suitable for early consumed wines [81]. MLF in tanks may simplify the control of the process; however, the use of oak wood during the MLF stage affects the chemical and sensory attributes of wines. In red wines, MLF container plays an important role on proanthocyanidin and anthocyanin concentration and evolution as oxygen in small quantities favors polymerization reactions among anthocyanins and tannins. Wines performing MLF in tanks present a higher total proanthocyanidin concentration (5.8 g/L wine) than that of those which accomplished MLF in medium-toasted barrels (4.9 g/L wine). The major wine glucosidic anthocyanin, malvidin, showed as well greater levels in wines carrying out their MLF (33 and 26 mg/L wine, respectively, for tank and barrel MLF). Regarding ellagitannin concentration, their content is strongly influenced by both barrel toasting and MLF container. For instance, in the case of medium-toasting barrels, castalagin was found at concentrations twofold times higher (19 mg/L wine) when MLF was performed in barrels [79]. Concerning sensory results, the MLF strengthens the organoleptic preference of wine when it takes place in barrels [79, 80, 82]. In white wines, total ellagitannin concentration varied from 7.8 to 17.4 mg castalagin equivalents/L for wines performing MLF in tanks and barrels, respectively [83].

4.3 Evolution of phenolic compounds during aging

The phenolic composition of wine changes along the wine aging process reflects in the color and astringency level of the final product. From 1978 to 2005 vintage for Cabernet Sauvignon wine, phenolic compounds, total tannins, and total anthocyanins varied from 1735 to 2903 mg/L, from 1.3 to 2.2 g/L, and from 15 to 123 mg/L, respectively [29]. In general, the relative anthocyanin content decreases upon aging although this chemical modification is associated with a very clear change in color. This characteristic is often used as a quality standard for aged wines. One of the main factors responsible for anthocyanin loss is the storage temperature [84]. The majority of red wines aged are in contact with oak wood, whether in form of barrels or in form of oak wood derivatives. As a consequence, their phenolic composition changes due to the addition of oak wood extracted compounds. These compounds include hydrolyzable tannins (*C*-glucosidic ellagitannins), aromatic carboxylic acids, and several aldehydes. Regarding wine-air interactions, barrel structure allows a controlled entrance of oxygen, which is essential to the polymerization and the slight oxidative reactions between different types of flavonoids, leading to a modification of the organoleptic properties of the wine. Indeed, wood can affect wine composition and, consequently, organoleptic properties through different mechanisms. On the one hand, wine compounds can be adsorbed onto wood surface. On the other hand, compounds, such as ellagitannins, can be extracted from wood to the wine due to the hydro-alcoholic nature of the latter. Ellagitannins can take part in oxidation reactions that may favor the polymerization reactions between flavanols and between flavanols and anthocyanins. Furthermore, they can directly react with these types of compounds giving rise to flavano-ellagitannins or anthocyano-ellagitannins [85]. The formation of flavano-ellagitannins and the *β*-1-*O*-ethylvescalagin in red wines aged in oak barrels has been reported. The ellagitannin concentrations fluctuated between 4 and 8 mg/L, being castalagin the ellagitannin with the highest concentration, followed by mongolicain A [86]. As a consequence, ellagitannins can modulate wine astringency and color through interactions with these compounds. Strong correlations have been observed between ellagitannin concentration and both antioxidant capacity and astringency sensation [5, 63, 78]. The amount of ellagitannins released into the wine depends on the content in the oak wood barrel, which in turn is dependent on several factors (Section 3). For instance, after 12-month aging with woods, the total ellagitannin level revealed a large diversity of concentrations ranging from 6.3 to 26.1 mg of ellagic acid/L wine. The wine with heavy toast woods and the wine with low toast woods presented, respectively, the less and the highest ellagitannin concentrations [78]. Storage with oak can also cause a decrease in anthocyanins, catechin, and epicatechin but an increase in total phenolic content and a stabilizing effect on color [77].

Besides the winemaking process, and oak wood aging, wine can be further exposed to oxygen during aging in the bottle, depending on the oxygen permeability of the closure. Because of the extremely low rates of oxygen ingress through a closure, this form of oxygen exposure has been referred to as nano-oxygenation. Oxygen transmission rates (OTR) of wine closures may vary widely depending on closure type and strongly influence the evolution of white and red phenolic composition and astringency during bottle aging [87].

5. Sensory impact of phenolic compounds in wine

"In-mouth" sensory properties of red wines encompass multiple interacting sensations such as acidity, sweetness, bitterness, retronasal aroma perception (flavor), viscosity, warmth, and astringency. Among these, the sensation of astringency and the taste of bitterness are related to phenolic compounds.

Bitterness perception is a taste recognition mediated by taste buds present in the tongue papillae. Each taste bud, consisting of approximately 50–100 taste receptor cells, is innervated by multiple taste fibers that transmit nervous signals to the brain [88]. In humans, each bitter receptor cell contains approximately 25 bitter G protein-coupled receptors encoded by a TAS2R gene family. It was evidenced by Soares and colleagues [89] that different phenolic compounds activate distinguished combination of TAS2Rs: epicatechin stimulated three receptors whereas procyanidin trimer only one. In general, in wines containing a large number of polyphenols, the taste of bitterness is attributed to flavan-3-ols and their polymers, although it is also able to be elicited by some flavonols [90], hydroxycinnamates, and benzoic acid derivatives [91].

While bitterness is a gustatory sense recognized by nervous signals, astringency is an oral sensation involving dryness and puckering [92]. According to the American Chemical Society, astringency refers to "the complex of sensations due to shrinking, drawing or puckering of the epithelium as a result of exposure to substances such as alums or tannins" [93]. It has been classically postulated that tannins possess the ability to interact with salivary proteins, with or without precipitation of the corresponding complexes [94]. In fact, the name of "tannins," originating from "tanning" of leather, has been used over decades for their capability of binding with proteins. The mechanisms of tannin-protein interactions involved different types of interaction such as the hydrophobic interactions, which are the predominant mechanisms involving the hydrophobic nature of the condensed tannin carbon skeleton and the apolar regions of the proteins. Together with hydrophobic interactions, some hydrogen interactions also occur between the carbonyl function of proline and the hydroxyl functions of phenols as well as some ionic interactions. This mechanism and thus the final astringency of wine are influenced by many factors, such as the structure and amount of condensed tannins in the wine as well as the composition of the wine matrix [95]. Indeed, the intensity and the quality of the astringency perception are influenced by the concentration of condensed tannins [96], their degree of polymerization [97], their percentage of galloylation [98], and prodelphinidins [99]. The matrix, on the other hand, impacts the perception according to its acidity [100], its ethanol concentration [101], and the presence of macromolecules such as polysaccharides. Fontoin and co-workers demonstrated that the astringency sensation of grape-seed oligomer tannins decreased with increasing pH and the percentage of ethanol [100]. For example, Cabernet Sauvignon wines having a mDP between 2.0 and 4.0 were characterized as mellow and slight astringent. Meanwhile, a rougher sensation (tannic) was perceived for wines with a mDP higher than 4.0. [102]. The analytical determination of the proanthocyanidin content and the type of subunit that is dominant in tannin chains might be a valuable tool for astringency estimation during wine aging [103]. Astringency intensity is influenced by the source of proanthocyanidin (seed or skin) and by well-defined proanthocyanidin fractions (oligomeric or polymeric). Polymeric seed fraction was perceived more astringent than the monomeric/oligomeric one, whereas polymeric skin fraction was characterized less tannic than the monomeric/oligomeric skin fraction, indicating a negative relationship between astringency intensity and % of prodelphinidins [104]. The presence of epigallocatechin units in the proanthocyanidins has been shown to lower the "coarse" perception through the increase of the degree of B-ring trihydroxylation. Furthermore, seed fraction with a higher proportion of galloyl group and a lower mDP was perceived to be bitterer than the skin fraction [105]. Both the inter-flavanoid bonds and stereochemistry of subunits impact bitterness sensation: catechin-(4–6)-catechin dimer was bitterer than both catechin-(4–8)-catechin and catechin-(4–8)-epicatechin [91].

Moreover, astringency sensation perceived always reduces with the increasing saliva volume flow rate. A linear correlation was found between protein concentration and tannin-binding affinity. The saliva proteins, including PRP family (acidic, basic, and glycosylated), α-amylase, statherin, histatins, and mucins, show diversified ability to interact with tannins [106]. Some proteins are specifically bound to astringents. For instance, tannins and alum precipitated PRPs, while acid and alum precipitated mucins [107, 108].

Ellagitannins (hydrolyzable tannins) impart an oral sensation described as astringent at relatively low threshold concentrations spanning from 0.2 to 6.3 mmols by means of the half-tongue test in bottled water (pH 4.5). Due to their lower astringent taste thresholds, hydrolyzable tannins are usually perceived as more astringent than condensed tannins (1.1 μM for both castalagin and vescalagin, compared to 410 μM for catechin, 930 μM for epicatechin, 240 μM for procyanidin B1, 190 μM for procyanidin B2, and 200 μM for procyanidin B3) [66, 69, 109]. Among ellagitannins, the monomers grandinin and roburin E exhibited an astringent mouthfeel at the lowest taste thresholds (0.2 μM), whereas values for dimers ranged between 2.9 and 6.3 μM. Thus, the *C*-glycation of castalagin and vescalagin seems to favor the astringent sensation, while dimerization seems to reduce it [66].

When the same concentration of ellagitannins and skin and seed tannins was tested, the ellagitannins were perceived softer [104].

Interaction of ellagitannins with salivary proteins has been poorly investigated up to now, probably because of their lower wine content compared to condensed tannins. Even if perceived as more astringent, ellagitannins have been noted as poorer protein precipitants than condensed tannins [94]. Soares and co-workers [110] stated that ellagitannins act as multidentate ligands cross-linking different salivary protein units, via their galloyl moieties. It is noteworthy to mention that these units are responsible for the antioxidant ability of hydrolyzable tannins; thus when complexed with salivary proteins, the antioxidant capacity of ellagitannins may be significantly impaired. At higher concentration levels, the main eight oak ellagitannins have also been observed to provide the wine with a bitter taste [66].

Apart from tannins, other polyphenolic compounds present in wine have been related with the overall perception of astringency sensation or bitterness. Very recently, some works have provided evidence about the interaction of anthocyanins and pyranoanthocyanins with salivary proteins. Indeed, malvidin-3-*O*-glucoside, the major anthocyanin of wine, has demonstrated to interact with acidic proline-rich proteins (aPRPs) showing dissociation constants (K_D) calculated by NMR of 1.88 mM [111] that can be compared to those obtained for procyanidins (dimers B1–4 and trimer C2) (between 0.4 and 8 mM) [112]. In addition, Paissoni and colleagues [113] tested the interaction with saliva proteins of the three representative of wine anthocyanins (glucosides, acetylated, and cinnamoylated) proving that cinnamoylated anthocyanins are the most reactive and also those that present the lowest perception threshold in wine model solutions. More recently, another work showed that pyranoanthocyanins (pyranomalvidin-3-glucoside, pyranolmalvidin-3-glucoside-catechol, and pyranomalvidin-3-glucoside-epicatechin) can also able to interact aPRPs with K_D even lower (more affinity) than for anthocyanins (between 0.87 and 1.73 mM) [114].

Concerning bitter taste, malvidin-3-*O*-glucoside has also demonstrated to stimulate one member of the bitterness receptor family (TAS2R7) at micromolar levels (12.6 μM) [89]. With regard to flavanols, the addition of quercetin-3-*O*-glucoside (0.25–2 g/L) to white and red wines resulted in a noticeable increase of astringency and bitterness evaluated by sensory analysis. In general, wines were described as smooth-tasting before the flavonol addition and became more astringent, rough, green, dry, bitter, and persistent in presence of quercetin-3-*O*-glucoside [33].

6. Conclusions

There is no doubt that wine is an extremely complex medium and that polyphenolic compounds play an essential role on its final sensory properties. There is an inestimable chemical diversity of polyphenols in both grapes and wines. Each family can be present in free or conjugated forms, with different hydroxylation levels and substitutions or even forming adducts between them. Starting with the raw material, phenolic compounds in grapes can vary substantially depending on several factors as ripening, viticulture practices, environmental conditions, and varietal or genetic differences. Such is the case that each cultivar may be considered as "exclusive" and consequently, the resulting wine too. Anthocyanins (mainly present in skins) and flavan-3-ols and condensed tannins (mainly present in skins and seeds) are the most abundant polyphenols in grapes. On the one hand, anthocyanins are responsible of the color of red wines, and their profile can be used as an analytical tool for authenticity certification. On the other hand, flavan-3-ols and condensed tannins are key compounds due to their implication on color stabilization and astringent and bitter properties. Finally, other phenolics as flavonols and hydroxycinnamic acids are mainly known for acting as copigments. Oak wood is commonly used during fermentation and/or aging of wines. The phenolic composition of oak wood will vary depending on species, geographical origin, and grain or wood processing. In quantitative terms, ellagitannins are the major phenolic constituents of oak wood, and their level and profile may affect the astringency and bitterness of wine. Winemaking produces important changes in polyphenolic composition. In fact, phenolics participate in several reactions such as copigmentation, cycloaddition, polymerization, and oxidation. Thus, new compounds as vitisins, ethyl-bridged anthocyanin-flavanol derivatives, or pyroanthocyanins are formed. Furthermore, wine aging in contact with oak wood affects the degree of complexity of phenolic compounds. In this sense, ellagitannins are actively engaged in oxidation reactions that favor the polymerization between flavanols and between flavanols and anthocyanins.

Overwhelming evidence has demonstrated that tannins (mostly condensed tannins), thanks to their ability to precipitate salivary proteins, are implicated on wine astringency. Astringency intensity, even if it is a multifaceted sensation complicated by a number of variables, is more influenced by the source of proanthocyanidin (seed or skin) and by well-defined proanthocyanidin fractions (oligomeric or polymeric). Up to now, ellagitannins' direct impact on astringency and bitterness sensation remains still unknown. One of the major limitations of the half-tongue test used to evaluate their sensory impact is the absence of contact between the ellagitannins and the entire oral cavity. Further studies are needed under wine conditions. Additionally, recent studies highlight that other phenolic compounds such as anthocyanins/pyranoanthocyanins or flavanols may also interact with salivary proteins and bitterness receptors. Thus, a new research line in the field of sensory properties linked to wine phenolic compounds sensory properties is opened.

Author details

Ruth Hornedo-Ortega[1], María Reyes González-Centeno[1,2], Kleopatra Chira[1,2], Michaël Jourdes[1] and Pierre-Louis Teissedre[1*]

1 Axe Qualité et Identité du Vin, Unité de Recherche Œnologie, Institut des Sciences de la Vigne et du Vin (ISVV), Université de Bordeaux, Villenave d'Ornon, France

2 Tonnellerie Nadalié, Ludon-Médoc, France

*Address all correspondence to: pierre-louis.teissedre@u-bordeaux.fr

References

[1] Vermerris W, Nicholson R. Phenolic compound biochemistry. In: Families of Phenolic. London: Springer; 2006. DOI: 10.1007/978-1-4020-5164-7

[2] Saltveit ME. Synthesis and metabolism of phenolic compounds. In: de la Rosa L, Alvarez-Parrilla E, Gonzalez-Águilar GA, editors. Fruit and Vegetable Phytochemicals Chemistry, Nutritional Value, and Stability. Hoboken, NJ, USA: Wiley-Blackwell; 2010. pp. 89-100. DOI: 10.1002/9780813809397.ch3

[3] Heras-Roger J, Alonso-Alonso O, Gallo-Montesdeoca A, Díaz-Romero C, Darias-Martín J. Influence of copigmentation and phenolic composition on wine color. Journal of Food Science and Technology. 2016;**53**:2540-2547. DOI: 10.1007/s13197-016-2210-3

[4] Li L, Sun B. Grape and wine polymeric polyphenols: Their importance in enology. Critical Reviews in Food Science and Nutrition. 2019;**59**:563-579. DOI: 10.1080/10408398.2017.1381071

[5] Chira K, Teissedre PL. Chemical and sensory evaluation of wine matured in oak barrel: Effect of oak species involved and toasting process. European Food Research and Technology. 2015;**240**:533-547. DOI: 10.1007/s00217-014-2352-3

[6] Cheynier V, Schneider R, Salmon J, Fulcrand H. Chemistry of wine. In: Mander L, Liu HW, editors. Comprehensive Natural Products II. Oxford: Elsevier; 2010. pp. 1119-1172

[7] Downey MO, Dokoozlian NK, Krstic MP. Cultural practice and environmental impacts on the flavonoid composition of grapes and wine. A review of recent research. American Journal of Enology and Viticulture. 2006;**57**:257-268

[8] Pinasseau L, Vallverdu Queralt A, Verbaere A, Roques M, Meudec E, Le Cunff L, et al. Cultivar diversity of grape skin polyphenol composition and changes in response to drought investigated by LC-MS based metabolomics. Frontiers in Plant Science. 2017;**8**:24. DOI: 10.3389/fpls.2017.01826

[9] Burns J, Mullen W, Landrault N, Teissedre P, Lean MEJ, Crozie A. Variations in the profile and content of anthocyanins in wines made from cabernet sauvignon and hybrid grapes. Journal of Agricultural and Food Chemistry. 2002;**50**:4096-4102. DOI: 10.1021/jf011233s

[10] Arapitsas P, Oliveira J, Mattivi F. Do white grapes really exist? Foodservice Research International. 2015;**69**:21-25. DOI: 10.1016/j.foodres.2014.12.002

[11] Guidoni S, Ferrandino A, Novello V. Effects of seasonal and agronomical practices on skin anthocyanin profile of Nebbiolo grapes. American Journal of Enology and Viticulture. 2008;**59**:22-29

[12] Pinelo-Jiménez M, Arnous A, Meyer ABS. Upgrading of grape skins: Significance of plant cell-wall structural components and extraction techniques for phenol release. Trends in Food Science and Technology. 2006;**17**:579-590. DOI: 10.1016/j.tifs.2006.05.003

[13] Ky I, Lorrain B, Kolbas N, Crozier A, Teissedre PL. Wine by-products: Phenolic characterization and antioxidant activity evaluation of grapes and grape pomaces from six different French grape varieties. Molecules. 2014;**19**:482-506. DOI: 10.3390/molecules19010482

[14] Kammerer D, Claus A, Carle R, Schieber A. Polyphenol screening of pomace from red and white grape varieties (*Vitis vinifera* L.)

by HPLC-DAD-MS/MS. Journal of Agricultural and Food Chemistry. 2004;**52**:4360-4367. DOI: 10.1021/jf049613b

[15] Arts IC, van de Putte B, Hollman PC. Catechin contents of foods commonly consumed in The Netherlands. 1. Fruits, vegetables, staple foods, and processed foods. Journal of Agricultural and Food Chemistry. 2000;**48**:1746-1751. DOI: 10.1021/jf000025h

[16] De Freitas VAP, Glories Y, Monique A. Developmental changes of procyanidins in grapes of red *Vitis vinifera* varieties and their composition in respective wines. American Journal of Enology and Viticulture. 2000;**51**:397-403

[17] Guendez R, Kallithraka S, Makris DP, Kefalas P. Determination of low molecular weight polyphenolic constituents in grape (*Vitis vinifera* sp.) seed extracts: Correlation with antiradical activity. Food Chemistry. 2005;**89**:1-9. DOI: 10.1016/j.foodchem.2004.02.010

[18] Rodríguez-Montealegre R, Peces R, Vozmediano J, Gascueña J, Romero E. Phenolic compounds in skins and seeds of ten grape *Vitis vinifera* varieties grown in a warm climate. Journal of Food Composition and Analysis. 2006;**19**:687-693. DOI: 10.1016/j.jfca.2005.05.003

[19] Yilmaz Y, Toledo RT. Major flavonoids in grape seeds and skins: Antioxidant capacity of catechin, epicatechin, and gallic acid. Journal of Agricultural and Food Chemistry. 2004;**52**:255-260. DOI: 10.1021/jf030117h

[20] He F, He JJ, Pan QH, Duan CQ. Mass-spectrometry evidence confirming the presence of pelargonidin-3-O-glucoside in the berry skins of cabernet sauvignon and pinot noir (*Vitis vinifera* L.). Australian Journal of Grape and Wine Research. 2010;**16**:464-468. DOI: 10.1111/j.1755-0238.2010.00107.x

[21] Vidal S, Hayasaka Y, Meudec E, Cheynier V, Skouroumounis G. Fractionation of grape anthocyanin classes using multilayer coil countercurrent chromatography with step gradient elution. Journal of Agricultural and Food Chemistry. 2004;**52**:713-719. DOI: 10.1021/jf034906a

[22] Vidal S, Meudec E, Cheynier V, Skouroumounis G, Hayasaka Y. Mass spectrometric evidence for the existence of oligomeric anthocyanins in grape skins. Journal of Agricultural and Food Chemistry. 2004;**52**:7144-7151. DOI: 10.1021/jf048939h

[23] Yonekura-Sakakibara K, Nakayama T, Yamazaki M, Saito K. Modification and stabilization of anthocyanins. In: Gould K, Davies K, Winefield C, editors. Anthocyanins Biosynthesis, Functions, and Applications. New York, USA: Springer; 2009. pp. 169-190. DOI: 10.1007/978-0-387-77335-3_6

[24] Castellarin SD, Di Gaspero G. Transcriptional control of anthocyanin biosynthetic genes in extreme phenotypes for berry pigmentation of naturally occurring grapevines. Biology. 2007;**7**:46. DOI: 10.1186/1471-2229-7-46

[25] Mazzuca P, Ferranti P, Picariello G, Chianese L, Addeo F. Mass spectrometry in the study of anthocyanins and their derivatives: Differentiation of *Vitis vinifera* and hybrid grapes by liquid chromatography/electrospray ionization mass spectrometry and tandem mass spectrometry. Journal of Mass Spectrometry. 2005;**40**:83-90. DOI: 10.1002/jms.778

[26] Mazza G, Fukumoto L, Delaquis P, Girard B, Ewert B. Anthocyanins, phenolics, and color of cabernet franc,

merlot, and pinot noir wines from British Columbia. Journal of Agricultural and Food Chemistry. 1999;**47**:4009-4017. DOI: 10.1021/jf990449f

[27] Picariello G, Ferranti P, Chianese L, Addeo F. Differentiation of *Vitis vinifera* L. and hybrid red grapes by matrix-assisted laser desorption/ionization mass spectrometry analysis of berry skin anthocyanins. Journal of Agricultural and Food Chemistry. 2012;**60**:4559-4566. DOI: 10.1021/jf300456k

[28] Smeriglio A, Barreca D, Bellocco E, Trombetta D. Proanthocyanidins and hydrolysable tannins: Occurrence, dietary intake and pharmacological effects. British Journal of Pharmacology. 2017;**174**:1244-1262. DOI: 10.1111/bph.13630

[29] Chira K. Structures moléculaires et perception tannique des raisins et des vins (Cabernet-Sauvignon, Merlot) du bordelais [PhD thesis]. University of Bordeaux; 2009

[30] Sun B, de Sa M, Leandro C, Caldeira I, Duarte FL, Spranger I. Reactivity of polymeric proanthocyanidins toward salivary proteins and their contribution to young red wine astringency. Journal of Agricultural and Food Chemistry. 2013;**61**:939-946. DOI: 10.1021/jf303704u

[31] Downey MO, Harvey JS, Robinson SP. Analysis of tannins in seeds and skins of shiraz grapes throughout berry development. Australian Journal of Grape and Wine Research. 2003;**9**:15-27. DOI: 10.1111/j.1755-0238.2003.tb00228.x

[32] Zeng L, Pons-Mercadé P, Richard T, Krisa S, Teissedre PL, Jourdes M. Crown Procyanidin tetramer: A procyanidin with an unusual cyclic skeleton with a potent protective effect against amyloid-β-induced toxicity. Molecules. 2019;**24**:1915. DOI: 10.3390/molecules24101915

[33] Ferrer-Gallego R, Brás NF, García-Estévez I, Mateus N, Rivas-Gonzalo JC, de Freitas V, et al. Effect of flavonols on wine astringency and their interaction with human saliva. Food Chemistry. 2016;**15**:358-364. DOI: 10.1016/j.foodchem.2016.04.091

[34] Downey MO, Harvey JS, Robinson SP. Synthesis of flavonols and expression of flavonol synthase genes in the developing grape berries of shiraz and chardonnay (*Vitis Vinifera L.*). Australian Journal of Grape and Wine Research. 2003;**9**:110-121

[35] Castillo-Muñoz N, Gómez-Alonso S, García-Romero E, Hermosín-Gutiérrez I. Flavonol profiles of *Vitis vinifera* red grapes and their single-cultivar wines. Journal of Agricultural and Food Chemistry. 2007;**55**:992-1002. DOI: 10.1021/jf062800k

[36] Mattivi F, Guzzon R, Vrhovsek U, Stefanini M, Velasco R. Metabolite profiling of grape: Flavonols and anthocyanins. Journal of Agricultural and Food Chemistry. 2006;**54**:7692-7702. DOI: 10.1021/jf061538c

[37] Wang H, Race EJ, Shrikhande AJ. Anthocyanin transformation in cabernet sauvignon wine during aging. Journal of Agricultural and Food Chemistry. 2003;**51**:7989-7994. DOI: 10.1021/jf034501q

[38] Zhu L, Zhang Y, Lu J. Phenolic contents and compositions in skins of red wine grape cultivars among various genetic backgrounds and originations. International Journal of Molecular Sciences. 2012;**13**:3492-3510. DOI: 10.3390/ijms13033492

[39] Boido E, García-Marino M, Dellacassa E, Carrau F, Rivas-Gonzalo JC, Escribano-Bailón MT. Characterisation and evolution of grape polyphenol profiles of *Vitis vinifera* L. cv. Tannat during ripening and vinification. Australian Journal of Grape and Wine Research. 2011;**17**:383-393

[40] Prodanov M, Vacas V, Hernández T, Estrella I, Amador B, Winterhalter P. Chemical characterisation of Malvar grape seeds (*Vitis vinifera* L.) by ultrafiltration and RP-HPLCPAD-MS. Journal of Food Composition and Analysis. 2013;**31**:284-292

[41] Darias-Martín J, Martín-Luis B, Carrillo-López M, Lamuela-Raventós R, Díaz-Romero C, Boulton R. Effect of caffeic acid on the color of red wine. Journal of Agricultural and Food Chemistry. 2002;**50**:2062-2067. DOI: 10.1021/jf010931+

[42] Adrian M, Jeandet P, Douillet-Breuil AC, Tesson L, Bessis R. Stilbene content of mature *Vitis vinifera* berries in response to UV-C elicitation. Journal of Agricultural and Food Chemistry. 2000;**48**:6103-6105. DOI: 10.1021/jf0009910

[43] Babazadeh A, Taghvimi A, Hamishehkar H, Tabibiazar M. Development of new ultrasonic–solvent assisted method for determination of trans-resveratrol from red grapes: Optimization, characterization, and antioxidant activity (ORAC assay). Food Bioscience. 2017;**20**:36-42. DOI: 10.1016/j.fbio.2017.08.003

[44] Németh G, Hegyi O, Dunai A, Kocsis L. Stilbenes in the different organs of *Vitis vinifera* cv. Merlot grafted on TK5BB rootstock. OENO One. 2017;**51**:323-328. DOI: 10.20870/oeno-one.2016.50.4.1068

[45] Guerrero RF, Puertas B, Fernández MI, Palma M, Cantos-Villar E. Induction of stilbenes in grapes by UV-C: Comparison of different subspecies of *Vitis*. Innovative Food Science & Emerging Technologies. 2010;**11**:231-238. DOI: 10.1016/j.ifset.2009.10.005

[46] Flamini R, De Rosso M, De Marchi F, Dalla Vedova A, Panighel A, Gardiman M, et al. An innovative approach to grape metabolomics: Stilbene profiling by suspect screening analysis. Metabolomics. 2013;**9**:1243-1253. DOI: 10.1007/s11306-013-0530-0

[47] Fernández-Marín MI, Guerrero RF, García-Parrilla MC, Puertas B, Richard T, Rodríguez-Werner MA, et al. Isorhapontigenin: A novel bioactive stilbene from wine grapes. Food Chemistry. 2012;**135**:1353-1359. DOI: 10.1016/j.foodchem.2012.05.086

[48] Feuillat F, Keller R. Variability of oak ood (*Quercus Robur L.*, *Quercus Petraea* Liebl.) anatomy relating to cask properties. American Journal of Enology and Viticulture. 1997;**48**:502-508

[49] Zhang B, Cai J, Duan CQ, Reeves MJ, He F. A review of polyphenolics in oak woods. International Journal of Molecular Sciences. 2015;**16**:6978-7014. DOI: 10.3390/ijms16046978

[50] Quideau S, Jourdes M, Lefeuvre D, Pardon P, Saucier C, Teissedre PL, et al. Ellagitannins-an underestimated class of bioactive plant polyphenols: Chemical reactivity of C-glucosidic ellagitannins in relation to wine chemistry and biological activity. In: Santos-Buelga C, Escribano-Bailon T, Lattanzio V, editors. Recent Advances in Polyphenol Research. Vol. 2. Singapore: Blackwell Publishing; 2010. pp. 81-137. DOI: 10.1002/9781444323375.ch4

[51] Hagerman AE. Hydrolyzable tannin structural chemistry. In: Ann EH, editor. Tannin Handbook. Oxford OH: Miami University; 2002. pp. 1-5. Available from: http://www.users.muohio.edu/hagermae/

[52] Jourdes M, Pouységu L, Deffieux D, Teissedre PL, Quideau S. Hydrolyzable tannins: gallotannins and ellagitannins. In: Ramawat K, Mérillon JM, editors. Natural Products. Berlin, Heidelberg: Springer; 2013. pp. 1975-2010. DOI: 10.1007/978-3-642-22144-6_65

[53] Cadahía E, Varea S, Muñoz L, Fernández de Simón B, García-Vallejo MC. Evolution of ellagitannins in Spanish, French, and American oak woods during natural seasoning and toasting. Journal of Agricultural and Food Chemistry. 2001;**49**:3677-3684. DOI: 10.1021/jf010288r

[54] Fernández de Simón B, Cadahía E, Conde E, García-Vallejo MC. Ellagitannins in woods of Spanish, French and American oaks. Holzforschung. 1999;**53**:147-150

[55] Fernández de Simón B, Cadahía E, Conde E, García-Vallejo MC. Evolution of phenolic compounds of Spanish oak wood during natural seasoning. First results. Journal of Agricultural and Food Chemistry. 1999;**47**:1687-1694. DOI: 10.1021/jf9805855

[56] Martínez J, Cadahía E, Fernández de Simón B, Ojeda S, Rubio P. Effect of the seasoning method on the chemical composition of oak heartwood to cooperage. Journal of Agricultural and Food Chemistry. 2008;**56**:3089-3096. DOI: 10.1021/jf0728698

[57] Masson G, Moutounet M, Puech JL. Ellagitannin content of oak wood as a function of species and of sampling position in the tree. American Journal of Enology and Viticulture. 1995;**46**:262-268

[58] Prida A, Puech JL. Influence of geographical origin and botanical species on the content of extractives in American, French, and East European oak woods. Journal of Agricultural and Food Chemistry. 2006;**54**:8115-8126. DOI: 10.1021/jf0616098

[59] Martínez-Gil AM, Cadahía E, Fernández de Simón B, Gutiérrez-Gamboa G, Nevares I, Alamo-Sanza M. *Quercus Humboldtii* (Colombian oak): Characterisation of wood phenolic composition with respect to traditional oak wood used in oenology. Ciência e Técnica Vitivinícola. 2017;**32**:93-101

[60] Martínez-Gil A, Del Alamo-Sanza M, Sánchez-Gómez R, Nevares I. Different woods in cooperage for oenology: A review. Beverages. 2018;**4**:94. DOI: 10.3390/beverages4040094

[61] Alañón ME, Castro-Vázquez L, Díaz-Maroto MC, Hermosín-Gutiérrez I, Gordon MH, Pérez-Coello MS. Antioxidant capacity and phenolic composition of different woods used in cooperage. Food Chemistry. 2011;**129**:1584-1590. DOI: 10.1016/j.foodchem.2011.06.013

[62] Jordao AM, Ricardo-da-Silva JM, Laureano O. Ellagitannins from Portuguese oak wood (*Quercus pyrenaica* Willd.) used in cooperage: Influence of geographical origin, coarseness of the grain and toasting level. Holzforschung. 2007;**61**:155-160. DOI: 10.1515/HF.2007.028

[63] Alañón ME, Castro-Vázquez L, Díaz-Maroto MC, Gordon MH, Pérez-Coello MS. A study of the antioxidant capacity of oak wood used in wine ageing and the correlation with polyphenol composition. Food Chemistry. 2011;**128**:997-1002. DOI: 10.1016/j.foodchem.2011.04.005

[64] Navarro M, Kontoudakis N, Gómez-Alonso S, García-Romero E, Canals JM, Hermosín-Gutíerrez I, et al. Influence of the botanical origin and toasting level on the ellagitannin content of wines aged in new and used oak barrels. Foodservice Research International. 2016;**87**:197-203. DOI: 10.1016/j.foodres.2016.07.016

[65] Puech JL, Feuillat F, Mosedale JR. The tannins of oak heartwood: Structure, properties, and their influence on wine flavor. American Journal of Enology and Viticulture. 1999;**50**:469-478

[66] Glabasnia A, Hofmann T. Sensory-directed identification of taste-active ellagitannins in American (*Quercus alba* L.) and European oak wood (*Quercus robur* L.) and quantitative analysis in Bourbon whiskey and oak-matured red wines. Journal of Agricultural and Food Chemistry. 2006;**54**:3380-3390. DOI: 10.1021/jf052617b

[67] Pracomtal G, Mirabel M, Teissier du Cros R, Monteau AC. Types of Oak Grain, Wine élevage in Barrel. USA: Practical Winery & Vineyard; 2014

[68] Del Alamo-Sanza M, Nevares I. Recent advances in the evaluation of the oxygen transfer rate in oak barrels. Journal of Agricultural and Food Chemistry. 2014;**62**:8892-8899. DOI: 10.1021/jf502333d

[69] Glabasnia A, Hofmann T. Identification and sensory evaluation of dehydro- and deoxy-ellagitannins formed upon toasting of oak wood (*Quercus alba* L.). Journal of Agricultural and Food Chemistry. 2007;**55**:4109-4118. DOI: 10.1021/jf070151m

[70] Chira K, Anguellu L, Da Costa G, Richard T, Pedrot E, Jourdes M, et al. Identification of new C-glycosidic ellagitannins formed upon oak wood toasting. Food Chemistry. (submitted-FOODCHEM-S-20-04464)

[71] Ribéreau-Gayon P, Glories Y, Maujean A, Dubourdieu D. Handbook of Enology, the Chemistry of Wine: Stabilization and Treatments. 2nd ed. Vol. 2. England: John Wiley & Sons, Ldt.; 2006. pp. 1-441

[72] Di Stefano R, Flamini R. High performance liquid chromatography analysis of grape and wine polyphenols. In: Flamini R, editor. Hyphenated Techniques in Grape and Wine Chemistry. Hoboken, NJ, USA: John Wiley & Sons; 2008. pp. 33-79. DOI: 10.1002/9780470754320

[73] Chira K, Schmauch G, Saucier C, Fabre S, Teissedre PL. Grape variety effect on proanthocyanidin composition and sensory perception of skin and seed tannin extracts from Bordeaux wine grapes (cabernet sauvignon and merlot) for two consecutive vintages (2006 and 2007). Journal of Agricultural and Food Chemistry. 2009;**57**:545-553. DOI: 10.1021/jf802301g

[74] Revilla E, García-Beneytez E, Cabello F, Martin-Ortega G, Ryan JM. Value of high-performance liquid chromatographic analysis of anthocyanins in the differentiation of red grape cultivars and red wines made from them. Journal of Chromatography. A. 2001;**915**:53-60. DOI: 10.1016/S0021-9673(01)00635-5

[75] Fang F, Li JM, Zhang P, Tang K, Wang W, Pan QH, et al. Effects of grape variety, harvest date, fermentation vessel and wine ageing on flavonoid concentration in red wines. Foodservice Research International. 2008;**41**:53-60. DOI: 10.1016/j.foodres.2007.09.004

[76] Saucier C. How do wine polyphenols evolve during wine ageing? Cerevisia. 2010;**35**:11-15. DOI: 10.1016/j.cervis.2010.05.002

[77] Gambuti A, Strollo D, Erbaggio A, Lecce L, Moio L. Effect of winemaking practices on color indexes and selected bioactive phenolics of Aglianico wine. Journal of Food Science. 2007;**72**:2347-2353. DOI: 10.1111/j.1750-3841.2007.00536.x

[78] Chira K, Teissedre PL. Extraction of oak volatiles and ellagitannins compounds and sensory profile of wine aged with French winewoods subjected to different toasting methods: Behaviour during storage. Food Chemistry. 2013;**140**:168-177. DOI: 10.1016/j.foodchem.2013.02.049

[79] González-Centeno MR, Chira K, Teissedre PL. Ellagitannin content,

volatile composition and sensory profile of wines from different countries matured in oak barrels subjected to different toasting methods. Food Chemistry. 2016;**210**:500-511. DOI: 10.1016/j.foodchem.2016.04.139

[80] González-Centeno MR, Chira K, Teissedre PL. Comparison between malolactic fermentation container and barrel toasting effects on phenolic, volatile, and sensory profiles of red wines. Journal of Agricultural and Food Chemistry. 2017;**65**:3320-3329. DOI: 10.1021/acs.jafc.6b05497

[81] Kyraleou M, Tzanakouli E, Kotseridis Y, Chira K, Ligas I, Kallithraka S, et al. Addition of wood chips in red wine during and after alcoholic fermentation: Differences in color parameters, phenolic content and volatile composition. OENO One. 2016;**50**:209-222. DOI: 10.20870/oeno-one.2016.50.4.885

[82] Chira K, González-Centeno MR, Teissedre PL. Wine maturation: Malolactic fermentation in barrel or steel tanks - what are the phenolic and sensorial differences? Wine & Viticulture Journal. 2018;**33**:22-27

[83] González-Centeno MR, Chira K, Teissedre PL. Use of oak wood during malolactic fermentation and ageing: Impact on chardonnay wine character. Food Chemistry. 2019;**278**:460-468. DOI: 10.1016/j.foodchem.2018.11.049

[84] He F, Mu L, Yan GL, Liang NN, Pan QH, Wang J, et al. Biosynthesis of anthocyanins and their regulation in coloured grapes. Molecules. 2010;**15**:9057-9091. DOI: 10.3390/molecules15129057

[85] Jourdes M, Lefeuvre D, Quideau S. C-glycosidic ellagitannins and their influence on wine chemistry. In: Quideau S, editor. Chemistry and Biology of Ellagitannins—An Underestimated Class of Bioactive Plant Polyphenols. 1st ed. Singapore: World Scientific; 2009. pp. 320-365. DOI: 10.1142/9789812797414_0009

[86] Rasines-Perea Z, Jacquet R, Jourdes M, Quideau S, Teissedre PL. Ellagitannins and flavano-ellagitannins: Red wines tendency in different areas, barrel origin and ageing time in barrel and bottle. Biomolecules. 2019;**9**:316. DOI: 10.3390/biom9080316

[87] Gambuti A, Rinaldi A, Ugliano M, Moio L. Evolution of phenolic compounds and astringency during aging of red wine: Effect of oxygen exposure before and after bottling. Journal of Agricultural and Food Chemistry. 2013;**61**:1618-1627. DOI: 10.1021/jf302822b

[88] Montmayeur JP, Matsunami H. Receptors for bitter and sweet taste. Current Opinion in Neurobiology. 2002;**12**:366-371. DOI: 10.1016/s0959-4388(02)00345-8

[89] Soares S, Kohl S, Thalmann S, Mateus N, Meyerhof W, De Freitas V. Different phenolic compounds activate distinct human bitter taste receptors. Journal of Agricultural and Food Chemistry. 2013;**61**:1525-1533. DOI: 10.1021/jf304198k

[90] Saenz-Navajas MP, Ferreira V, Dizy M, Fernández-Zurbano P. Characterization of taste-active fractions in red wine combining HPLC fractionation, sensory analysis and ultra-performance liquid chromatography coupled with mass spectrometry detection. Analytica Chimica Acta. 2010;**673**:151-159. DOI: 10.1016/j.aca.2010.05.038

[91] Peleg H, Noble AC. Perceptual properties of benzoic acid derivatives. Chemical Senses. 1995;**20**:393-400. DOI: 10.1093/chemse/20.4.393

[92] Breslin PAS, Gilmore MM, Beauchamp GK, Green BG. Psychophysical evidence that oral astringency is a tactile sensation.

Chemical Senses. 1993;**18**:405-417. DOI: 10.1093/chemse/18.4.405

[93] ASMT. Standard definitions of terms relating to sensory evaluation of materials and products. In: Annual Book of ASTM Standards. Philadelphia: American Society for Testing and Materials, Philadelphia; 1989. pp. 19-22

[94] Obreque-Slier E, López-Solís R, Peña-Neira A, Zamora-Marín F. Tannin–protein interaction is more closely associated with astringency than tannin–protein precipitation: Experience with two oenological tannins and a gelatin. International Journal of Food Science and Technology. 2010;**45**:2629-2636. DOI: 10.1111/j.1365-2621.2010.02437.x

[95] Kennedy JA, Ferrier J, Harbertson JF, des Gachons CP. Analysis of tannins in red wine using multiple methods: Correlation with perceived astringency. American Journal of Enology and Viticulture. 2006;**57**:481-485

[96] Harbertson JF, Kilmister RL, Kelm MA, Downey MO. Impact of condensed tannin size as individual and mixed polymers on bovine serum albumin precipitation. Food Chemistry. 2014;**160**:16-21

[97] Chira K, Jourdes M, Teissedre PL. Cabernet sauvignon red wine astringency quality control by tannin characterization and polymerization during storage. European Food Research and Technology. 2012;**234**:253-261. DOI: 10.1007/s00217-011-1627-1

[98] Ferrer-Gallego R, García-Marino M, Hernández-Hierro JM, Rivas-Gonzalo JC, Escribano-Bailón MT. Statistical correlation between flavanolic composition, colour and sensorial parameters in grape seed during ripening. Analytica Chimica Acta. 2010;**660**:22-28. DOI: 10.1016/j.aca.2009.09.039

[99] Schwarz B, Hofmann T. Is there a direct relationship between oral astringency and human salivary protein binding? European Food Research and Technology. 2008;**227**:1693-1698. DOI: 10.1007/s00217-008-0895-x

[100] Fontoin H, Saucier C, Teissedre PL, Glories Y. Effect of pH, ethanol and acidity on astringency and bitterness of grape seed tannin oligomers in model wine solution. Food Quality and Preference. 2008;**19**:286-291. DOI: 10.1016/j.foodqual.2007.08.004

[101] Demiglio P, Pickering GJ. The influence of ethanol and pH on the taste and mouthfeel sensations elicited by red wine. Journal of Food, Agriculture and Environment. 2008;**6**:143-150

[102] Chira K, Pacella N, Jourdes M, Teissedre PL. Chemical and sensory evaluation of Bordeaux wines (cabernet-sauvignon and merlot) and correlation with wine age. Food Chemistry. 2011;**126**:1971-1977. DOI: 10.1016/j.foodchem.2010.12.056

[103] Basalekou M, Kyraleou M, Pappas C, Tarantilis P, Kotseridis Y, Kallithraka S. Proanthocyanidin content as an astringency estimation tool and maturation index in red and white winemaking technology. Food Chemistry. 2019;**299**:125-135. DOI: 10.1016/j.foodchem.2019.125135

[104] Chira K, Liming Z, Le Floch A, Péchamat L, Jourdes M, Teissedre PL. Compositional and sensory characterization of grape proanthocyanidins and oak wood ellagitannin. Tetrahedron. 2015;**71**:2999-3006. DOI: 10.1016/j.tet.2015.02.018

[105] Brossaud F, Cheynier V, Noble AC. Bitterness and astringency of grape and wine polyphenols. Australian Journal of Grape and Wine Research. 2001;**7**:33-39

[106] Soares S, Mateus N, de Freitas V. Interaction of different polyphenols

with bovine serum albumin (BSA) and human salivary α-amylase (HSA) by fluorescence quenching. Australian Journal of Grape and Wine Research. 2007;**55**:6726-6735

[107] Lee CA, Ismail B, Vickers ZM. The role of salivary proteins in the mechanism of astringency. Journal of Food Science. 2012;**77**:C381-C387. DOI: 10.1111/j.1750-3841.2012.02644.x

[108] Meyerhof W, Born S, Brockhoff A, Behrens M. Molecular biology of mammalian bitter taste receptors. A review. Flavour and Fragrance Journal. 2011;**26**:260-268. DOI: 10.1002/ffj.2041

[109] Hufnagel JC, Hofmann T. Quantitative reconstruction of the nonvolatile sensometabolome of a red wine. Journal of Agricultural and Food Chemistry. 2008;**56**:9190-9199. DOI: 10.1021/jf801742w

[110] Soares S, Brandão E, García-Estevez I, Fonseca F, Guerreiro C, Ferreira-da-Silva F, et al. Interaction between ellagitannins and salivary proline-rich proteins. Journal of Agricultural and Food Chemistry. 2019;**67**:9579-9590. DOI: 10.1021/acs.jafc.9b02574

[111] Ferrer-Gallego R, Soares S, Mateus N, Rivas-Gonzalo J, Escribano-Bailón MT, de Freitas V. New anthocyanin–human salivary protein complexes. Langmuir. 2015;**31**:8392-8401. DOI: 10.1021/acs.langmuir.5b01122

[112] Cala O, Pinaud N, Simon C, Fouquet E, Laguerre M, Dufourc EJ, et al. NMR and molecular modeling of wine tannins binding to saliva proteins: Revisiting astringency from molecular and colloidal prospects. The FASEB Journal. 2010;**24**:4281-4290. DOI: 10.1096/fj.10-158741

[113] Paissoni MA, Waffo-Teguo P, Ma W, Jourdes M, Rolle L, Teissedre PL. Chemical and sensorial investigation of in-mouth sensory properties of grape anthocyanins. Scientific Reports. 2018;**8**:17098. DOI: 10.1038/s41598-018-35355-x

[114] García-Estévez I, Cruz L, Oliveira J, Mateus N, de Freitas V, Soares S. First evidences of interaction between pyranoanthocyanins and salivary proline-rich proteins. Food Chemistry. 2017;**228**:574-581. DOI: 10.1016/j.foodchem.2017.02.03

Chapter 2

Emerging Technologies to Increase Extraction, Control Microorganisms, and Reduce SO_2

Antonio Morata, Iris Loira, Buenaventura Guamis, Javier Raso, Juan Manuel del Fresno, Carlos Escott, María Antonia Bañuelos, Ignacio Álvarez, Wendu Tesfaye, Carmen González and Jose Antonio Suárez-Lepe

Abstract

This chapter reviews the main non-thermal technologies with application in enology and their impact in: the extraction of phenolic compounds from grapes, the elimination of indigenous microorganisms, and the subsequent effect in SO_2 reduction. The technologies are physical processes with null or low repercussion in temperature and therefore gentle with sensory quality of grapes. High hydrostatic pressure (HHP), ultra high pressure homogenization (UHPH), pulsed electric fields (PEFs), electron-beam irradiation (eBeam), ultrasound (US), and pulsed light (PL) have interesting advantages and some drawbacks that are extensively reviewed highlighting the potential applications in current technology.

Keywords: non-thermal technologies, wine, grape, HHP, UHPH, PEF, eBeam, ultrasound, PL

1. Emerging non-thermal technologies for grape and must processing

Traditionally, must extraction for winemaking is different in whites and reds. White grapes are pressed and optionally destemmed, especially if they are going to be cold soaked. Later, the grape juice is cleaned by settling before fermentation (**Figure 1**). Red wines need maceration to extract polyphenolic compounds (tannins and pigments). Consequently, the grapes are crushed and later simultaneously macerated and fermented in a tank for several days until a suitable polyphenolic content is reached (**Figure 1**). Emerging non-thermal technologies can help to speed up the extraction of phenolic compounds and aromatic molecules from skins, to eliminate or reduce wild microorganisms from grapes thus facilitating new fermentation biotechnologies, and some of them can also destroy or reduce the activity of oxidative enzymes [1–3]. Since the use of non-thermal technologies does not increase significantly the temperature, usually, their application does not affect negatively the sensory quality of the wine. Additionally, the elimination of microorganisms and control of enzymes helps to reduce the SO_2 content in wines [1].

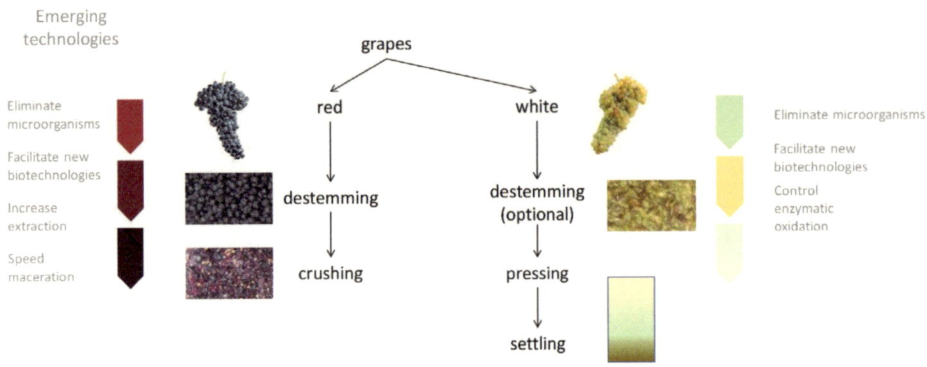

Figure 1.
Red and white winemaking and main steps where to apply emerging technologies and potential advantages.

The emerging technologies that showed the greatest potential for increasing the extraction of skin compounds and controlling indigenous microorganisms are as follows: high hydrostatic pressure (HHP) [4–6], ultra high pressure homogenization (UHPH) [7–9], pulsed electric fields (PEFs) [10–15], electron-beam irradiation (eBeam) [16–18], ultrasound (US) [15, 19, 20], and pulsed light (PL) [21–23]. Most of these technologies are either approved or under evaluation as new enological practices in the International Organization of Vine and Wine (OIV) regulations [24–26]. The treatment conditions and the most significant effects on microbial loads, grape extraction, and enzyme inactivation are described in **Table 1**.

Wine freshness is a complex sensory perception in which acidity is strongly involved, but it is also represented by a certain aromatic profile in which floral and fruity esters predominate, as well as a bluish red color in young wines with no oxidation or a pale yellow color in white wines without browning [29, 30]. The acidity makes the wines to be perceived more refreshing and this is currently well appreciated by consumers. Moreover, fresh aromas such as citric, fruity or floral add complexity and elegance to the wine. New emerging non-thermal technologies can eliminate or strongly reduce the wild microorganisms in grapes, thus facilitating the use of new biotechnologies such as fermentation with non-*Saccharomyces* yeasts which help modulate acidity and aroma [29, 30]. Currently, acidity can be increased in several ways; one efficient way is by producing lactic acid during alcoholic fermentation with the non-*Saccharomyces* yeast *Lachancea thermotolerans* (Lt) [31–35]. The use of Lt during fermentation allows reducing pH values by 0.2–0.5 units, producing lactic acid from sugars and without significant side effects [34, 35]. Moreover, some strains can also have positive sensory impacts by producing fruity or floral esters [34]. Concerning fruity or floral aroma, several non-*Saccharomyces* as *Torulaspora delbrueckii* (Td), *Wickerhamomyces anomalus* (Wa), *Metschnikowia pulcherrima* (Mp), *Hanseniaspora vineae* (Hv), and *Hanseniaspora/Kloeckera* spp., *Lachancea thermotolerans* or *Candida stellata* (Cs) have demonstrated their ability to modulate or influence wine aroma during fermentation by producing fermentation esters or by developing enzymatic activities that release varietal precursors of aromatic molecules [30]. Some of these species are also able to produce stable pyranoanthocyanins or promote the formation of polymeric pigments, thus improving and stabilizing wine color [36–39]. Among them, *Schizosaccharomyces pombe* (Sp) has interesting properties such as the formation of vitisin A-type pyranoanthocyanins because of its high production of pyruvate.

The use of non-*Saccharomyces* yeasts has the main drawback of being normally little competitive against *Saccharomyces*, so their successful implantation in the must is not easy and they cannot express their metabolic properties or enzymatic

Technology	Microbial inactivation	Extraction of anthocyanins	Extraction of polyphenols	Inactivation of oxidative enzymes	Processing flow and temperature	Reference
HHP	Yeast elimination: 400 MPa/10 min Lactic acid bacteria resistance even at 550 MPa	Increased	Improved	unclear	Until 3 t/h Room temperature or even refrigeration	[4–6, 27]
UHPH (EU Patent)	300 MPa/continuously yeast and bacteria elimination. Sporulated bacteria depending on in-valve temperature	Increased	Not described	>90%	Available 10,000 L/h pumps, higher rates in multi-modular systems. 20–25°C. In valve 98°C less than 0.2 s	[8, 28]
PEF	Depending on dose until 5 log cycles of bacteria not sporulated	4–10 kV/cm Increased 10–20%	4–10 kV/cm Increased 10–20%	Not described	Extraction: Flow 20 t/h Specific energy <10 kJ/kg ΔT^a < 3°C Inactivation: until 2000 L/h Specific energy >50 kJ/kg 20–25°C for less than 1 s	[11–15]
eBeam	1–2-log reductions for yeast and bacteria with 1 kGray, and elimination with 10 kGray	70% higher when 10 KGray dose is used	10–20% with 10 kGray	Not described	Increase 3–5°C at 10 kGray	[16]
US	Low effectivity, at high intensity thermal effects	Treatment and 3 days of maceration achieved similar contents to 8 days of conventional maceration	Treatment and 3 days of maceration yielded double content compared to 8 days of conventional maceration	Not described	ΔT^a: 2–3°C	[19, 20]
PL	Effective against yeast and bacteria. 1–3-log reductions	Slightly higher, but some final degradation due to oxidizing effects	Not described	Not described	ΔT^a: 2–3°C	[21]

Table 1.
Main features and effects of the emerging non-thermal technologies used in grape processing.

activities. Most non-*Saccharomyces* usually have lower fermentative power than *S. cerevisiae* or slower fermentation kinetics (e.g., Sp), which reduces the possibilities of being well implanted [40]. New emerging non-thermal technologies facilitate the use of non-*Saccharomyces* yeasts by eliminating or strongly reducing wild yeasts from the grapes [2, 6, 8, 14, 16, 21].

2. High hydrostatic pressure (HHP)

HHP was developed at an industrial scale in the last century by Gec Alston Company in Europe, even when their basis and initial prototypes where defined by Hite in 1899 [41]. Technological difficulties delayed the production of industrial devices able to work at industrial scale until the 80s of XX century. HHP technique uses the hydrostatic pressure transmitted by a liquid, usually water, to pressurize a liquid or solid food in order to eliminate microorganisms or modify some food properties (**Figure 2**). In HHP systems, the pressures are applied by a liquid and therefore homogeneous pressurization is produced on the whole surface of the food. The low-pressure pump is used to quickly fill the vessel with water. It is important to reach a high filling ratio to reduce the volume of water necessary and therefore decrease the dead-times. When air is removed and the vessel is completely filled with water, the secondary high pressure pump is used to continue introducing water to increase the pressure. It is normally necessary to introduce an additional 4% volume of water to reach the working pressures of 400–600 MPa needed to eliminate the microorganisms.

Typical pressure ranges to eliminate microorganisms by HHP are 300–400 MPa for yeasts and molds, 400 MPa for Gram-negative bacteria, and 500–600 MPa for Gram-positive during processing times of 3–10 min [42]. Sporulated bacteria cannot be controlled by HHP because pressures around 1000 MPa are needed, which is not feasible for industrial equipment. The application of 200 MPa for 10 min on the grapes can decrease the indigenous yeast populations, with 400 MPa often being enough to eliminate the yeasts from the grapes [5, 6]. However, even the treatment of 550 MPa for 10 min may not be able to completely eliminate bacterial populations [5]. The elimination of indigenous yeasts facilitates the implantation of inoculated non-*Saccharomyces* yeasts and the better expression of their metabolic activities [6]. Furthermore, the elimination of both wild yeasts and bacteria helps to reduce the levels of sulfites needed for their control.

HHP can be considered a non-thermal technology because the adiabatic compression heat is around 2–3°C/100 MPa, so pressurization at 500 MPa increases

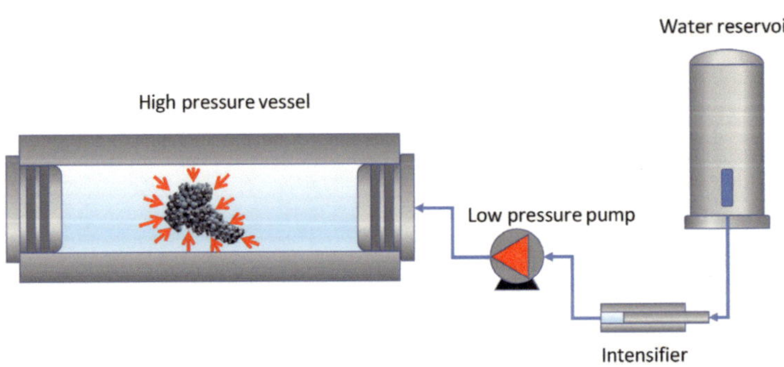

Figure 2.
High hydrostatic pressure system. The pressure vessel is filled with water by means of a low-pressure, high flow rate pump and then increased to working pressure by an intensifier.

the temperature by only 10–15°C. In addition, heat exchangers are incorporated in the pressurization vessel to either refrigerate or to apply thermo-pressurization. Moreover, pressure forces are unable to modify covalent bonds, so most of the molecules responsible for sensory properties (pigments, aromas and flavor substances) remain unaffected during HHP processing. Therefore, HHP can be considered a gentle technique in terms of preserving sensory quality. Moreover, even when grapes are pressurized with intense treatments (500–600 MPa), no heat treatment markers such as Maillard compounds are observed in the HHP-processed musts.

The antimicrobial effect of pressure mainly affects the microbial envelopes: cell wall, cell membrane, and nuclear membrane in yeasts. When microorganisms are pressurized and subsequently returned to atmospheric pressure, all these microbial structures are affected by the poration or completely broken, making the cells non-viable. At the same time, foods are unaffected or slightly affected by their texture and appearance. The pressure only produces a homogeneous volume reduction that can be quantified as 4% and the external appearance of the grapes remains unaltered even at pressures of 550 MPa (**Figure 3**).

The external aspect does not show the effect of the HHP processing but on a molecular scale both membranes and cell wall are altered, causing the migration of pigments and phenols from the grape skin to the pulp (**Figure 4**). Even anthocyanin staining of seeds is observed after HHP treatment [5]. This migration of phenolic compounds produces a faster extraction of these molecules and probably of the aromatic precursors from the skins to the pulp, facilitating grape maceration. Anthocyanin extraction may increase by 20–80% in HHP-processed grapes compared to controls [4, 5], and the wines produced have a significantly higher color intensity and tannin content [5].

HHP processing can be used as a gentle technique to control microorganisms in grapes, thus facilitating the implantation of inoculated starters, especially when they have weaker performance than *S. cerevisiae*. In addition, the intense effect on cell walls helps to increase extraction by reducing conventional maceration times in red winemaking [3].

Currently, industrial devices equipped with horizontal vessels of 55–525 L are available for HHP, able to work up to 600 MPa, and to process up to 3000 kg/h [27]. To reduce dead-times, HHP machines are equipped with a high number of intensifiers: 4–16 depending on the vessel capacity [27, 43]. The most important companies in HHP technology are Hiperbaric [27] and Avure [43].

Figure 3.
External appearance of Vitis vinifera *L. cv. Tempranillo grapes processed by HHP at 200, 400, and 550 MPa for 10 min.*

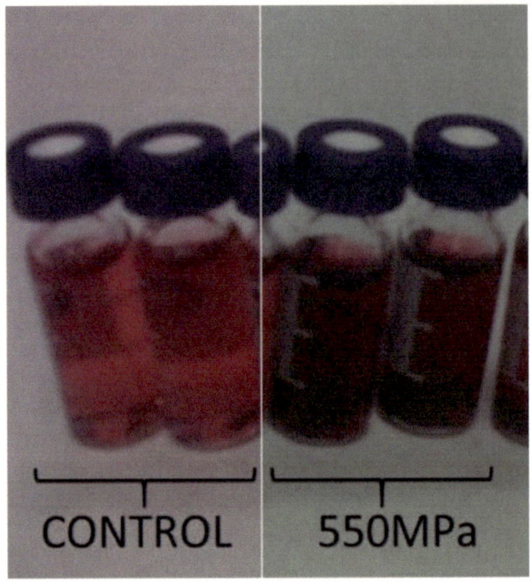

Figure 4.
Color of the grape juice extracted without maceration: unprocessed (control) or pressurized at 550 MPa for 10 min.

3. Ultra high pressure homogenization (UHPH)

High pressure homogenization (HPH) is a continuous treatment in which a food liquid is pumped at high pressure and later depressurized when the fluid passes through a special valve. It is normally called HPH when pressurization occurs at 100–200 MPa (**Figure 5A**) and ultra high pressure homogenization (UHPH) when the pressure range is higher than 200 MPa (**Figure 5B**) [7, 9].

The "UHPH sterilization system" is a novel process, patented in Europe (EP2409583B1) by UAB (Autonomous University of Barcelona), extended to a lot of countries and exclusively exploited and manufactured by Ypsicon Advanced

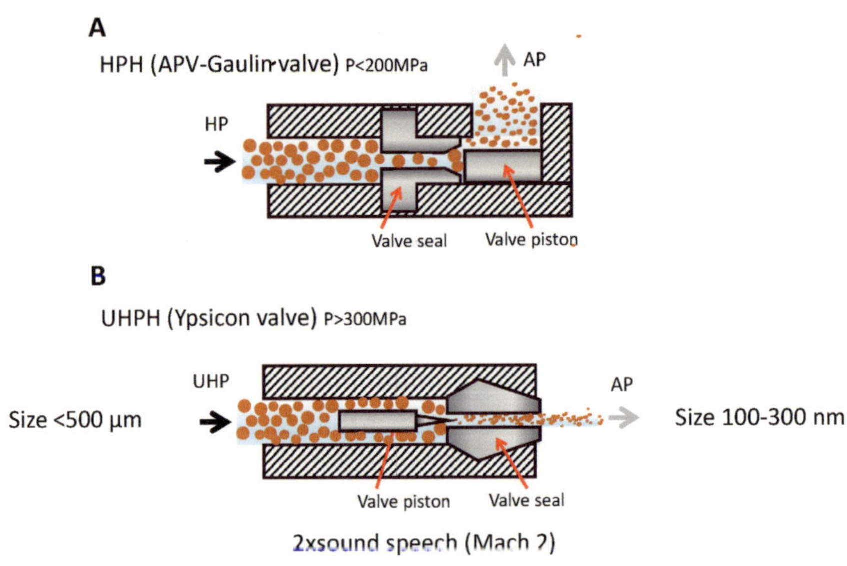

Figure 5.
Valve components in HPH (A) and UHPH (B) systems (Ypsicon [28]). HP: high pressure, UHP: ultra high pressure, AP: atmospheric pressure.

Technologies (Barcelona, Spain) (www.ypsicon.com). It consists of a continuous device capable of working from 200 to 400 MPa and applying shear, impact, cavitation, and turbulence forces in a special valve at high speed (Mach 2). As a consequence, the particles size is reduced from 100 to 300 nm, microorganisms are destroyed, enzymes inactivated and stable emulsions are produced without additives, and consuming less energy than thermal treatments.

UHPH is highly efficient in controlling microorganisms. The antimicrobial effect is produced by the strong impact forces together with the shear stresses and the complementary effect of local cavitation and friction [7–9]. This process produces intense heating in the in-valve time but during a really short period of time of 0.02 and 0.2 s for the global residence time. The temperature in the valve can reach 100°C when the inlet temperature is 20°C, being reduced to 25°C after the valve [8] by adiabatic-expansion chilling. Even when high temperatures are reached instantaneously in the valve, as a whole it can be considered a gentle technology with no thermal effect on sensory degradation. After the UHPH process, no formation of thermal markers such as furfural or 5-hydroxymethylfurfural is observed, probably because of the very low residence time. UHPH processing using suitable in-valve temperatures can produce sterilization capable of destroying even sporulated bacteria. Because of this feature, it can be considered a gentle alternative to UHT, since in this thermal technique a temperature of 140°C for 3–4 s is necessary. In contrast, UHPH only requires a total processing time of less than 0.2 s.

UHPH can be applied to liquids containing particles, but the particle size must be lower than 500 μm. The average particle size range at the valve outlet is 100–300 nm. The valve design is a critical point, as well as the performance of the UHPH process, but especially the antimicrobial effect depends on the geometry and materials of the valve piston and seal. Especially, efficient designs are manufactured by Ypsicon [44] (**Figure 6**). UHPH systems are currently available with flow rates of up to 10,000 L/h [28]. The processing rate can be increased in a modular way by using several systems working in parallel. Some UHPH pumps can work up to 400 MPa continuously with a pressure oscillation of 1 MPa.

Concerning the elimination of microorganisms, UHPH has proven to be highly effective with 6-log reductions for *Saccharomyces* and non-*Saccharomyces* yeasts

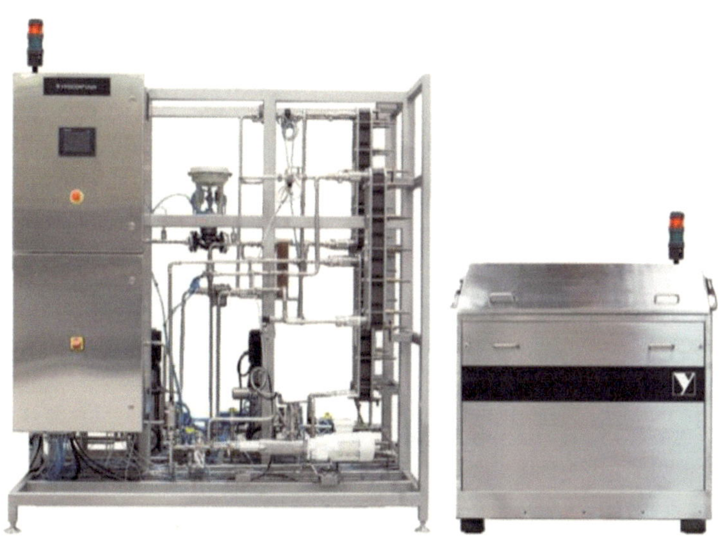

Figure 6.
UHPH industrial machine (Ypsicon [28]).

Figure 7.
Color evolution in the control and UHPH musts after 2 days at room temperature in the absence of SO_2. White must of Vitis vinifera *L. cv. Muscat.*

and 4-log for aerobic and lactic acid bacteria in the must [8]. All these wild microorganisms remained undetected in the must after UHPH processing at 300 MPa (inlet temperature 20°C, in-valve 98°C, outlet 26°C, and in-valve time 0.02 s) [8]. Therefore, UHPH is a powerful technology for eliminating indigenous microorganisms and facilitating the use of modern biotechnologies, such as the use of non-*Saccharomyces* or yeast-bacteria co-inoculations [2]. Simultaneously, microbial control facilitates the reduction of SO_2 content.

UHPH technology has also shown high efficiency in enzyme destruction. The intense impact and shear forces that the fluid undergoes when pumped through the valve produce a molecular depolymerization that reduces colloidal particles, microorganisms, and enzymes to small fragments. In the case of cells and spores, it causes microbial death, and in the case of enzymes, it causes denaturalization and inactivation. In musts processed by UHPH, a reduction in oxidase activity higher than 90% has been observed for polyphenol oxidase (PPO) enzymes [8]. In addition, the fragmentation effect of colloidal particles can increase the nutrient availability for alcoholic fermentation, which can have positive impact on the production of fermentative aroma [8].

When measuring the color intensity in white grape musts, the value was lower in UHPH than in the unprocessed controls. This is an indication of a paler color that can be correlated with low oxidation by PPO enzymes [8]. The same results have been observed when white musts were kept at room temperature under oxidation conditions and without SO_2: the UHPH musts remained pale and the controls quickly browned (**Figure 7**). UHPH is a key technology for reducing sulfites in must by controlling oxidative enzymatic activities which lead to browning and aroma degradation.

4. Pulsed electric fields (PEFs)

Pulsed electric fields (PEFs) are a non-thermal technology with a high potential in the extraction of phenolic and aromatic compounds from grapes [11–13] and also in the elimination of microorganisms [10, 14]. This technology uses high intensity voltages (10–40 kV) producing strong electric fields of 1–30 kV/cm. However, it can be considered a non-thermal technology because these high intensity fields are applied in ultra-short periods of a few microseconds which usually produce a temperature increase of only a few degrees Celsius.

Typical pulses are produced as bipolar square waves because of their higher efficiency compared to other types of waves. Exponential decay waves can also be used, but the energy transferred and the effectiveness of electroporation is lower for the same field intensity.

PEFs are applied to food products in a treatment chamber using two electrodes built with inert metals. The location of the electrodes is usually in two consecutives sections of the pipeline separated by an isolating section. In this design, the electrodes from the treatment chamber, keeping the same diameter of the pipeline through which the crushed grapes are pumped. The force lines of the electric field are tangential to the flow direction.

The effect of PEFs on the cell is the electroporation which increases the permeability of the cell membrane in microorganisms and plant cells. This increase in permeability is due to the formation of pores. The electric field strength required to produce the electroporation depends on the size of the cells, with intensities below 10 kV/cm and specific energies below 10 kJ/kg being sufficient to produce this effect in grape skin cells. However, higher electric fields (>10 kV/cm) and specific energies (>50 kJ/kg) are required for the electroporation of microbial cells [12].

In wine technology, electroporation is a powerful tool to increase the extraction of pigments and tannins, thus reducing maceration times (**Figure 8**). The skin contact time when crushed grapes are processed by PEFs can be reduced by 2–3 times in comparison with the unprocessed grapes, also allowing to finish the fermentation in the absence of solids which allows a cleaner and more controlled fermentation. The low temperature increase protects the aroma compounds and facilitates the preservation of the varietal aroma.

Moreover, PEF can be applied continuously to the crushed grapes during the pumping from the crusher to the tank (**Figure 9**). Industrial devices can process more than 10 t/h. PEF industrial treatments decrease maceration times, thus increasing the availability of the fermentation facilities. In addition, the equipment is moderate in size, it requires little space in the winery, and the energy inputs required are low compared to other traditional techniques.

The use of PEFs to control microorganisms in grapes needs higher field intensities due to the smaller size of the microbial cells [45, 46]. These field intensities produce an inactivation ranging from 0.6 to 4.94 log cycles for several wine spoilage yeasts and lactic acid bacteria [46]. PEFs can be used as a powerful non-thermal tool to control indigenous microorganisms, thus helping to decrease SO_2 levels.

Figure 8.
*Color intensity and degree of pigment extraction in musts obtained from untreated (right) and PEF-treated (left) grapes (*V. vinifera *L. cv. Garnacha) after 1 h of maceration.*

Figure 9.
PEF experimental unit to process the crushed grapes at a flow of 3 t/h to improve the polyphenols extraction in the maceration-fermentation stage of red winemaking.

5. eBeam or β-irradiation

Electron-beam irradiation or beta-irradiation is a technology that uses low-dose ionizing radiation to eliminate microorganisms in food and also to delay ripening by extending shelf-life. eBeam irradiation and X-rays use ionizing radiation but they are not produced by radioactive materials. The radiation is generated by an electron accelerator that is a switch-on/off electronic technology that can be disconnected when is not in use. The electrons are generated in a heavy metal-doped cathode and then increase their speed by being accelerated by radio frequency fields in a high vacuum cavity (**Figure 10**). The maximum energy allowed is 10 MeV to avoid

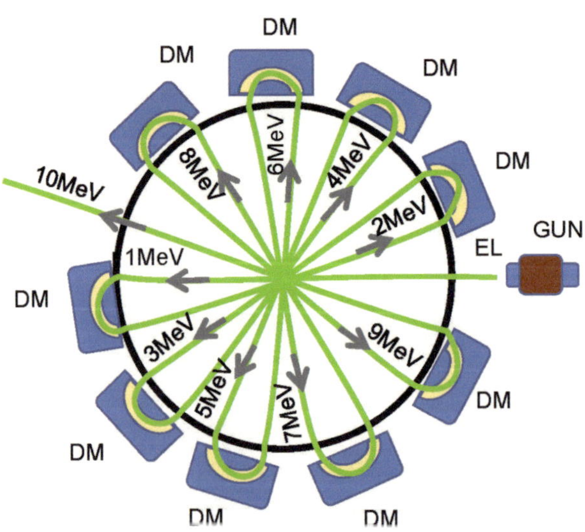

Figure 10.
Radiofrequency cavity in a rodothron™-type ebeam accelerator. GUN: low energy electron producing cathode, EL: electromagnetic lens, DM: deflector magnets.

the production of unstable nuclei in food. This maximum value is less than the 14 MeV necessary to generate radioactive isotopes. The energy of the radiation is related to its penetration power and it can be estimated as 3 mm depth in water per MeV. Therefore, 3 cm of water-like density products can be treated at 10 MeV.

The energy of 10 MeV makes the electrons capable of breaking the atomic and molecular bonds, thus generating ions and free radicals that can react with other molecules and produce secondary ions and radicals [47, 48]. During food irradiation, water is mainly affected by the formation of hydroxyl ($^•OH$), hydrogen ($H^•$), superoxide ($HO_2^•$) radicals or peroxide (H_2O_2) [48]. These free radicals may react with several molecules or biopolymers in the cells, including DNA, making them unviable. The irradiation dose is measured in kiloGray (1 kGy = 1000 J/kg). Food irradiation applications can be classified according to dose: low dose (<1 kGy) with disinfection applications, medium dose (1–10 kGy) with antimicrobial effect to extend shelf-life, and high dose (10–60 kGy) for sterilization purposes [48].

The dose must be checked after irradiation by locating radio-chromic dosimeters (**Figure 11**) on the treated foods in order to verify if the scheduled and applied doses correspond to actual values in the food. It is essential to check at all depths to ensure that the eBeam radiation is suitable for the entire material. Radiochromic dosimeters are formed by a transparent film, inside an aluminum envelope, containing a radiochromic pigment that is colored when is irradiated, and color intensity depends on received dose. After treatment, the irradiation dose can be measured spectrophotometrically. Additionally, a radiochromic sticker is added (**Figure 11**) to each food package to verify that it has been properly treated. It is often difficult or impossible to visually detect if foods have been treated, and this sticker is necessary to distinguish treated from untreated samples.

Irradiation can be considered a non-thermal technology with slight temperature increments of a few degrees in conventional treatments. Moreover, it can be applied in refrigerated foods. The effect on sensory quality is gentle, thus many molecules with sensory repercussion as pigments or flavor compounds remain unaffected. The oxidative effect of free radicals generated during the irradiation can produce some oxidative processes in some molecules.

When the electron beam is produced, it must scan from left to right to generate a treatment surface. Electron beam must move at high frequency, usually 100 Hz, to

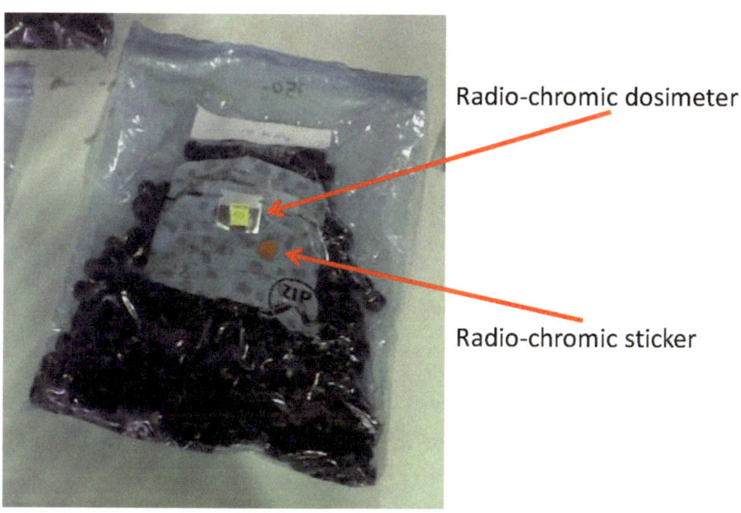

Figure 11.
Grape sample with both, radiochromic dosimeter and sticker.

treat each irradiated section several times in a second. The electron beam is moved by using intense electromagnetic fields. When the food is moved below the treatment plane, all food volume is irradiated. The received dose depends on the speed at which the food moves below the irradiation section that is controlled by using a belt conveyor (**Figure 12**).

The external appearance of foods after irradiation frequently remains unaffected; sometimes, a brighter outer aspect can be observed (**Figure 13A**) [16]. Irradiation doses of 0.5–1 kGy produce 1 log reductions in yeast and lactic acid bacteria in grapes [16] without modifications in the external appearance and firmness [16]; similar effects have been observed in other fruits [49, 50]. Irradiation dose of 10 kGy produces 6-log reductions in yeasts and 3-log in bacteria [16]; this dose decreases firmness and softens the texture of grapes, thus enhancing the extraction of pigments in juice (**Figure 13B**) [16].

Irradiation can also be considered non-thermal technology with high efficiency to eliminate indigenous microorganisms from grapes. This reduction in the population of microorganisms allows a better implantation of selected yeasts and the reduction in SO_2 doses. Depending on the irradiation dose (10 kGy), better

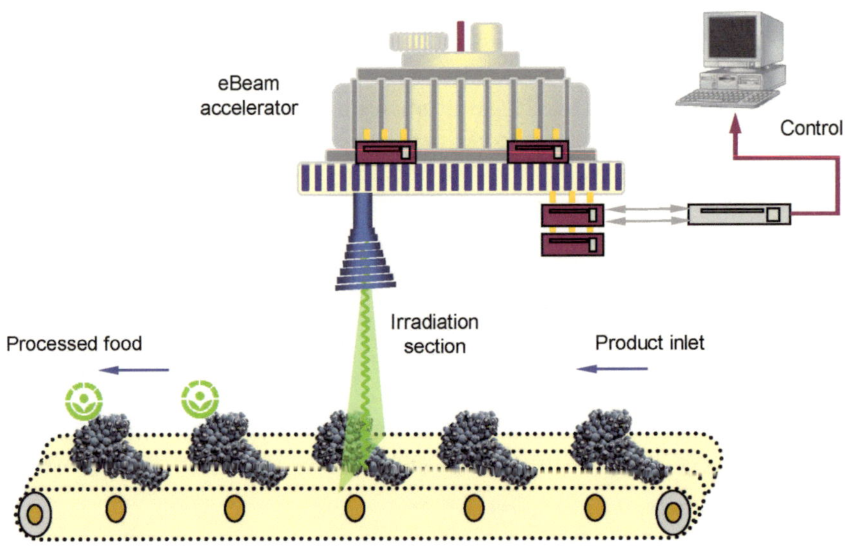

Figure 12.
Electron beam accelerator and detail of the irradiation process.

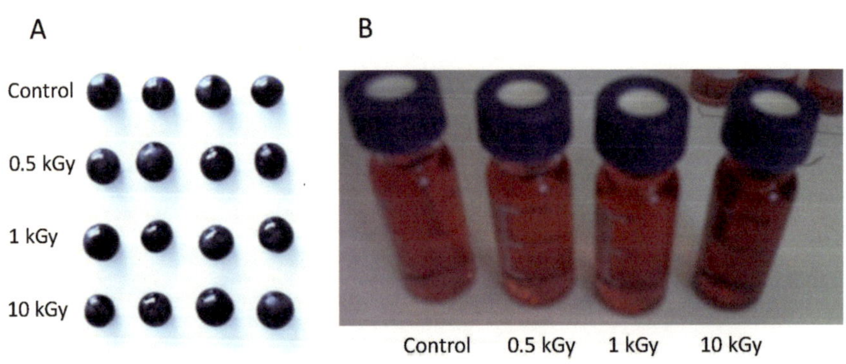

Figure 13.
(A) External appearance of grapes irradiated at 0.5, 1, and 10 kGy. (B) Color of juice from the irradiated grapes.

extraction of pigments and polyphenols can be observed with subsequent improvements on the maceration processes.

6. Ultrasound

Ultrasound (US) is sonic waves with a frequency range of 20–100 kHz producing cavitation phenomena with locally extreme temperature and pressure. Cavitation is generated by compression-decompression cycles producing the formation and implosion of gas bubbles [51] (**Figure 14**). This phenomenon produces intense agitation and dispersive effects that help to disrupt vegetal tissues, depolymerizing cell walls and favoring a better extraction. US efficiency in extraction processes improves when the frequency is closed to 30 kHz [52].

US technology can be used in winemaking for continuous processing of crushed grapes. As a consequence, there is a weakening of the cell wall and an increase in the extraction of tannins, pigments, and aromatic compounds [15, 19, 20, 52]. This effect can also be enhanced with the use of pectolytic enzymes [53]. When enzymes are used as the sole extraction technique, tannin concentration is 13% higher while, after US treatment, there is an increase of 16%. The initial use of enzymes followed by the subsequent application of US is especially synergistic increasing color intensity (18%) and tannin content (30%) [53].

Antimicrobial effect of US is quite reduced, and the intensity and time needed normally produce significant increments of temperature. Therefore, it is difficult to consider US as a non-thermal technology. However, US produces synergistic antimicrobial effects when applied together with conventional thermal technologies or emerging non-thermal processes.

Industrial devices are currently available to process crushed grapes increasing extraction and reducing maceration times. The sonication device has a tubular structure to increase sonication surface, normally with a polygonal section to better dispose of the sonoplates (**Figure 15**). Currently, this technology is developed by several companies; among them, Agrovin inside a H2020 European project, has developed the Perseo™ system [54] with 50 kW of power and 8 cavitation cells to process up to 10 t/h [52, 54]. With this technology, it is possible to reduce maceration times from 7 to 2–3 days with similar contents of anthocyanins and tannins; moreover, the aromatic fraction is in the same time enhanced. Prof. Emilio Celotti

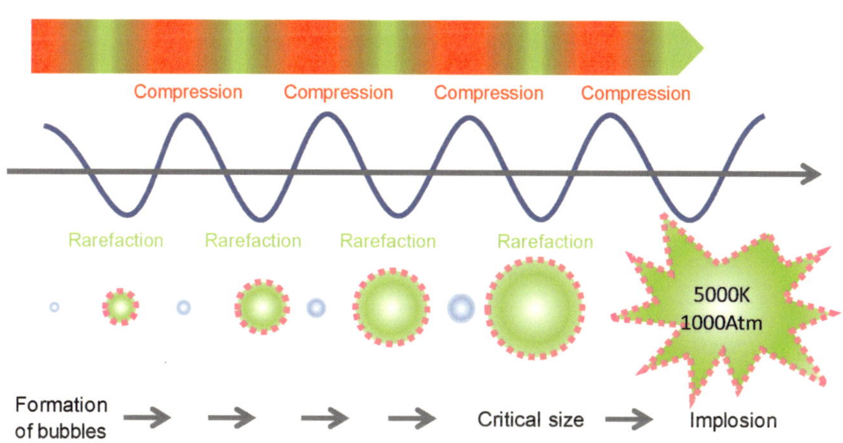

Figure 14.
Implosion of bubbles and cavitation produced by alternative compression-rarefaction effects generated by US waves.

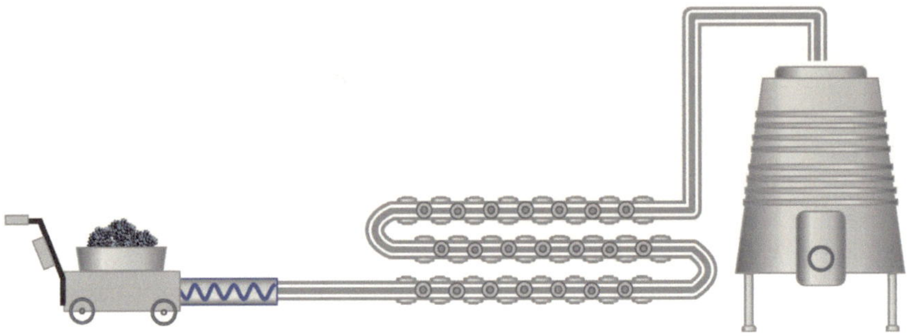

Figure 15.
Cavitation cells disposed in a tubular structure with the sonoplates to apply the US waves.

(Udine University, Italy) is also working in the use of US to improve the maceration process and has developed a prototype to favor a faster extraction of polyphenols, aroma, and precursors [55, 56].

7. Pulsed light

Pulsed light (PL) is the use of wide spectrum (170–2600 nm) high intensity light [57] applied in short flashes during a few microseconds. The spectrum includes ultraviolet, visible, and near infrared radiation. PL spectrum is quite similar to the solar radiation, but having lower wavelengths from the UV (<320 nm), which, in the case of sun light, these are filtered by the ozone layer in the atmosphere. The intensity of light can reach values 10^5 folds higher than the sun radiation at sea level [58]. PL technology uses very short energization times, for example, a flash lamp applying energy of 300 J during 300 μs produces a peak intensity of 1 MW which produces ca. 1 kW/cm^2 on the irradiation surface. PL was initially used in Japan in the 1970s; later, in 1988, it was developed by a Californian company PurePulse Technologies Inc., but the applications in the food industry increased after it was approved by FDA for food processing [59].

The antimicrobial effect of PL is due to dimerization of DNA pyrimidines promoted by the 254 nm UV radiation but also due to the localized instantaneous heating producing membrane and cell wall breakage [57]. The effect on membrane

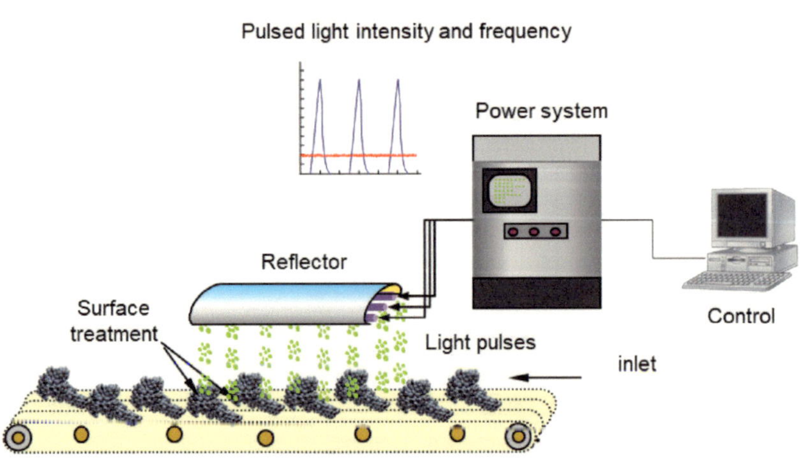

Figure 16.
Pulsed light system to treat grape surface for the elimination of microorganisms in a continuous flow.

and cell walls can be observed by electronic microscopy, and it is associated to a higher concentration of eluted protein [60]. These effects affect vegetative and spore forms of microorganisms. *Saccharomyces* and non-*Saccharomyces* yeasts and lactic acid bacteria are controlled in grapes at low temperature and without effect in external appearance and sensory quality of food [60] (**Figure 16**).

Currently, PL devices not only for continuous food treatment, but also for the sterilization of food packaging are available [61]. The PL equipment is formed by an electronic module containing both control and monitoring systems, high voltage components for the flashing lamps, cooling system, and, finally, the optical unit with the flash lamps. PL is another non-thermal technology with powerful applications to sterilize the grape surface, thus enabling the control of microbial loads and, therefore, facilitating the use of selected starters and the reduction of SO_2 levels in wines.

8. Conclusions

Emerging non-thermal technologies open new possibilities in winemaking technology, generally facilitating at the same time the control of indigenous microorganisms and the use of new biotechnologies such as the fermentation with non-*Saccharomyces* yeasts or the use of yeast-bacteria co-inoculations. Most of them (HHP, UHPH, PEF, eBeam, and US) also facilitates a faster extraction of phenolic compounds from the grape skins, including not only pigments and tannins, but also aromatic and flavor compounds, thus reducing the maceration times. Several of them (UHPH, PEF, US, and PL) can be applied in a continuous mode when the crushed grape or must is pumped to the fermentation tank, increasing the processing yield and reducing the dead times. Some of these technologies, such as UHPH, produce an intense inactivation of oxidative enzymes, preserving better the sensory quality and strongly reducing the SO_2 needs. Finally, most of these technologies have low energetic requirements, so the running costs are moderated.

Acknowledgements

This research was funded by Ministerio de Ciencia, Innovación y Universidades grant number [RTI2018-096626-B-I00] and by European Regional Development Fund (ERDF) through the National Smart Growth Operational Programme FEDER INTERCONECTA grant number [EXP-00111498/ITC-20181125], project: FRESHWINES.

Author details

Antonio Morata[1*], Iris Loira[1], Buenaventura Guamis[2], Javier Raso[3], Juan Manuel del Fresno[1], Carlos Escott[1], María Antonia Bañuelos[4], Ignacio Álvarez[3], Wendu Tesfaye[1], Carmen González[1] and Jose Antonio Suárez-Lepe[1]

1 enotecUPM, Universidad Politécnica de Madrid, Spain

2 Departament de Ciència Animal i dels Aliments, Facultat de Veterinària, Centre d'Innovació, Recerca i Transferència en Tecnologia dels Aliments (CIRTTA), TECNIO, XaRTA, Universitat Autònoma de Barcelona, Bellaterra, Spain

3 Food Technology, Facultad de Veterinaria, Instituto Agroalimentario de Aragón-IA2, Universidad de Zaragoza-CITA, Zaragoza, Spain

4 Departament de Biotecnología-Biología Vegetal, ETSIAAB, Universidad Politécnica de Madrid, Spain

*Address all correspondence to: antonio.morata@upm.es

References

[1] Suárez-Lepe JA, Morata A. Levaduras para vinificación en tinto. Vol. 8. Madrid. Spain: AMV Ediciones; 2015. pp. 275-291

[2] Morata A, Loira I, Vejarano R, González C, Callejo MJ, Suárez-Lepe JA. Emerging preservation technologies in grapes for winemaking. Trends in Food Science and Technology. 2017;**67**:36-43

[3] Morata A, González C, Tesfaye W, Loira I, Suárez-Lepe JA. Maceration and fermentation: New technologies to increase extraction. In: Red Wine Technology. Amsterdam, The Netherlands: Elsevier; 2019. pp. 35-49

[4] Corrales M, García AF, Butz P, Tauscher B. Extraction of anthocyanins from grape skins assisted by high hydrostatic pressure. Journal of Food Engineering. 2009;**90**:415-421

[5] Morata A, Loira I, Vejarano R, Bañuelos MA, Sanz PD, Otero L, et al. Grape processing by high hydrostatic pressure: Effect on microbial populations, phenol extraction and wine quality. Food and Bioprocess Technology. 2015;**8**:277-286

[6] Bañuelos MA, Loira I, Escott C, Del Fresno JM, Morata A, Sanz PD, et al. Grape processing by high hydrostatic pressure: Effect on use of non-*Saccharomyces* in must fermentation. Food and Bioprocess Technology. 2016;**9**:1769-1778

[7] Zamora A, Guamis B. Opportunities for ultra-high-pressure homogenization (UHPH) for the food industry. Food Engineering Reviews. 2015;7(2):130-142

[8] Loira I, Morata A, Bañuelos MA, Puig-Pujol A, Guamis B, González C, et al. Use of ultra-high pressure homogenization processing in winemaking: Control of microbial populations in grape musts and effects in sensory quality. Innovative Food Science and Emerging Technologies. 2018;**50**:50-56

[9] Comuzzo P, Calligaris S. Potential applications of high pressure homogenization in winemaking: A review. Beverages. 2019;**5**:56

[10] Wu Y, Mittal GS, Griffiths MW. Effect of pulsed electric field on the inactivation of microorganisms in grape juices with and without antimicrobials. Biosystems Engineering. 2005;**90**:1-7

[11] López N, Puértolas E, Condón S, Álvarez I, Raso J. Effects of pulsed electric fields on the extraction of phenolic compounds during the fermentation of must of Tempranillo grapes. Innovative Food Science and Emerging Technologies. 2008;**9**:477-482

[12] Puértolas E, López N, Condón S, Álvarez I, Raso J. Potential applications of PEF to improve red wine quality. Trends in Food Science and Technology. 2010;**21**:247-255

[13] Puértolas E, Saldaña G, Álvarez I, Raso J. Experimental design approach for the evaluation of anthocyanin content of rosé wines obtained by pulsed electric fields. Influence of temperature and time of maceration. Food Chemistry. 2011;**126**:1482-1487

[14] Saldaña G, Álvarez I, Condón S, Raso J. Microbiological aspects related to the feasibility of PEF technology for food pasteurization. Critical Reviews in Food Science and Nutrition. 2014;**54**:1415-1426

[15] Maza M, Álvarez I, Raso J. Thermal and non-thermal physical methods for improving polyphenol extraction in red winemaking. Beverages. 2019;5(3):47. DOI: 10.3390/beverages5030047

[16] Morata A, Bañuelos MA, Tesfaye W, Loira I, Palomero F, Benito S, et al.

Electron beam irradiation of wine grapes: Effect on microbial populations, phenol extraction and wine quality. Food and Bioprocess Technology. 2015;**8**:1845-1853

[17] Gupta S, Padole R, Variyar PS, Sharma A. Influence of radiation processing of grapes on wine quality. Radiation Physics and Chemistry. 2015;**111**:46-56

[18] Błaszak M, Nowak A, Lachowicz S, Migdał W, Ochmian I. E-beam irradiation and ozonation as an alternative to the sulphuric method of wine preservation. Molecules. 2019;**24**(18):3406. DOI: 10.3390/molecules24183406

[19] Bautista-Ortín AB, Jiménez-Martínez MD, Jurado R, Iniesta JA, Terrades S, Andrés A, et al. Application of high-power ultrasounds during red wine vinification. International Journal of Food Science and Technology. 2017;**52**(6):1314-1323. DOI: 10.1111/ijfs.13411

[20] Gómez PE, Jurado R, Iniesta JA, Bautista-Ortín AB. High power ultrasounds: A powerful, non-thermal and green technique for improving the phenolic extraction from grapes to must during red wine vinification. 41st World Congress of Vine and Wine. BIO Web of Conferences. 2019;**12**:02001. DOI: 10.1051/bioconf/20191202001

[21] Escott C, Vaquero C, del Fresno JM, Bañuelos MA, Loira I, Han S, et al. Pulsed light effect in red grape quality and fermentation. Food and Bioprocess Technology. 2017;**10**:1540-1547. DOI: 10.1007/s11947-017-1921-4

[22] Escott C, Loira I, Morata A, Bañuelos MA, Suárez-Lepe JA. In: Morata A, Loira I, editors. Wine Spoilage Yeasts: Control Strategy, Yeast - Industrial Applications. Vol. 4. London, UK: IntechOpen; 2017. DOI: 10.5772/intechopen.69942

[23] Wiktor A, Mandal R, Singh A, Singh AP. Pulsed light treatment below a critical Fluence (3.82 J/cm2) minimizes photo-degradation and browning of a model phenolic (Gallic acid) solution. Food. 2019;**8, 9**:380. DOI: 10.3390/foods8090380

[24] OIV-OENO 594A-2019. Reduction of indigenous microorganisms in grapes and musts by discontinuous high pressure processes (high hydrostatic pressure – HHP). 2019. Available from: http://www.oiv.int/public/medias/6821/oiv-oeno-594a-2019-en.pdf

[25] OENO-MICRO 16-594B Et5. PROVISIONAL DRAFT RESOLUTION. Elimination of wild microorganisms in musts by continuous High Pressure Processes (Ultra High Pressure Homogenization – UHPH)

[26] OIV-OENO 616-2019. Treatment of crushed grapes with ultrasound to promote the extraction of their compounds. 2019. Available from: http://www.oiv.int/public/medias/6826/oiv-oeno-616-2019-en.pdf

[27] Available from: https://www.hiperbaric.com/. 2019

[28] Available from: https://www.ypsicon.com/. 2019

[29] Morata A, Loira I, Del Fresno JM, Escott C, Bañuelos MA, Tesfaye W, et al. Strategies to improve the freshness in wines from warm areas. In: Morata A, Iris L, editors. Advances in Grape and Wine Biotechnology. London, UK: InTech; 2019

[30] Morata A, Escott C, Bañuelos MA, Loira I, del Fresno JM, González C. Suárez-Lepe JA contribution of non-*Saccharomyces* yeasts to wine freshness. A review. Biomolecules. 2020;**10**:34. DOI: 10.3390/biom10010034

[31] Comitini F, Gobbi M, Domizio P, Romani C, Lencioni L, Mannazzu I, et al.

Selected non-*Saccharomyces* wine yeasts in controlled multistarter fermentations with *Saccharomyces cerevisiae*. Food Microbiology. 2011;**28**:873-882

[32] Gobbi M, Comitini F, Domizio P, Romani C, Lencioni L, Mannazzu I, et al. *Lachancea thermotolerans* and *Saccharomyces cerevisiae* in simultaneous and sequential co-fermentation: A strategy to enhance acidity and improve the overall quality of wine. Food Microbiology. 2013;**33**:271-281

[33] Hranilovic A, Gambetta JM, Schmidtke L, Boss PK, Grbin PR, Masneuf-Pomarede I, et al. Oenological traits of *Lachancea thermotolerans* show signs of domestication and allopatric differentiation. Scientific Reports. 2018;**8**:14812-14825

[34] Morata A, Loira I, Tesfaye W, Bañuelos MA, González C, Suárez Lepe JA. *Lachancea thermotolerans* applications in wine technology. Fermentation. 2018;**4**:53

[35] Morata A, Bañuelos MA, Vaquero C, Loira I, Cuerda R, Palomero F, et al. Bi Yang, *Lachancea thermotolerans* as a tool to improve pH in red wines from warm regions. European Food Research and Technology. 2019;**245**:885-894. DOI: 10.1007/s00217-019-03229-9

[36] Morata A, Loira I, Suárez Lepe JA. In: Morata A, Loira I, editors. Influence of Yeasts in Wine Color, Grape and Wine Biotechnology. Vol. 13. IntechOpen; 2016. DOI: 10.5772/65055

[37] Escott C, Morata A, Ricardo-da-Silva JM, Callejo MJ, González C, Suárez-Lepe JA. Effect of *Lachancea thermotolerans* on the formation of polymeric pigments during sequential fermentation with *Schizosaccharomyces pombe* and *Saccharomyces cerevisiae*. Molecules. 2018;**23**:2353

[38] Escott C, del Fresno JM, Loira I, Morata A, Tesfaye W, González MC, et al. Formation of polymeric pigments in red wine through sequential fermentation of flavanol-enriched musts with non-*Saccharomyces* yeasts. Food Chemistry. 2018;**239**:975-983

[39] Morata A, Escott C, Loira I, del Fresno JM, González C, Suárez-Lepe JA. Influence of *Saccharomyces* and non-*Saccharomyces* yeasts in the formation of pyranoanthocyanins and polymeric pigments during red wine making. Molecules. 2019;**24**:4490

[40] Morata A. Enological repercussions of non-*Saccharomyces* species in wine biotechnology. Fermentation. 2019;**5**:72

[41] Morata A. In: Morata A, editor. High hydrostatic pressures, Nuevas tecnologías de conservación de alimentos. Vol. 2. Madrid, Spain: AMVEdiciones; 2010. p. 15

[42] Smelt JPPM. Recent advances in the microbiology of high pressure processing. Trends in Food Science and Technology. 1998;**9**(4):152-158

[43] Available from: https://www.avure-hpp-foods.com/. 2019

[44] UAB European Patent registered in 2010 (European Patent EP2409583B1). Validated in (EU, USA, China, Japan, Australia, India, Brazil, México, etc.). Manufactured by Ypsicon Advance Technologies

[45] Puértolas E, López N, Condón S, Raso J, Álvarez I. Pulsed electric fields inactivation of wine spoilage yeast and bacteria. International Journal of Food Microbiology. 2009;**130**:49-55. DOI: 10.1016/j.ijfoodmicro.2008.12.035

[46] González-Arenzana L, Portu J, López R, López N, Santamaría P, Garde-Cerdán T, et al. Inactivation of wine-associated microbiota by continuous pulsed electric field treatments. Innovative Food Science and Emerging Technologies. 2015;**29**:187-192

[47] Clemmons HE, Clemmons EJ, Brown EJ. Electron beam processing technology for food processing. In: Pillai SD, Shayanfar S, editors. Electron Bean Pasteurization and Complementary Food Processing Technologies. UK: Woodhead Publishing; 2015. pp. 11-25

[48] Lung H-M, Cheng Y-C, Chang Y-H, Huang H-W, Yang BB, Wang C-Y. Microbial decontamination of food by electron beam irradiation. Trends in Food Science and Technology. 2015;**44**(1):66-78

[49] Zhao M, Moy J, Paul RE. Effect of gamma-irradiation on ripening papaya pectin. Postharvest Biology and Technology. 1996;**8**:209-222

[50] D'Innocenzo M, Lajolo FM. Effect of gamma irradiation on softening changes and enzyme activities during ripening of papaya fruit. Journal of Food Biochemistry. 2001;**25**:425-438

[51] Chemat F, Khan MK. Applications of ultrasound in food technology: Processing, preservation and extraction. Ultrasonics Sonochemistry. 2011;**18**(4):813-835

[52] Morata A, González C, Tesfaye W, Loira I, Suárez-Lepe JA. Maceration and fermentation: New technologies to increase extraction. In: Morata A, editor. Red Wine Technology. Amsterdam, The Netherlands: Elsevier; 2019. pp. 35-49

[53] Osete Alcaraz A, Bautista-Ortín AB, Ortega-Regules AE, Gomez-Plaza E. Combined use of pectolytic enzymes and ultrasounds for improving the extraction of phenolic compounds during vinification. Food and Bioprocess Technology. 2019;**12**(8):1330-1339. DOI: 10.1007/s11947-019-02303-0

[54] Available from: https://www.agrovin.com/team/perseo/

[55] Celotti E, Ferraretto P. Recent application of ultrasound in winemaking: From the maceration to the wine aging. In: Oral Communication as Invited Speaker "2nd International Conference on Ultrasound-Based Applications: From Analysis to Synthesis". 6th–8th June. Caparica – Lisbon: Proceedings book; 2016. pp. 89-90

[56] Celotti E, Ferraretto P. Studies for the ultrasound application in winemaking for a low impact enology. Oral Communication, 39th World Congress of Vine and Wine, Bento Gonçalves, Brazil, 24-28 October. Book of Abstracts, ISBN-979-10-91799-62-1; 2016. pp. 104-106

[57] Morata A. In: Morata A, editor. Pulsed light, Nuevas tecnologías de conservación de alimentos. Vol. 9. Madrid, Spain: AMVEdiciones; 2010. p. 127

[58] Takeshita K, Shibato J, Sameshima T, Fukunaga S, Isobe S, Arihara K, et al. Damage of yeast cells induced by pulsed light irradiation. International Journal of Food Microbiology. 2003;**85**(1-2):151-158

[59] Oms-Oliu G, Martín-Belloso O, Soliva-Fortuny R. Pulsed light treatments for food preservation. A review. Food and Bioprocess Technology. 2010;**3**:13-23. DOI: 10.1007/s11947-008-0147-x

[60] Escott C, Vaquero C, del Fresno JM, Bañuelos MA, Loira I, Han SY, et al. Pulsed light effect in red grape quality and fermentation. Food and Bioprocess Technology. 2017;**10**:1540-1547

[61] Available from: http://www.claranor.com/

Chapter 3

Management of Wine Aroma Compounds: Principal Basis and Future Perspectives

Antia G. Pereira, Maria Fraga, Paula Garcia-Oliveira, Maria Carpena, Cecilia Jimenez-Lopez, Catarina Lourenço-Lopes, Lillian Barros, Isabel C.F.R. Ferreira, Miguel Angel Prieto and Jesus Simal-Gandara

Abstract

Wine's aroma is defined by volatile and non-volatile compounds that contribute to its make-up. The complex variety of volatile compounds, coming from grapes, interact with other non-volatile substances of the wine as precursors of wine's aroma, known as primary aromas, which give the aroma of the young wine. The volatile compounds present in the skin and in the grape juice change according to the grape variety. Most of wine volatile compounds responsible for aroma are linked to sugars and they initially form odorless glycosides. Through the process of hydrolysis, they are reverted into an aromatic form. Chemical reactions among these compounds occur during the fermentation and in the first months of a wine's existence, triggering fast and multiple modifications in wine's aroma at this point. As wine ages and matures, changes and development in aroma will continue to take place but at a slower and more gradual pace. The study of the compounds responsible for aroma and flavor, as well as their correlation with the wine quality, is ongoing. Improving the knowledge of wine aromatic compounds could increase the risk of its potential adulteration; however, consumers prefer wine for its natural origin, so this scenario is unlikely in the future.

Keywords: wine, aroma, compounds, grapes, volatile

1. Introduction

1.1 Wine, quality and aroma

Since ancient times, wine has had a fundamental role in diet for several reasons. It is a good method of liquid conservation because its composition prevents the development of microorganisms and rot, it presents properties against certain diseases and provides health benefits, as long as its consumption is moderate. Wine is rich in antioxidants, mainly phenolic compounds, which are natural bioactive compounds related to its diverse properties. Numerous epidemiological studies have associated regular and moderate consumption of wine with a lower incidence of mortality and morbidity from cardiovascular diseases in European countries, mainly in the Mediterranean [1].

The general term "quality" is a complex idea, which is not simpler when applied to wine. Throughout history, different institutions and laws have tried to regulate this concept. The International Organization of Vine and Wine (OIV) considers wine quality as the set of characters that differentiate one wine from another, being one of these characters the consumer taste [2]. According to this definition, the quality of a wine is relative and may vary over time, depending on several factors, such as the country of consumption, its habits, or new advances in the wine industry, among others.

However, the Regulation (EEC) No 823/87 has established a more permanent concept of wine quality, through the definition of the following requirements:

- Delimitation of the production area.
- Distribution of varieties within the wine area.
- Cultivation systems and uses.
- Vinification methods.
- Minimum natural volumetric alcoholic strength.

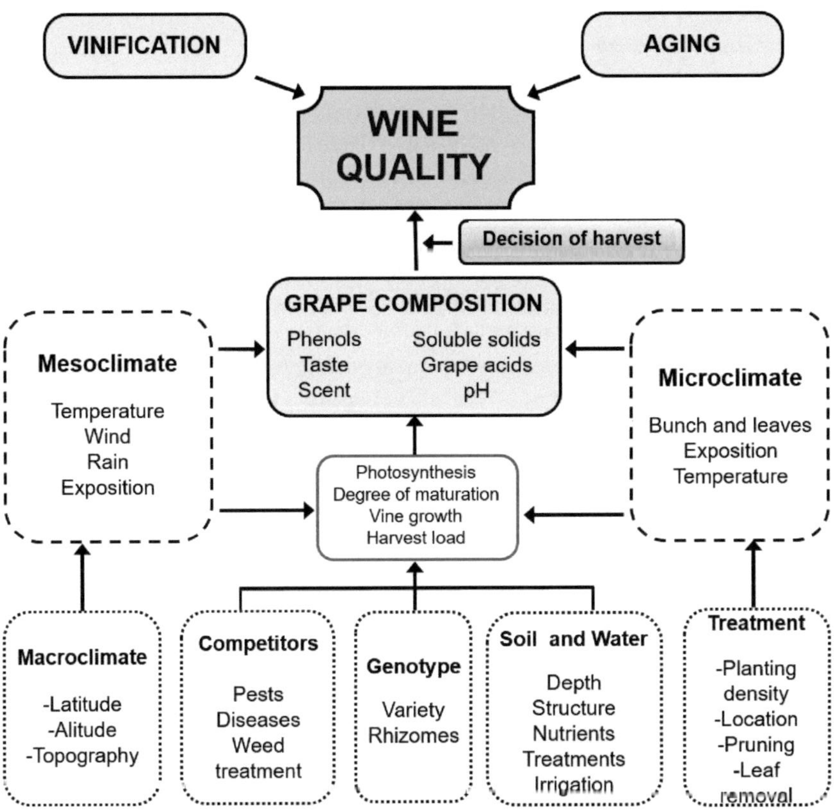

Figure 1.
Environmental and viticultural factors that influence the composition of the wine. Adapted from [3].

- Yield per hectare.
- Analysis and evaluation of sensory characteristics.

Therefore, the quality of a wine will depend on many factors such as soil, type of grapes, presence of grape diseases, weather conditions, flavor complexity, alcoholic strength, balance, longevity, etc. (**Figure 1**) [4, 5]. These factors will also be important in the development and accumulation of the aroma compounds.

1.2 Accumulation of aroma compounds

The aroma of wine is complex, a particular wine contains more than 800 volatile compounds. When wine is ingested, these volatile compounds are dissolved in the palate and pharynx mucosa. Then, they reach the pituitary gland through the Eustachian tube. The perception of this odorous stimulus is the aroma of a wine and it cannot be evaluated without tasting it [6]. The concentration of these compounds may range from a few nanograms per liter to a few hundred micrograms or milligrams. They conform to a complex mixture of many different chemical compounds immersed in a hydroalcoholic solution [7].

Different factors determine the wine aroma such as the grape metabolism, the grape variety, the graft, the year, the weather conditions, the location of the vineyard, the region, the soil, cultural practices, the pre-fermentative biochemical phenomena (oxidations, hydrolysis) and the fermentative metabolism of microorganisms (conservation, aging) [8]. Since wine production is a long process that involves different stages, aromatic compounds from grapes can be transformed or revealed and many can be produced during alcoholic fermentation. In the case of aging wines in wood barrels, many reactions will occur in the wine forming new aroma compounds that will interact with the wood extracted aroma compounds.

The ripening of grapes also plays an important role since many compounds of interest are synthesized at the skin, such as terpenoids, linalool, alpha-terpineol, and geraniol, which will influence the final chemical composition and therefore, the accumulation of wine's aroma. The concentration of volatile compounds in grapes increases in advanced stages of ripening, when sugar accumulation in berries slows down. However, this cumulative process varies from red grapes to white grapes. In red ones, the maximum amount of varietal volatile compounds is reached at maturity and remains constant in the following weeks; whereas, in white grapes varietal volatiles content cannot be related to a maturity stage, because of its concentration changes during ripening [9].

1.3 Main factors affecting aroma quality

As mentioned before, different factors influence the wine quality: type of soil, grape variety, grape diseases, weather, balance, longevity, etc. (**Figure 1**) [10]. Thermal variation is an important factor since it affects grape ripening. Slight thermal variations along grape development will suppress grapes acidity, while stronger thermal fluctuations will yield grapes with better balance. Not just thermal variations but hours of sunshine exposition are another parameter to consider. Environmental conditions affect the grape taste and quality and they also influence the quantity and availability of nutrients in the soil. A deficiency of nutrients affects negatively to the development of vines and to the production and the quality of the wine. Regarding soil characteristics, granularity and texture are two important factors. Stony soils are considered favorable for wine quality, but unfavorable for the plant production capacity due to its lower fertility. Clay soils are related to aromatic, well colored and, often thick, wines [11].

The variety of graft inserted in the vine is also important since the combinations of different varieties allow the obtaining of wines with greater complexity and creativity. Fruit physicochemical characteristics and, therefore wine properties define each variety [12].

Grape production, transformation and certification techniques may be considered, too. A good production is defined by several parameters. Grapes might be whole, healthy and mature. They should be grown following respectful cultivation and be manufactured using proper mechanization procedures. In addition, using appropriate technologies, oenological assessments should be performed for keeping high levels of quality along the control systems. As a result of all this chain of processes, a specific wine will be produced and therefore all these parameters will define its brand, image, authenticity and origin, among other aspects. All these factors lead to a new term "terroir", which is used to express the distinctive characteristics that encompass the physical and biological environment and the applied wine practices that allow producing this wine in this situation [2].

1.4 Management of aroma in vineyards

One of the main factors that influence the evolution of the aroma is management in vineyards. Viticulture practices, which include management of vineyard cultivation, are controlled by human beings and, may be used to change the quality of the wine. These practices involve very varied factors as for example those based on decisions affecting the orientation vineyards' rows, their training system, density, the calendar for pruning, trimming, fungicide treatments, or the way in which soil surface is managed, which includes its tillage, the manipulation of the canopy structure, the nitrogen fertilization [13]. Additionally, vineyard cultivation may be affected by sunlight and water, among others, which will determine the final quality of berries since they are very sensitive to the microclimate. All the factors could change year by year, so the potential aroma of the wine depends on the year.

The amount of light absorbed by the vine leaf area determines the rate of photosynthesis, hence its metabolism. Therefore, light influences the development of grape aroma compounds. This relation is complex, as direct sunlight induces beneficial changes in photosynthetic pigment levels, but can cause stress by dehydration or temperature increase. Excessive sunlight exposure may burn the grapes' skin, resulting in a bitter flavor and affecting the quality of the wine [9]. Several studies reflect that sun exposure plays an important role in the variations of the volatile composition of grapes [14–16]. Some authors [17, 18] reported that high levels of terpene precursors are produced in over-ripe grape. The concentration of free and bound terpenes can also be increased by exposure of grapes to the sun [19, 20].

Canopy management is also important. Limiting canopy growth by removing leaves and vine shoots allows the increment of glucose production in the fruit, improving fruit's sunlight exposure. Nowadays, different ways for canopy structure management have been developed. The most common one is the plucking of leaves. It improves the microclimate of the clusters, reduces grape rot (caused by different pathogenic bacteria, yeast and, fungi, such as *Botrytis cinerea*) and provides better fruit maturation since less shadow is generated and enhances the pass of the sunlight. Another technique is head trimming which consists of cutting growing shoot tips in order to remove non-photosynthetic leaves. This technique reduces transpiration and induces the lignification of the plant, balances the development of branches and increases airflow and insulation within the foliage [13]. Additionally, the removal of leaves may increase the content of total and phenol-free glycosides (one group of compounds that influence the aroma) [21]. In fact, wines obtained from defoliated grapes usually present higher fruity notes [22, 23]. Leaf-plucked grapes show the highest values of terpene and norisoprenoids compounds in Tempranillo, Merlot, and Gewürztraminer varieties where the synthesis of precursors is higher in warmest years and in grapes with more exposure to the sun [13].

As it has been previously described, soil and weather have an important effect on wine aroma, and this is mostly mediated by water. In general terms, water deficit has been proved to improve wine quality, enhance color, flavor and/or aroma [24]. However, no definitive scientific results can prove that the lack of water has such direct and positive impact on the quality since the effect of water depends on the species. For instance, during the maturation of Riesling grapes, the accumulation of terpene precursors is lower under irrigation [25]. In Tempranillo grapes, a positive correlation between irrigation and aroma potential at harvest has been detected [26, 27]. In Agiorgitiko grapes, precursors' synthesis is promoted by the limited availability of water [28]. There are several studies about the impacts of irrigation on fruit composition and wine sensory properties. However, the majority of these studies have been conducted in arid regions where irrigation is considered necessary and only a few have been carried out in humid regions where irrigation is mostly not necessary [29].

Among the nutritive compounds of must, the amino acids are of great importance. They can serve as precursors of esters and also may play an indirect role in supporting cellular metabolic activities [30]. Consequently, wine growers can also modify soil fertilization to change the aromatic profile of the resulting wine. The nitrogen soil composition can lead to excessive vine vigor and susceptibility to gray rot, but can also enhance aroma expression since yeasts use that same nitrogen [31]. Nevertheless, when juice presents a sufficient amount of nitrogen, the addition of amino acids does not improve the volatile composition of wine, therefore the excess of the amino acids will probably be consumed by other cellular processes [30].

As wine quality strongly depends on the grape quality, it is necessary to process healthy grapes at the correct ripeness stage in order to obtain high-quality wines. For this reason, winegrowers have to be especially careful in the prevention of parasite attacks on the grapevine. Fungicide treatments are applied to prevent diseases such as botrytis, powdery mildew or downy mildew, which cause serious problems in the production. However, these fungicides can pass to wines at trace levels and can affect the aroma, and also the safety of the wine [9]. In addition, pesticide residues on grapes can be transferred to the must and influence the selection and development of yeast strains. On the other hand, yeast can reduce the levels of the pesticides in the wine or adsorb them on the dregs of the wine. The persistence of pesticides depends on several factors such as the chemical characteristics of the active ingredients, photodegradation, thermodegradation, co-distillation and enzymatic degradation [32]. However, more studies are needed to orient the preferences of winegrowers for one or another fungicide.

2. Determination of aroma compounds

The analysis of the food aroma can be performed through tastings or instrumental methods, being necessary to correlate the latter with the former (**Figure 2**). Traditionally, sensory analysis was used to carry out the aroma determination. However, this method has the disadvantage (**Table 1**) of being subjective, and so,

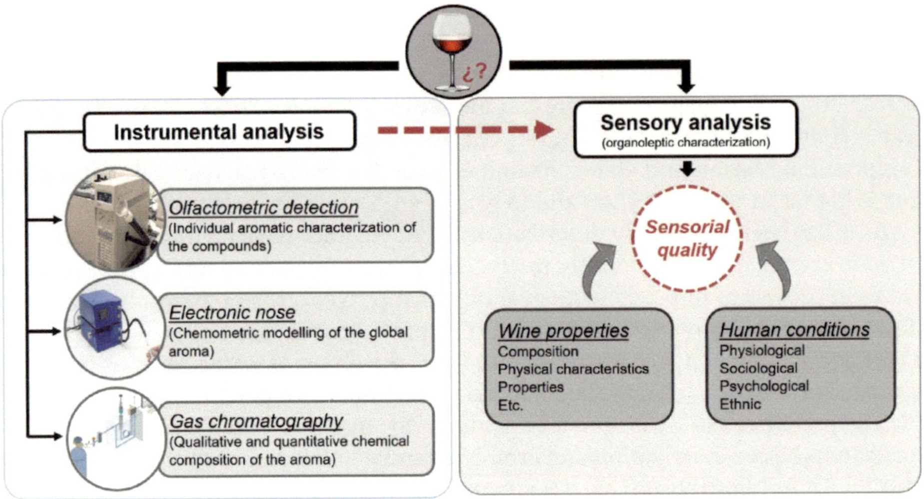

Figure 2.
Different techniques that can be applied to wine aroma analyses.

Advantages		Disadvantages	
Sensory	**Instrumental**	**Sensory**	**Instrumental**
Direct	Objective	Subjective	Indirect
Few material	Low quantity of sample	High quantity of sample	Expensive equipment
More realistic	One person	Several persons	Need of sensory analysis for realistic results
	Repeatable	Non repeatable	
	Can be toxic or no innocuous	Innocuous	

Table 1.
Advantages and disadvantages of each type of analysis methodology.

numerous regulations and protocols were elaborated to standardize the methodology and aromatic descriptors. The obtained results were significant but never distinguishable by the subjectivity of the taster. In this regard, technological advances that allow the investigation of the compounds responsible for the aroma and aromatic profile of a wine have been carried out. Nonetheless, there is no analytical method that allows interpreting the perceptions received during the tasting [33].

As mentioned before, the aroma of wines is complex and more than 800 volatile compounds have been identified, some of them at the ng/L level. Because of that complexity, the determination of aromatic compounds is not simple and it is difficult to isolate a specific aroma character. However, the development of techniques such as gas chromatography coupled to mass spectrophotometry or electric nose has allowed to identify hundreds of them [34].

Sensory methods and gas chromatography are analytical methodologies often slow and expensive. Developing new analysis techniques to solve these limitations is of great interest. This could be the case of the electronic nose (e-nose), characterized by its high sensitivity, speed, low cost and minimal, or no sample preparation [35]. Therefore, there are complementary tools that in combination will provide a lot of information regarding the evaluation of aroma [33].

2.1 Sensory analysis

The sensory analysis includes wine tasting, its sensory estimation and appreciation, and its description. In this process, the subjectivity of the individual towards the sensations, emotions and memories that can awaken on him a certain smell (affective aspect) is very involved. Subjectivity is present because two stages coexist in the process: chemistry (biochemical reactions between the aroma/taste molecules and our papillae) and psychology (electrical signals processed in the brain that assigns a description) [36].

To avoid subjectivity in the obtained results, it is necessary to standardize the process and establish a series of standards. Various levels such as CEE, OIV, ISO, etc., have described regulations to carry out the tasting [37].

Even though sensory analysis does not provide information on the chemical composition of the aroma, it is useful for the search of certain sensory alterations, detection of adulterations or the typification of the product according to its origin and variety. However, the incorporation of instrumental techniques (olfactometric detection, electronic nose, etc.) requires sensory analysis for calibration and interpretation [37].

2.2 Instrumental analysis

Despite the complexity of the human smell and the study of wine's aroma, various techniques have provided valuable and complementary information on different aspects of aromas. Those include gas chromatography (qualitative and quantitative chemical composition of the aroma), olfactometric detection (individual aromatic characterization of compounds) and electronic nose (chemometric modeling of the global aroma). This analysis allows the quantitative determination and the qualitative evaluation of the aromatic compounds, as well as an instrumental interpretation of the overall aroma. It is necessary for the winemaker to know these techniques when seeking answers to their needs [35].

One of these techniques is gas chromatography (GC), which allows the identification and quantification of the volatile compounds of the wine. This system has been improved through stages of extraction and concentration of the volatile compounds, chromatographic separation and detection systems, such as mass spectrometry (MS) or flame ionization (FID) [38].

There is another technique coupled to the chromatographic analysis called olfactometric detection or sniffing (gas chromatography-olfactometry, GCO), which allows to determine the aromatic profile of a wine. In this assay, a person smells the compounds after they have been separated and eluted from the column. Thus, each peak is associated with an olfactory descriptor. Aromas are classified as active or inactive based on the value of aromatic activity (OAV) (This value allows measuring the aromatic activity of a compound (C) in a specific matrix as far as its activity is above the detection threshold (DT)). Thus, active aromas possess numerical OAV greater than one [39].

$$OAV = \frac{C}{DT} \qquad (1)$$

This assay presents some limitations, for example, the evaluation of the compound β-damascenone, which is overestimating due to its very low detection threshold with GCO, its wide range and its dependence on the composition of the medium [34]. Nevertheless, this technique has been one of the most used in the last decade for overall identification of odor-active compounds in all types of wine (white and red musts, young and aged wines, sweet wines) made from different varieties of grapes such as Chardonnay, Riesling, Gewürztraminer, Merlot, Cabernet Sauvignon, Grenache, Tempranillo, Zalema, Palomino Fino, Touriga Nacional, Aragonez or Trincadeira. It can also be applied to study the sensory profiles of wines produced with sound and sour rot affected grapes and compare them to understand the role of sour rot in the odor nuances of wines [40].

Recently, the electronic nose has been introduced in the wine industry. It consists of an instrument equipped with chemical sensors and a chemometric model recognition program, capable of identifying and comparing individual or complex odors. As its main objective is to obtain results comparable to those from the human olfactory system, the aim of this method is to relate the perceived aroma with a response that, after being stored in memory, will serve as a model in further analysis. It has been displayed as a useful tool due to the advantages it offers: short analysis time in chromatography position (5–10 min), continuous control, it is a non-destructive method and it does not require qualified personnel. However, it is limited by the effectiveness of the detectors [41]. Most of its applications are related to the discrimination of wines to prevent their adulteration or detection of disagreeable odors, but only a few of them consider the identification of the quality of wine aromas. Despite all, this system allows a good classification of typical red and white wine aromas [42].

Finally, along the scientific literature, it has been described as an innovative technique of aroma determination. It consists of an array of conducting polymer sensors coupled to a selective solid-phase micro-extraction (SPME) fiber. This assay allows carrying an analysis of the principal components, differentiating the aromas of the sample, even for wines with very similar sensory characteristics. Moreover, the response is fast and consistent. The selective adsorption of the fiber provides a better distinction, increasing the concentration of the minor compounds of an aroma [43, 44].

3. Types of aroma

Despite the massive amount of aromatic compounds present in wine, not all of them contribute to the perceived aroma, since some of them are found in concentrations below the perception threshold. Compounds that exceed this concentration are called active compounds. Aromas can be classified in different ways according to the parameter considered. In this way, a classification can be made based on its presence (basic, subtle and special) or according to its origin or sequence of wine production as described in **Figure 3**. In the last case, the classification based in the sequence of wine production allows to differentiate the aromas accordingly to its process as primary, secondary and post-fermentative aroma [9, 45], facilitating the inference of wine makers. The following paragraphs will describe the different stages of wine production and the aromas originated during the process.

3.1 Primary aroma

The primary aroma is formed by the varietal aromatic constituents. Three large groups of compounds can be distinguished: the free varietal aroma, the precursors of varietal origin (non-volatile or non-odorous precursors and odorous volatile compounds) and the substances that are formed from the precursors [46, 47].

3.1.1 Free varietal aroma

This type of aroma distinguishes the different varieties of grapes. Although relatively few studies have been able to identify a compound responsible for the

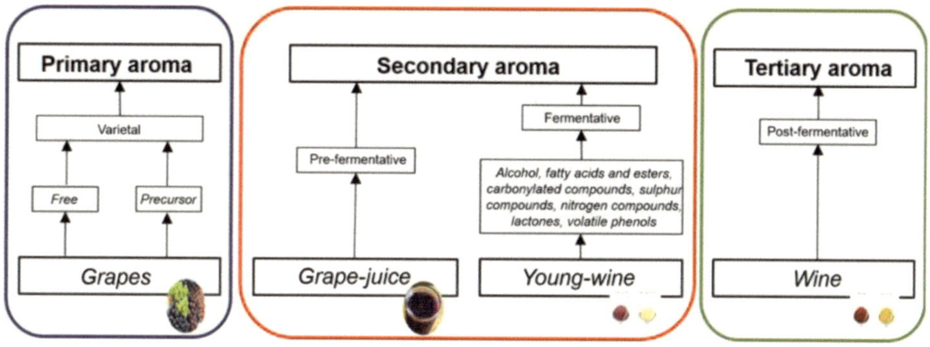

Figure 3.
Classification of wine aroma according to its origin along wine production.

Compound	Chemical formula	Chemical structure	Olfactory descriptor
3-Isobutyl-2-methoxypyrazine	$CH_2CH(CH_3)_2$	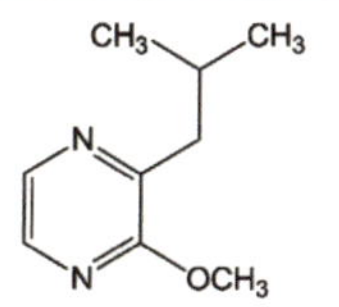	Green peppers
2-Methoxy-3-sec-butyl-pyrazine	$CH(CH_3)$ CH_2CH_3	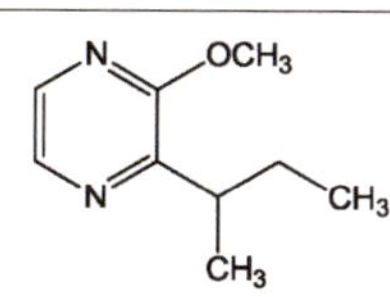	Green peppers
2-Methoxy-3-(2-methylpropyl) pyrazine	$CH(CH_3)_2$	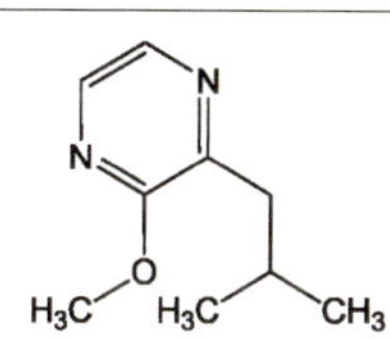	Green peppers

Table 2.
Principal pyrazines found in wines.

varietal character, notable difference between varieties have been reported. This difference is therefore due to the overall aromatic profile [48]. Among the compounds that determine the free varietal aroma, determined by volatile substances linked to the aromatic typicity of the variety, two chemical families are distinguished: pyrazines and terpenes [46].

3.1.1.1 Pyrazines

These compounds (**Table 2**) were first identified in Cabernet Sauvignon grape variety. They are nitrogen compounds derived from the catabolism of some amino acids such as leucine, isoleucine, valine and glyoxal. They are usually found in concentrations below the perception threshold. Its concentration has a positive correlation with the herbaceous note of some wines such as Cabernet Sauvignon and Sauvignon Blanc [46, 49]. The concentration of pyrazines has been estimated in different grape varieties, such as Sauvignon Blanc (3 ng/L), Semillon (2 ng/L), Cabernet Sauvignon (2–24 ng/L) [50]. Regarding wine, pyrazines content has been estimated at about 1 ng/L in white wines, while in red wines, the concentration reached 10 ng/L [51]. Several factors have been described to influence the pyrazines content in grapes, especially grape variety and maturation. The degree of grapes' ripeness influences their content, being pyrazines' content inversely proportional to this factor. Thus, its content decreases appreciably from summer and disappears practically under optimal conditions of maturation. Other factors have been described to influence pyrazine content, such as temperature and irradiation of vineyard [46, 50]. The soil also plays a significant role in pyrazine levels. A higher amount of pyrazines has been found in vineyards grown in limestone and clay soils than in sandy soils [49].

3.1.1.2 Terpenes

Within this group, very abundant in the plant kingdom, are monoterpenes (formed by 10 carbon atoms), sesquiterpenes (15 carbon atoms) and the corresponding alcohols and aldehydes [52].

They are the most studied odoriferous compounds found in *Vitis vinifera*, having identified around 70 compounds in both wines and grapes. They can be found as free forms or as odorless precursors, mainly glycosylates. The most odoriferous monoterpenes are monoterpenic alcohols like linalool, α-terpineol, nerol, geraniol, citronellol and ho-trienol (**Figure 4**), which provides floral aromas (rose, lily, citronella, linden, etc.). These floral attributes are characteristic of white wines [49, 53]. In addition, the content in terpenoles is determined by the state of the grape since they are very sensitive to the attack of *Botrytis cinerea* [54].

Both monoterpenoles and sesquiterpenes are synthesized from isopentyl pyrophosphate (IPP) and dimethylalkyl pyrophosphate (DMAPP). IPP and DMAPP precursors are produced through the cytosolic mevalonic acid (MVA) pathway (from three molecules of acetyl-CoA) or through 2-C-methyl-D-eryritol-4-phosphate plastidial (MEP) pyruvate and glyceraldehyde-3-phosphate [48].

Regarding the location of these compounds in grapes, more than 50% are concentrated in the solid part (pulp and skin), reaching concentrations of 90% of geraniol and nerol in these parts. In contrast, half of the linalool is in the juice; thus,

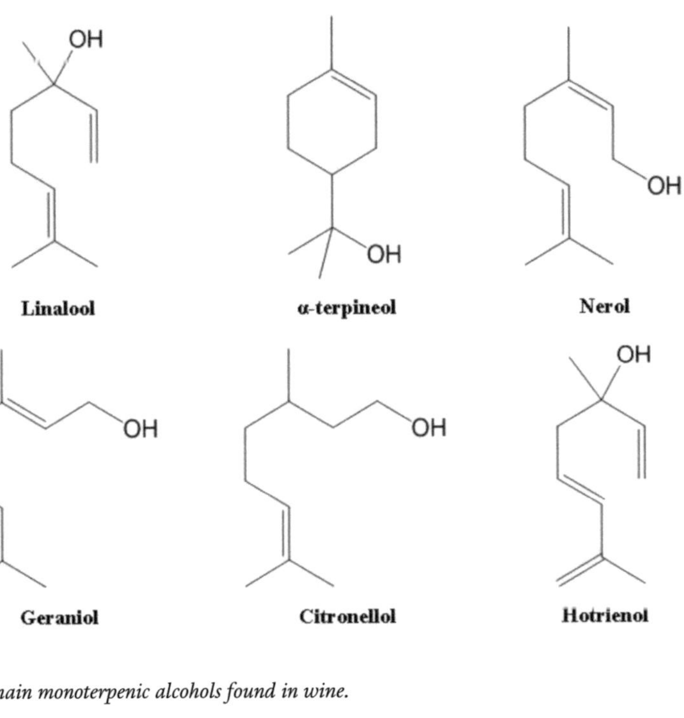

Figure 4.
Structure of the main monoterpenic alcohols found in wine.

processes that involve an increase in the exchange of solid and liquid parts (e.g. maceration) have important implications on the final aromatic characteristics of the wine [46].

3.1.2 Precursors of varietal origin

Although some precursors do not possess odoriferous characteristics, they can give rise to odoriferous substances. These are monoterpenes, diols or terpene polyols, fatty acids, carotenoids, glycosylated precursors of aroma and volatile phenols [55].

3.1.2.1 Monoterpenes

This group of compounds is one of the most studied in wine and includes a wide variety of compounds. The formation of these molecules is mainly due to the oxidative metabolism of linalool in grapes [56]. Monoterpenols are sensitive to the hydration and oxidation reactions that occur during winemaking and cause the transformation of one into another [57, 58]. Some yeasts can increase the content of this type of compound. This is the case of *Issatchenkia* spp. [59].

3.1.2.2 Diols

These compounds are characterized by transforming at relatively acidic pH, such as those found in musts or wines. Some of these compounds thus obtained are fragrant, but may be the cause of transfers of strange aromas to the wines [46]. Red grapes are not very rich in these compounds, but their action on the aroma is not improbable if the olfactory threshold is taken into account, since it is very low in some of its derivatives, either isolated or well mixed [8].

3.1.2.3 Carotenoids

Carotenoid content decreases throughout ripening, with a higher content in the grapes exposed to the shade than those exposed to the sun. In grapes grown in high altitude, the content is also lower, due to the low temperatures and higher humidity [60–62]. Its content in grapes ranges between 15 and 2.000 μg/kg. Lutein and β-carotene stand out as the most abundant, as well as neoxanthin, flavoxanthin and others in smaller quantities [63].

Carotenoids are not found in grape juices and in wines made without maceration, as they are degraded during the breaking of the grape and the vinification. Light and oxidases are capable of degrading carotenoids into smaller fragments, more soluble and more fragrant. Among the compounds that are formed in the decomposition of carotenoids, norisoprenoids are worth mentioning because they have low perception thresholds that make them play an important role in the aroma of wine. This degradation can be direct or with an intermediate step that is the formation of glycoconjugates, which can then release their volatile aglycone during fermentation through enzymatic and acid hydrolysis processes [64, 65].

3.1.2.4 Glycosylated precursors

All grape varieties have the same glycosylated derivatives, being Moscatel varieties the most concentrated. They appear in greater content in the skin than in the pulp or juice. These compounds are four types of glycosides: one monoglycoside (β-D-glucopyranoside) and three diglucosides

(*O*-α-L arabinofuranosyl-β-D-glucopyranoside, 6-*O*-α-L rhamnosyl-β-D-glucopyranoside and 6-*O*-β-D apiosyl-β-D-glucopyranoside) [66].

Although grapes contain enzymes (β-glycosidases) capable of releasing some of these compounds during winemaking, under normal winemaking conditions, glycosylated precursors have a poor effect on the development of the aroma since its optimal activity is developed in other pH values [67].

3.1.2.5 Volatile phenols

These compounds can be responsible for the originality or wine aroma defects. Phenolic compounds are non-odorous compounds, which can be submitted to different enzymatic reactions, transforming into volatile phenols and contributing to unpleasant aromas of pharmacy, smoke, forest, leather or pepper. These compounds are characteristic of carbonic maceration wines, although they are present in other wines [68].

Among phenolic acids, caffeic, ferulic and *p*-coumaric are found in the pulp and skin in the form of tartaric esters (caftaric acid, feruloyl tartaric acid) and coutaric (*p*-coumaroyl tartaric). Its olfactory threshold is very low and they strongly decrease during the ripening of the grape, with large variations within the same variety [46].

During vinification, free acids and then volatile phenols are formed. The process of hydrolysis causes the appearance of other phenolic compounds such as vanillin, methyl vanilla or homovainyl alcohol [69].

3.2 Secondary aroma

3.2.1 Pre-fermentative aroma

This type of aroma is developed during the winemaking process, more specifically, between the harvest of the grapes and the beginning of the fermentation. The mechanical processes performed during this stage, such as grape transport, crushing, de-stemming, pressing and even carbonic maceration have a great influence in the pre-fermentative aroma. All these processes involve grape cellular rupture, allowing enzymes to come into contact with the aroma precursors. Thus, the availability of aroma precursors is directly proportional to the degree of grape rupture. Moreover, the mechanical process also favors the incorporation of oxygen, leading to enzymatic oxidation reactions. Throughout the process of crushing grapes, relatively large amounts of aldehydes and alcohols of 6 carbon atoms are formed. In some cases, these compounds give the smell of cut leaves, bitter taste and low olfactory threshold. The formation of C6 aroma compounds varies depending on the ripeness of the grape, although the greatest potential occurs before the ripening date, where grape lipids levels are higher and also depends on the type of variety [70].

3.2.2 Fermentative aroma

Fermentation is the main phase in the transformation of grape juices into wine, with two biological transformations occurring at this stage: alcoholic (yeast transform sugars into alcohol and other secondary products) and malolactic fermentation (lactic acid bacteria cause the degradation of malic acid in lactic acid) [71].

While malolactic fermentation subtly modifies the aroma of wine, alcoholic fermentation is of great importance in the aroma since it is responsible for the winey note that constitutes the common aromatic base for all wines. Moreover, volatile constituents formed during fermentation (**Table 3**) will quantitatively represent most of the aroma constituents [72]. Malolactic fermentation is not interesting in all situations. For example, in white wines it is interesting to avoid it since this type of fermentation leads to a loss of freshness and varietal aromas, which are desirable characteristics in this type of wine.

3.2.2.1 Alcohols

Main alcohols synthesized in alcoholic fermentation are 2- and 3-methylbutanol, 1-propanol, 2-methylpropanol, 1-butanol, 1-pentanol, 2-phenylethanol, 3-methylthio-propanol, tyrosol and tryptophol, being its average total content between 400 and 500 mg/L. These compounds are produced at the metabolism of amino acids, so their concentration depends on the nitrogen content of the grape juice, the yeast species and the factors that influence their development [73].

3.2.2.2 Fatty acids and esters

Together with alcohols, these compounds are the main markers of the fermentative aroma. Fatty acids are formed as a result of the interruption of the process of elongation of acetyl-CoA by hydrolysis. Regarding esters, most of them are formed enzymatically, by the initial activation of fatty acids combined with coenzyme A (CoA) [74]. These compounds are usually found in low concentrations, but above the threshold of perception. They are also pleasant aromas (fruit mainly) except ethyl acetate (over 100 mg/L) [75].

3.2.2.3 Carbonylated compounds

These compounds arise from the metabolism of yeasts by decarboxylation of the α-ketone acids formed in the biosynthesis of higher alcohols. They remain in the wine in small quantities, due to the strong reducing activity of yeast during fermentation. Sulfur dioxide and alcohols react to the pH of the wine with the carbonylated compounds, forming sulphydric combinations [69].

3.2.2.4 Sulfur compounds

These are compounds that generally provide intense and unpleasant odors. Within this group are thiols, sulfides and thioesters. They can be classified according to their weight in compounds of low molecular weight (sulfur functional group predominantly on odor) or high molecular weight (complex participation in aroma due to its analogy with higher alcohols) [69].

3.2.2.5 Nitrogen compounds

The most abundant nitrogen compounds are acetamides from primary amines and their corresponding amines. Acetamides have a smell similar to uric acid. They are produced by *Bacterium manitopeum* [76].

3.2.2.6 Lactones

They are compounds related to hydroxy acids because they are obtained by intramolecular esterification of these compounds: 4-hydroxy acids lead to γ-lactones and 5-hydroxy acids to δ-lactones; both are oxygenated heterocycles [69].

Compound	Molecular formula	Chemical structure	Olfactory descriptor	ODT (mg/l)
A. **Alcohols, acids and esters**				
Ethanol	C_2H_6O	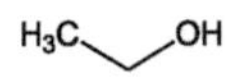	Alcohol	0.150
Ethyl acetate	$C_4H_8O_2$	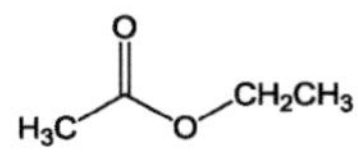	Fruity	0.605
Isoamyl alcohol	$C_5H_{12}O$	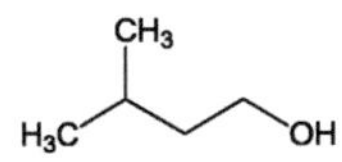	Almond Cream, fusel oil	30
Isoamyl acetate	$C_7H_{14}O$	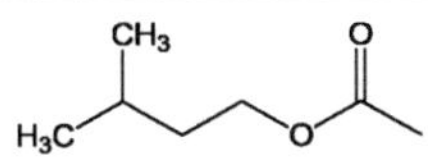	Banana	0.030
Chain R = H 2-phenylethanol Chain R = Acetate; 2-Phenylethyl acetate	$C_8H_{10}O$	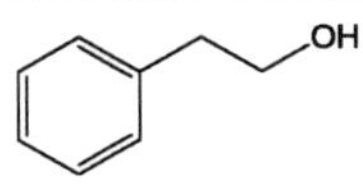	Roses	14
Acetic acid	$C_2H_4O_2$	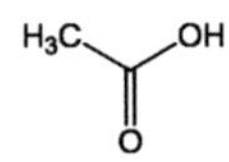	Vinegar	2.3
Chain R = H; Hexanoic acid Chain R = Ethyl; Ethyl octanoate		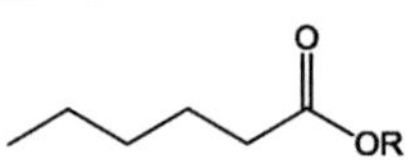	Sweat, sour Pineapple	0.420 0.014
Chain R H; Octanoic acid Chain R = Ethyl; Ethyl octanoate		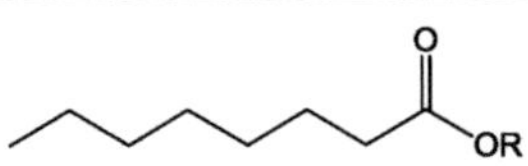	Butter Fruit	2.200 0.005
Chain R = H; 3-methylbutanoic acid Chain R = Ethyl; Ethyl 3-methylbutanoate		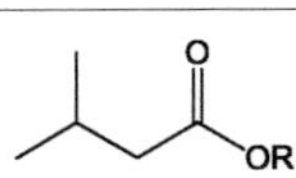	Cheese Fruit	0.250 0.003
Phenylacetic acid		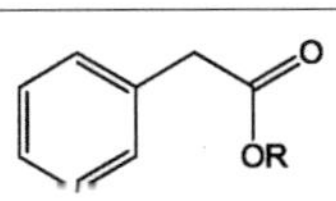	Animal Rose	1 1.8
Chain R = H; lactic acid Chain R = Ethyl; Ethyl lactate		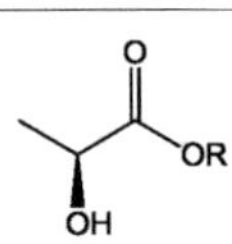	Sour milk Milk	0.065 150
Chain R = H; Succinic acid Chain R = R = Ethyl; Ethyl succinate		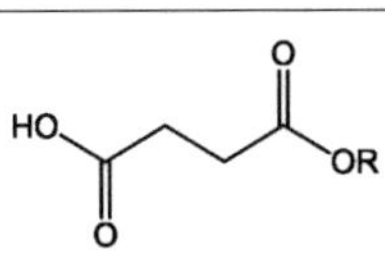	Cantaloupe	No data 1.2
B. **Carbonylated compounds**				
Acetaldehyde	C_2H_4O	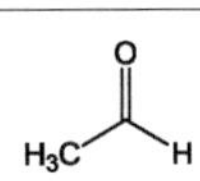	Last apple	0.090
Phenylacetaldehyde	C_8H_8O	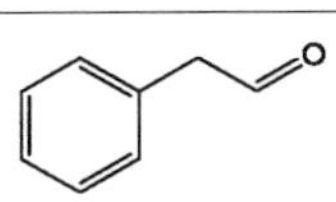	Floral	No data

Compound	Molecular formula	Chemical structure	Olfactory descriptor	ODT (mg/l)
Diacetyl	$C_4H_6O_2$	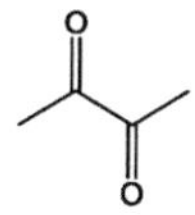	Butter	0.100
Acetoin	$C_4H_8O_2$	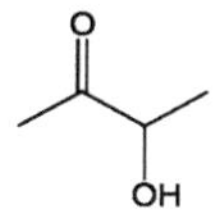	Butter	150
C. **Sulfur compounds**				
Hydrogen sulfide	H_2S	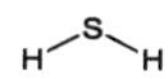	Rotten egg	0.016
Methanethiol	CH_4S	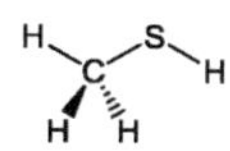	Light taste	1
Methional	C_4H_8OS	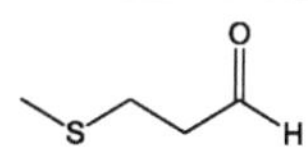	Cooked potatoes	0.200
Methylthiophene-3-one	C_5H_6OS	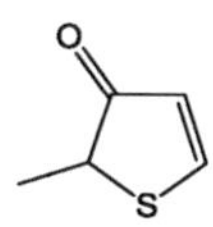	Breadcrumbs	No data
D. **Nitrogen compounds**				
2-phenylethylamine	$C_8H_{11}N$	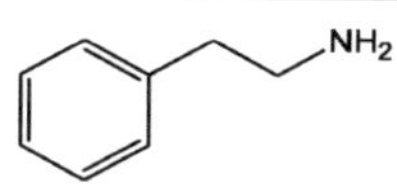	Rot	No data
N-(2-phenylethyl)-acetamide	$C_{10}H_{13}NO$	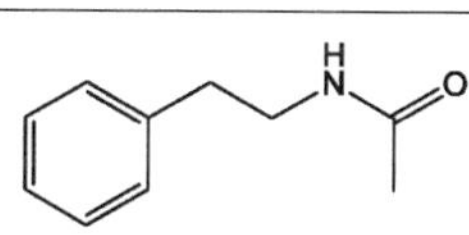	Similar to uric acid	6
E. **Lactones**				
γ-nonalactone	$C_9H_{16}O_2$	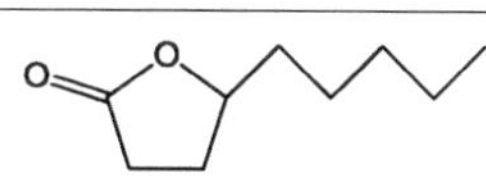	Coconut nut	0.025
Δ-decalactone	$C_{10}H_{18}O_2$	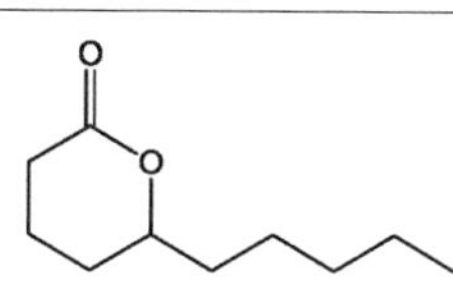	Peach	0.110
γ-butyrolactone	$C_4H_6O_2$	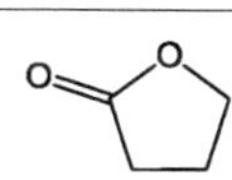	Coconut nut	0.400
F. **Volatile phenols**				
4-vinylphenol	C_8H_8O	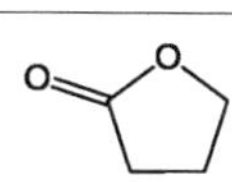	Nail	0.375
4-vinylguaiacol	$C_9H_{10}O$	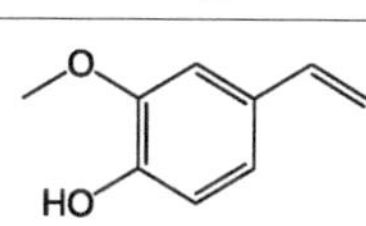	Clove, roasted almonds	0.006

Compound	Molecular formula	Chemical structure	Olfactory descriptor	ODT (mg/l)
4-ethylphenol	$C_8H_{10}O$	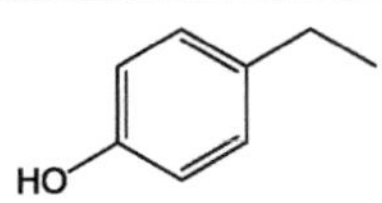	Medicinal horse manure	0.620
4-ethylguaiacol	$C_9H_{12}O_2$	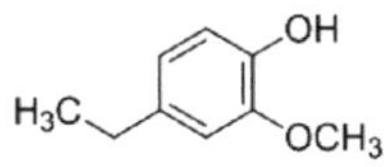	Wood	0.05

Olfactory detection threshold (ODT).

Table 3.
Relevant aroma compounds produced during wine fermentation [77, 78].

The most abundant lactones areγ-butyrolactone, 4-carbethoxy-8-butyrolactone and 4-ethoxy-γ-butyrolactone.

3.2.2.7 Volatile phenols

The main volatile phenols produced by *Saccharomyces cerevisiae* yeasts are 4-vinylphenol and 4-vinylguaiacol. Within this group is also 4-ethylphenol, which is produced by yeasts belonging to the genus *Brettanomyces/Dekkera*. It is produced by enzymatic decarboxylation of *p*-coumaric and ferulic acids during alcoholic fermentation [69].

3.3 Post-fermentative aroma

This type of aroma includes all volatile compounds formed throughout the aging of the wine, giving rise to what is called "bouquet". During the conservation of the wine, physicochemical and biological reactions transform the aromatic compounds produced in previous stages, so post-fermentative aroma suffers deep changes [79]. Depending on the type of aging, two types of bouquet can be distinguished:

- The bouquet of oxidation takes place in wooden barrels. It is produced by the synthesis of acetaldehydes and/or acetals, as well as by the extraction of a series of nuances from wood and the diffusion of oxygen through the wood pores [79].

- The bouquet of reduction is characteristic of aging in the bottle, determined by the interaction between the wine compounds synthesized throughout the fermentation [80].

During this stage, the fruity aroma of the wines disappears progressively, evolving towards a more complex aroma. These reactions are slow and limited, sensibly manifesting in the wines after several years.

In this type of wines, there are compounds given by wood such as furans (furfural, 5-methylfurfural, 5-hydroxymethylfurfural) which are responsible for the aromas of roasted almonds [81], phenolic aldehydes (vanilla), phenyl ketones (vanilla aroma), oxygenated heterocycles (caramel aroma), volatile phenols (toasted smell and burnt wood) [82], 4-vinylphenol (carnation aroma), 4-vinylguaiacol (clove aroma), whiskey-lactones (coconut aroma) [83].

However, some compounds are transformed. During aging, the concentration of monoterpenic alcohols in wine from grapes (linalool, geraniol and citronellol) decreases. On the other hand, the concentration of isomeric oxides, whose olfactory thresholds are higher than the starting alcohols is increased. Hence, there is a loss of varietal aromatic load [84].

4. Future perspectives

The complexity of wine's aroma has been considered more an art than a scientific fact since little knowledge of the mechanisms involved in the process is available. However, in the last decades, this situation is changing as more and more wine growers turn to science to improve their production. In the nearly future, wine-related genomic, proteomic and metabolomic research will be applied to *Vitis vinifera* cultivars (the important ones from a commercial point of view), to other *Vitis* species, as well as to the yeast strains employed in fermentation [9]. Nowadays, scientific research helps, for example, to understand the impact of wine microorganisms and their derivatives on varietal aromas. This allows winemakers to generate different wines with different characteristics (including aromatic ones) from the same matrix (same grapes), using different species of microorganisms.

The aromatic quality of wines in the near future will depend on climate change and the effect of greenhouse gases. The increment of these gases will carry an average raise on global surface temperatures, evaporative demand, and the frequency and intensity of drought. Studies about changes in the volatile composition, phenolic content and antioxidant activity of wines produced at elevated CO_2 concentrations are very limited, so it will be necessary to optimize the quality of wine in a future scenario of climate change [9]. Climate change will also allow to cultivate vineyards in not cultivable areas until now. For instance, the viticulture sector of United Kingdom has undergone a rapid growth (148% during 2004–2013), in part attributed to a warmer temperature in the areas of England and Wales (13–15°C), deemed suitable for cool-climate viticulture [85].

There is a future concern about that wines in the future could being "manipulated" by using chemical additives to add enhance complexity and additional aromas to wine such as created a manufactured perfume. In fact, there have already been cases. In 2004, a South-African winery was found to have added illegal flavoring to their Sauvignon blanc to enhance the aroma [86]. Increasing the knowledge about the mechanisms and compounds responsible for wine aroma helps to develop new methods to allow the improvement of the quality of common wines to resemble high-quality wines. However, any artificial manipulation of wine aroma is currently forbidden, and there is a significant number of winemakers and consumers that prefer more natural wines, without or with minimum chemical manipulation. The wine is widely consumed around the world by its nice sensory taste and its natural origin from fermented grape juice. Change this view could be dangerous to the wine industry and can lead to a significant noteworthy wine devaluation towards by the wine consumers and connoisseurs.

5. Conclusions

Despite the plenty of advances for understanding how the grape berry is developed and which chemical components are important for wine aroma and flavor, there is still too much to do. The advances that are being carried out in different fields such as the determination of aromas and their evolution will allow to continue expanding this knowledge. Nowadays, the quality of wines has improved due to many of the mechanisms involved in the quality of a wine are known. Thus, manipulation of grapes in vineyards through different production practices in order to improve wine's quality is possible.

Acknowledgements

The research leading to these results was funded by FEDER under the program Interreg V Spain-Portugal (POPTEC, ref. 0377-Iberphenol-6-E); by MICINN supporting the Ramón&Cajal grant for M.A. Prieto (RYC-2017-22891); by Xunta de Galicia and University of Vigo supporting the post-doctoral grant of M. Fraga-Corral (ED481B-2019/096), the pre-doctoral grants for A.G. Pereira (ED481A-2019/0228) and P. García-Oliveira (ED481A-2019/295); by Axudas Conecta Peme (Xunta de Galicia) supporting the IN852A 2018/58 NeuroFood Project and AlgaMar (www.algamar.com) for supporting the pre-doctoral grant for C. Lourenço-Lopes; NANOEATERS Project (0181_NANOEATERS_01_E) for supporting the pre-doctoral work of C. Jimenez-Lopez; EcoChestnut Project (Erasmus+ KA202) for supporting the work of M. Carpena; Ibero-American Program on Science and Technology (CYTED - AQUA-CIBUS, P317RT0003) for financial support. This project has received funding from the Bio Based Industries Joint Undertaking (JU) under grant agreement No 888003 UP4HEALTH Project (H2020-BBI-JTI-2019), the JU receives support from the European Union's Horizon 2020 research and innovation program and the Bio Based Industries Consortium.

Author details

Antia G. Pereira[1,2], Maria Fraga[1,2], Paula Garcia-Oliveira[1,2], Maria Carpena[1], Cecilia Jimenez-Lopez[1,2], Catarina Lourenço-Lopes[1], Lillian Barros[2], Isabel C.F.R. Ferreira[2], Miguel Angel Prieto[1*] and Jesus Simal-Gandara[1*]

1 Nutrition and Bromatology Group, Analytical and Food Chemistry Department, Faculty of Food Science and Technology, University of Vigo, Ourense, Spain

2 Centro de Investigação de Montanha (CIMO), Instituto Politécnico de Bragança, Bragança, Portugal

*Address all correspondence to: mprieto@uvigo.es and jsimal@uvigo.es

References

[1] Gutiérrez Maydata A. Vino, Polifenoles y Protección a la Salud. Revista Cubana de Alimentación y Nutrición. 2002;**16**(2):134-141

[2] OIV—Organización Internacional de la Viña y el Vino. Código Internacional de Prácticas Enológicas. 2016

[3] Jackson DI, Lombard PB. Environmental and management practices affecting grape composition and wine quality—A review. American Journal of Enology and Viticulture. 1993;**44**:409-430

[4] Noble AC. Wine analysis and production. Food Quality and Preference. 1996;7(2):145

[5] Wang R, Sun Q, Chang Q. Soil types effect on grape and wine composition in Helan Mountain area of Ningxia. PLoS One. 2015;**10**(2):e0116690

[6] CSIC. Curso de Análisis Sensorial De Alimentos. Vol. 1. Madrid, Spain: CSIC; 2011. p. 79

[7] Flanzy C. Oenologie, fondements scientifiques et technologiques. In: Sciences et Techniques Agroalimentaires. France: Tec & Doc Lavoisier; 1998. ISBN: 2-7430-0243-3

[8] Ribéreau-Gayon P, Dubourdieu D, Donèche B, Lonvaud A. Traité d'oenologie - Tome 1 - 6e éd. - Microbiologie du vin. Vinifications. France: DUNOD; 2012. ISBN: 2100588745

[9] González-Barreiro C, Rial-Otero R, Cancho-Grande B, Simal-Gándara J. Wine aroma compounds in grapes: A critical review. Critical Reviews in Food Science and Nutrition. 2015;**55**:202-218

[10] Franco E. Aspectos vitícolas en la calidad del vino. ACE: Revista de Enología. 2013;**137**:48-56

[11] Início - Vine to Wine CircleVine to Wine Circle | Portal da vinha e do vinho Jorge Böhm PLANSEL. Available from: http://www.vinetowinecircle.com/

[12] Ruiz VS. Avances en Viticultura en el Mundo. Revista Brasileira de Fruticultura. 2011;**33**(SPE1):131-143

[13] Hernandez-Orte P, Concejero B, Astrain J, Lacau B, Cacho J, Ferreira V. Influence of viticulture practices on grape aroma precursors and their relation with wine aroma. Journal of the Science of Food and Agriculture. 2015; **95**(4):688-701

[14] Bureau SM, Razungles AJ, Baumes RL. The aroma of Muscat of Frontignan grapes: Effect of the light environment of vine or bunch on volatiles and glycoconjugates. Journal of the Science of Food and Agriculture. 2000;**80**(14):2012-2020

[15] Belancic A, Agosin E, Ibacache A, Bordeu E, Baumes R, Razungles A, et al. Influence of sun exposure on the aromatic composition of chilean Muscat grape cultivars Moscatel de Alejandria and *Moscatel rosada*. American Journal of Enology and Viticulture. 1997;**48**: 181-186

[16] Zhang H, Fan P, Liu C, Wu B, Li S, Liang Z. Sunlight exclusion from Muscat grape alters volatile profiles during berry development. Food Chemistry. 2014;**164**:242-250

[17] Bayonove C, Cordonni R. Researches on aroma of Muscat. 1. Development of volatile constituants during maturation of Muscat-of-Alexandria. Annales de Technologie Agricole. 1970;**19**:79-93

[18] Mencarelli F, Tonutti P. Sweet, Reinforced and Fortified Wines: Grape Biochemistry, Technology and

Vinification. Italy: John Wiley & Sons, Ltd; 2013. ISBN: 9780470672242

[19] Harney KR. Influence of fruit microclimate on monoterpene levels of gewürztraminer. American Journal of Enology and Viticulture. 1996;**40**:149-154

[20] Reynolds AG. Managing Wine Quality: Viticulture and Wine Quality. Cambridge: Woodhead; 2010. ISBN: 9781845694845

[21] Zoecklein BW, Wolf TK, Duncan SE, Marcy JE, Jasinski Y. Effect of fruit zone leaf removal on total glycoconjugates and conjugate fraction concentration of Riesling and Chardonnay (*Vitis vinifera* L.) grapes. American Journal of Enology and Viticulture. 1998;**49**:259-265

[22] Reynolds AG, Wardle DA, Dever M. Vine performance, fruit composition, and wine sensory attributes of Gewrztraminer in response to vineyard location and canopy manipulation. American Journal of Enology and Viticulture. 1996;**47**(1):77-92

[23] Mozzon M, Savini S, Boselli E, Thorngate JH. The herbaceous character of wines. Italian Journal of Food Science. 2016;**28**(2):190-207

[24] Ou C, Du X, Shellie K, Ross C, Qian MC. Volatile compounds and sensory attributes of wine from Cv. Merlot (*Vitis vinifera* L.) Grown under differential levels of water deficit with or without a kaolin-based, foliar reflectant particle film. Journal of Agricultural and Food Chemistry. 2010; **58**(24):12890-12898

[25] McCarthy M, Coombe B. Water status and wine grape quality. Acta Horticulturae. 1985;**171**:447-456

[26] Escalona JM, Flexas J, Schultz HR, Medrano H. Effect of moderate irrigation on aroma potential and other markers of grape quality. In: Proceedings of the Acta Horticulturae; 1999

[27] Baeza P, Sánchez-de-Miguel P, Centeno A, Junquera P, Linares R, Lissarrague JR. Water relations between leaf water potential, photosynthesis and agronomic vine response as a tool for establishing thresholds in irrigation scheduling. Scientia Horticulturae (Amsterdam). 2007;**114**(3):151-158

[28] Koundouras S, Marinos V, Gkoulioti A, Kotseridis Y, Van Leeuwen C. Influence of vineyard location and vine water status on fruit maturation of nonirrigated cv. Agiorgitiko (*Vitis vinifera* L.). Effects on wine phenolic and aroma components. Journal of Agricultural and Food Chemistry. 2006;**54**(14):5077-5086

[29] Reynolds AG. Viticultural and vineyard management practices and their effects on grape and wine quality. In: Managing Wine Quality: Viticulture and Wine Quality. Canada: Woodhead Publishing Limited; 2010. pp. 365-444. ISBN: 9781845694845

[30] González-Marco A, Jiménez-Moreno N, Ancín-Azpilicueta C. Influence of nutrients addition to nonlimited-in nitrogen must on wine volatile composition. Journal of Food Science. 2010;**75**:206-211

[31] Lacroux F, Tregoat O, Van Leeuwen C, Pons A, Tominaga T, Lavigne-Cruège V, et al. Effect of foliar nitrogen and sulphur application on aromatic expression of *Vitis vinifera* L. cv. Sauvignon blanc. Journal International des Sciences de la Vigne et du Vin. 2008;**42**(3):125-132

[32] Caboni P, Cabras P. Pesticides' influence on wine fermentation. In: Advances in Food and Nutrition Research. Vol. 59. Academic Press; 2010. pp. 43-62

[33] Muñoz-González C, Rodríguez-Bencomo JJ, Moreno-Arribas MV,

Pozo-Bayón MÁ. Beyond the characterization of wine aroma compounds: Looking for analytical approaches in trying to understand aroma perception during wine consumption. Analytical and Bioanalytical Chemistry. 2011;**401**(5):1497-1512

[34] Barbe JC, Pineau B, Silva Ferreira AC. Instrumental and sensory approaches for the characterization of compounds responsible for wine aroma. Chemistry & Biodiversity. 2008;**5**(6): 1170-1183

[35] Guasch J, Busto O, Mestres M. Nariz electrónica. Aplicaciones enológicas. ACE: Revista de Enología. 2011;(126):3

[36] Robinson AL, Boss PK, Solomon PS, Trengove RD, Heymann H, Ebeler SE. Origins of grape and wine aroma. Part 2. Chemical and sensory analysis. American Journal of Enology and Viticulture. 2014;**65**:25-42

[37] Torrens J. El análisis del aroma en el control de calidad de los vinos. ACE Rev. Enol. 2000;(5):1

[38] Wang M, Zeng L, Lu S, Shao M, Liu X, Yu X, et al. Development and validation of a cryogen-free automatic gas chromatograph system (GC-MS/FID) for online measurements of volatile organic compounds. Analytical Methods. 2014;**6**:9424-9434

[39] del Pozo Bayón M. Descifrando las claves químicas que explican el aroma del vino. ACE Rev. Enol. 2011;**127**:1-14

[40] Barata A, Campo E, Malfeito-Ferreira M, Loureiro V, Cacho J, Ferreira V. Analytical and sensorial characterization of the aroma of wines produced with sour rotten grapes using GC-O and GC-MS: Identification of key aroma compounds. Journal of Agricultural and Food Chemistry. 2011; **59**(6):2543-2553

[41] Busto O. La nariz electrónica: una nueva herramienta para analizar el aroma. Ponencias de Tecnologia. 2002; **18**(2):209-217

[42] Lozano J, Santos JP, Aleixandre M, Sayago I, Gutiérrez J, Horrillo MC. Identification of typical wine aromas by means of an electronic nose. IEEE Sensors Journal. 2006;**6-1**:173-178

[43] Guadarrama A, Fernández JA, Íiguez M, Souto J, De Saja JA. Discrimination of wine aroma using an array of conducting polymer sensors in conjunction with solid-phase micro-extraction (SPME) technique. Sensors and Actuators B: Chemical. 2001;**77** (1–2):401-408

[44] Pino JA, Queris O. Analysis of volatile compounds of pineapple wine using solid-phase microextraction techniques. Food Chemistry. 2010;**122** (4):1241-1246

[45] Tésis. Oliveira JM. Aroma varietais e de fermentação determinantes da tipicidade das castas loureiro e alvarinho. 2001

[46] Flanzy C. Oenologie, fondements scientifiques et technologiques. In: Sciences et Techniques Agroalimentaires. France: Lavoisier; 2003. ISBN: 2-7430-0243-3

[47] Zhu F, Du B, Li J. Aroma compounds in wine. In: Grape and Wine Biotechnology. Croatia: InTech; 2015

[48] Robinson AL, Boss PK, Solomon PS, Trengove RD, Heymann H, Ebeler SE. Origins of grape and wine aroma. Part 1. Chemical components and viticultural impacts. American Journal of Enology and Viticulture. 2014;**65**:1-24

[49] Ribéreau-Gayon P, Glories Y, Maujean A, Dubourdieu D. Handbook of Enology, The Chemistry of Wine: Stabilization and Treatments. 2nd ed. Bordeaux: Wiley; 2006. ISBN: 9780470010396

[50] Sidhu D, Lund J, Kotseridis Y, Saucier C. Methoxypyrazine analysis and influence of viticultural and enological procedures on their levels in grapes, musts, and wines. Critical Reviews in Food Science and Nutrition. 2015;**55**(4): 485-502

[51] Godelmann R, Limmert S, Kuballa T. Implementation of headspace solid-phase-microextraction-GC-MS/MS methodology for determination of 3-alkyl-2-methoxypyrazines in wine. European Food Research and Technology. 2008; **227**:449-461

[52] Blouin J, Guimberteau G. In: Féret E, editor. Maturation et Maturité des Raisins. France: Merignac; 2012

[53] Panighel A, Flamini R. Applications of solid-phase microextraction and gas chromatography/mass spectrometry (SPME-GC/MS) in the study of grape and wine volatile compounds. Molecules. 2014;**19**(12):21291-21309

[54] Boidron J. Relation entre les substances terpéniques et la qualité du raisin. Role du *Botrytis cinerea*. Annales de Technologie Agricole. 1978;**27**:141-145

[55] Ferreira V, Lopez R. The actual and potential aroma of winemaking grapes. Biomolecules. 2019;**9**(12):818

[56] Ilc T, Werck-Reichhart D, Navrot N. Meta-analysis of the core aroma components of grape and wine aroma. Frontiers in Plant Science. 2016; **7**:1472

[57] Ribereau-Gayon P, Dubourdieu D, Doneche B, Lonvaud A. Handbook of Enology: The Microbiology of Wine and Vinifications. 2nd ed. West Sussex, England: John Wiley & Sons; 2006. ISBN: 9780470010365

[58] Moreno García J. Proteomic and metabolomic study of wine yeasts in free and immobilized formats, subjected to different stress conditions. Thesis. Cordoba, Soain: Universidad de Córdoba; 2017

[59] Capozzi V, Garofalo C, Chiriatti MA, Grieco F, Spano G. Microbial terroir and food innovation: The case of yeast biodiversity in wine. Microbiological Research. 2015;**181**:75-83

[60] Bouard J, Guimberteau G, editor. La Viticulture à l'aube du IIIe Millénaire. Bourdeaux: Vigne et Vin Publications Internationales; 1996

[61] Giovanelli G, Brenna OV. Evolution of some phenolic components, carotenoids and chlorophylls during ripening of three Italian grape varieties. European Food Research and Technology. 2007;**225**(1):145-150

[62] Van Leeuwen C, Destrac-Irvine A. Modified grape composition under climate change conditions requires adaptations in the vineyard. OENO One. 2017;**51**(2):147-154

[63] Razungles A, Bayonove C, Cordonnier R, Sapis J. Grape carotenoids: Changes during the maturation period and localization in mature berries. American Journal of Enology and Viticulture. 1988;**39**(1):44-48

[64] Mendes-Pinto MM. Carotenoid breakdown products the-norisoprenoids-in wine aroma. Archives of Biochemistry and Biophysics. 2009; **483**(2):236-245

[65] Crupi P, Coletta A, Antonacci D. Analysis of carotenoids in grapes to predict norisoprenoid varietal aroma of wines from apulia. Journal of Agricultural and Food Chemistry. 2010; **58**(17):9647-9656

[66] Di Stefano R, Maggiorotto G. Actes du symposium internacional. In: Proceedings of the Connaissance Aromatique des Cépages et Qualité des Vins. Montpellier; 1993

[67] Michlmayr H, Nauer S, Brandes W, Schümann C, Kulbe KD, Del Hierro AM, et al. Release of wine monoterpenes from natural precursors by glycosidases from *Oenococcus oeni*. Food Chemistry. 2012;**135**(1):80-87

[68] Guzzon R, Malacarne M, Larcher R, Franciosi E, Toffanin A. The impact of grape processing and carbonic maceration on the microbiota of early stages of winemaking. Journal of Applied Microbiology. 2020;**128**(1):209-224

[69] Bayonove C, Baumes R, Crouzet J, Günata Z. In: Flanzy C, editor. Enología: Fundamentos Científicos y Tecnológicos. Madrid: Mundi-Prensa; 2003

[70] Cordonnier R, Bayonove CL. Etude de la phase préfermentaire de la vinification: Extraction et formation de certains composés de l'arôme; cas des terpenols, des aldehydes et des alcools en C_6 *. OENO One. 1981;**15**(4):269-286

[71] Bleve G, Tufariello M, Vetrano C, Mita G, Grieco F. Simultaneous alcoholic and malolactic fermentations by *Saccharomyces cerevisiae* and *Oenococcus oeni* cells co-immobilized in alginate beads. Frontiers in Microbiology. 2016;**7**:943

[72] Bordiga M, Nollet LML. Food Aroma Evolution: During Food Processing, Cooking, and Aging. Boca Raton: CRC Press; 2019

[73] Baumes RL. Enología: Fundamentos Científicos y Tecnológicos. Aromas. Madrid: Mundi-Prensa; 2003

[74] Lambrechts MG, Pretorius IS. Yeast and its importance to wine aroma—A review. South African Journal for Enology and Viticulture. 2019;**21**(1):97-129

[75] Jackson RS. Wine Tasting (Third Edition)—A Professional Handbook. London: Elsevier; 2017. ISBN: 9780128018262

[76] Mendes-Ferreira A, Barbosa C, Lage P, Mendes-Faia A. The impact of nitrogen on yeast fermentation and wine quality. Ciência e Técnica Vitivinícola. 2011;**26**(1):17-32

[77] Bordiga M, Nollet LML, Bertrand E. Food processing, cooking, and aging: A practical case study. In: Food Aroma Evolution. United Kingdom: Taylor Francis Group; 2019

[78] Domínguez AM, Eduardo A. Gas chromatography coupled with mass spectrometry detection for the volatile profiling of *Vitis vinifera* CV. Carménère wines. Journal of the Chilean Chemical Society. 2010;**55**(3):385-391

[79] Picard M, Tempere S, de Revel G, Marchand S. A sensory study of the ageing bouquet of red bordeaux wines: A three-step approach for exploring a complex olfactory concept. Food Quality and Preference. 2015;**42**:110-122

[80] Ugliano M. Oxygen contribution to wine aroma evolution during bottle aging. Journal of Agricultural and Food Chemistry. 2013;**61**(26):6125-6136

[81] Pérez-Coello MS, Díaz-Maroto MC. Volatile compounds and wine aging. In: Wine Chemistry and Biochemistry. Madrid: Springer; 2009. ISBN: 9780387741161

[82] Fernández De Simón B, Cadahía E, Jalocha J. Volatile compounds in a Spanish red wine aged in barrels made of Spanish, French, and American Oak Wood. Journal of Agricultural and Food Chemistry. 2003;**51**(26):7671-7678

[83] Skouroumounis GK, Sefton MA. Acid-catalyzed hydrolysis of alcohols and their β-D-glucopyranosides. Journal of Agricultural and Food Chemistry. 2000;**48**(6):2033-2039

[84] Oliveira I, Ferreira V. Modulating fermentative, varietal and aging aromas of wine using non-Saccharomyces

yeasts in a sequential inoculation approach. Microorganisms. 2019;**7**(6):164

[85] Nesbitt A, Kemp B, Steele C, Lovett A, Dorling S. Impact of recent climate change and weather variability on the viability of UK viticulture - combining weather and climate records with producers' perspectives. Australian Journal of Grape and Wine Research. 2016;**22**:324-335

[86] Robinson J, Harding J. The Oxford Companion to Wine. 3rd ed. Oxford: Oxford University Press; 2014. ISBN: 9780198609902

Chapter 4

Formation of Aromatic and Flavor Compounds in Wine: A Perspective of Positive and Negative Contributions of Non-*Saccharomyces* Yeasts

Liliana Godoy, Andrea Acuña-Fontecilla and Daniela Catrileo

Abstract

Wine is a complex matrix that involves compounds of different chemical nature, with volatile compounds being primarily responsible for the aromatic quality of the wine. The formation of these volatile compounds is mainly due to yeasts' metabolism during alcoholic fermentation. Several studies in the microbiology field have reported that *Saccharomyces cerevisiae* is responsible for alcoholic fermentation, influencing the sensory quality of the wine and affecting the metabolic activity of other genera and species of yeasts, called non-*Saccharomyces*, which would positively affect sensory quality. Non-*Saccharomyces* yeasts, considered until recently as undesirable or spoilage yeasts, can improve the chemical composition and aroma profile of the wine. The activity of these yeasts is considered essential for the final wine aroma profile. Thus, the metabolism of these microorganisms could be a decisive factor that strongly influences the aroma of the wine, impacting on its quality. However, there are few studies that explain the impact of non-*Saccharomyces* yeasts on the final wine aroma profile. This chapter summarizes relevant aspects and pathways involved in the synthesis of aromatic compounds by non-*Saccharomyces* yeasts as well as studies at the genetic and transcriptional level associated with their formation.

Keywords: wine, non-*Saccharomyces* yeasts, fermentation, aroma, flavor

1. Introduction

The study of the yeasts involved in the wine fermentation process has shown that the main yeasts genera are *Saccharomyces, Candida, Debaryomyces, Hanseniaspora, Kloeckera, Pichia* and *Torulaspora* [1]. Despite this large number of genera involved in this process, it has been identified that the species responsible for alcoholic fermentation is *Saccharomyces cerevisiae*.

Starter cultures of *S. cerevisiae* are currently used by winemakers in order to homogenize the microbiota and to prevent unwanted yeast species from prevailing in the fermentation process. These cultures quickly position themselves against the rest of the yeasts, thus ensuring the quality of the final product without being conditioned by the other microorganism species present in the fermentation.

IntechOpen

Due to the importance of the aromatic products obtained during the fermentation process, numerous works have been carried out correlating the strain of *S. cerevisiae* used versus the aroma of the wine obtained [2]. Thereby, several studies carried out in the field of wine microbiology have reported that not only *S. cerevisiae* has an effect on the sensory quality of the wine, but also the metabolic activity of other genera and species of yeast, called non-*Saccharomyces*, would positively affect sensory quality.

In this way, the sequential action of these different genera and yeast species contributes to the wine aroma and flavor, determining the final sensory quality. The wine aroma and flavor are mainly obtained by many volatile compounds formed during the alcoholic fermentation, including alcohols, esters, organic acids, phenols, thiols, monoterpenes and norisoprenoids.

In this context, *Candida stellata* and *Kloeckera apiculata* stand out for their high glycerol production. This compound provides sweetness and fullness in wines, but the perception of these sensations depends on the concentration and on the wine [3].

Candida colliculosa stands out for its production of acetaldehyde and n-propanol, which can have a positive influence on the quality of the wine. Likewise, other non-*Saccharomyces* species possess ß-glucosidase activity, an enzyme that can hydrolyze aromatic precursors [2, 4]. In this way, the initial activity of these yeasts in the must is considered essential for the final wine aroma profile, because they are responsible for different reactions in the development of a wide range of volatile and nonvolatile products. As mentioned before, partially, it has been reported that the use of non-*Saccharomyces* yeasts in wine fermentation improves several parameters associated with the final wine quality, such as the increase in glycerol content [5], aromatic complexity [6], acidity [7] and anthocyanin content [8].

However, despite the aromatic potential of non-*Saccharomyces* yeasts, most of them have a low fermentative capacity, due to their low tolerance to alcohol, being unable to finish the fermentation. These characteristics have limited their use in the industry, despite their potential.

Currently, one of the strategies is the use of mixed cultures of non-*Saccharomyces* species with *S. cerevisiae* strains. This represents a useful tool that allows taking advantage of the sensory qualities of non-*Saccharomyces* species and the fermentative fitness of *S. cerevisiae*, favoring the sensory complexity and, therefore, the quality of the wine obtained [9]. Studies of mixed cultures of *Candida cantarelli* and *S. cerevisiae* reported that the use of sequential inoculation of these yeasts contributes to the improvement of the sensory characteristics of Syrah variety wine [10]. Likewise, Jolly et al. [4] observed that the aromatic profile of Chenin Blanc wines was improved with mixtures of *Candida pulcherrima* and *S. cerevisiae*. García et al. [11] reported similar observations for Chardonnay wines. Regarding fermentation of musts with a high concentration of sugar, it has been reported that the use of *T. delbrueckii* and *S. cerevisiae* reduced the volatile acidity and improved the analytical profile of the wine [12]. Also, the combined use of *Debaryomyces vanriji* and *S. cerevisiae* increased the concentration of geraniol [13]. An increase in varietal thiols was observed in cofermentation with *Pichia kluyveri* and *S. cerevisiae* [14]. Clemente-Jimenez et al. [15], using sequential inoculum of *Pichia fermentans* and *S. cerevisiae,* observed an increase in the concentration of specific aromatic components.

These results show that there is a huge potential for the application of non-*Saccharomyces* yeasts in oenology and strong evidence that their use contributes to the production of specific volatile compounds improving the aromatic composition of wines (**Table 1**).

Species	Metabolites		References
	Increase	Decrease	
Saccharomyces cerevisiae	Acetaldehyde Ethyl esters (caprylate)		[16, 17]
Torulaspora delbrueckii	Fruity esters Ethyl propanoate Ethyl isobutanoate Ethyl dihydrocinnamate Thiols Terpenes Glycerol 3-methylthio-1-propanol 4-MSP	Acetic acid Acetaldehyde Ethanol Higher alcohols	[5, 6, 18–30]
Kluyveromyces spp	Lactic acid Esters Monoterpenic alcohols 2-phenylethanol Carboxylic acids Ketones Furans Isoamyl acetate	Acetic acid	[31, 32]
K. marxianus	Polygalacturonases 2-phenylethanol Phenethyl acetate Ethyl acetate		[33–35]
K. lactis	Monoterpenoids		[36–40]
Hanseniaspora spp	Acetic acid Acetate ester Ethyl acetate Sulfur compounds Hydrogen sulfide		[16, 41]
H. uvarum	Acetic acid		[42]
H. guilliermondii	Acetate ester		[43]
H. vineae	Acetate and ethyl ester 2-phenylethyl acetate β-damascenone Isoamyl acetate Phenylacetaldehyde	2-phenylethanol	[44–48]
Metschnikowia pulcherrima	Free terpenes Linalool Geraniol Nerol Citronerol Alpha-terpineol Biogenic amines (histamine, tyramine and putrescine) Acetate esters β-damascenone Higher alcohols (isobutanol and phenylethanol)	C6 alcohols	[48–54]
Brettanomyces bruxellensis	Volatile phenols (4-ethylphenol) 2-acetyl-3,4,5,6-tetrahydropyridine 2-acetyl-1,2,5,6-tetrahydropyridine 2-ethyl-3,4,5,6-tetrahydropyridine Isoamyl alcohol Isoamyl acetate Esters		[55–60]

Species	Metabolites		References
	Increase	Decrease	
Schizosaccharomyces spp	H2S Acetaldehyde Pyruvic acid 2, 3-butanediol Acetoin Acetic acid	Esters Higher alcohols Gluconic acid	[61–70]

Table 1.
*Metabolites produced in wine by non-*Saccharomyces *yeasts in mixed fermentations compared to fermentations with* S. cerevisiae.

2. Wine aromas produced by non-*Saccharomyces* yeasts

The formation of aromatic compounds has been extensively studied in *S. cerevisiae*. In this regard, higher alcohols are synthesized from amino acids by transamination and decarboxylation reactions (**Figure 1**). Permeases of amino acids participate in these reactions, which are encoded by the *GAP1, BAP2* and *MEP2* genes. Subsequently, the transamination reactions are carried out by enzymes encoded by the *BAT1* and *BAT2* genes, which code for branched-chain amino acid transaminases, and the *ARO8* and *ARO9* genes that code for aromatic amino acid aminotransferases, which catalyze the transfer of amines between amino acids and their respective α-keto acid. Subsequently, the decarboxylation reactions of the α-keto acid occur to form the respective aldehydes, where the *PDC1*, *PDC5*, *PDC6*, *THI3* and *ARO10* genes are responsible for coding for enzymes with decarboxylase activity and, finally, dehydrogenases act, which reduce aldehydes to alcohols, a reaction that is carried out by alcohol dehydrogenases, encoded by the *ADH1–7* and *SFA1* genes, and aryl alcohol dehydrogenases, encoded by *AAD* genes [71–73].

Other important compounds are acetate esters, and their synthesis occurs by condensation between higher alcohols and acetyl-CoA (**Figure 2**). This reaction is carried out by acetyltransferases, encoded by the *ATF1* and *ATF2* genes. The ethyl esters are produced by condensation between ethanol and acyl CoA, a reaction mediated by acyltransferases encoded by the genes *EHT1*, *EEB1* and YMR210W, encoding for a monoacylglycerol lipase [74].

Likewise, it has been reported that *S. cerevisiae* participates in the primary release of aromas through the activity of glucosidase enzymes [76].

2.1 *Torulaspora delbrueckii*

Among the non-*Saccharomyces* yeasts, *T. delbrueckii* has gained interest in the vitiviniculture industry because it modifies the aromatic properties of final wines in a very positive way, producing higher levels of fruity esters, thiols and terpenes and lower amounts of higher alcohols, thus respecting the initial character of the grape [6, 18, 19]. Also, *T. delbrueckii* typically produces low concentrations of acetic acid [12], one of the main quality parameters in wine production. It has also been reported that *T. delbrueckii* produces wines with higher levels of glycerol [5] and, consequently, with lower concentrations of ethanol [20]. This is currently a relevant feature because as a consequence of climate change, an increase in sugar concentration in the must has been observed, resulting in wines with higher alcohol content.

During alcoholic fermentation, the ethanol production is usually higher than 12% (v/v), so the associated microorganisms must have resistance mechanisms for this compound. In practice, the phenotype of ethanol resistance among wine yeasts

DOI: http://dx.doi.org/10.5772/intechopen.92562

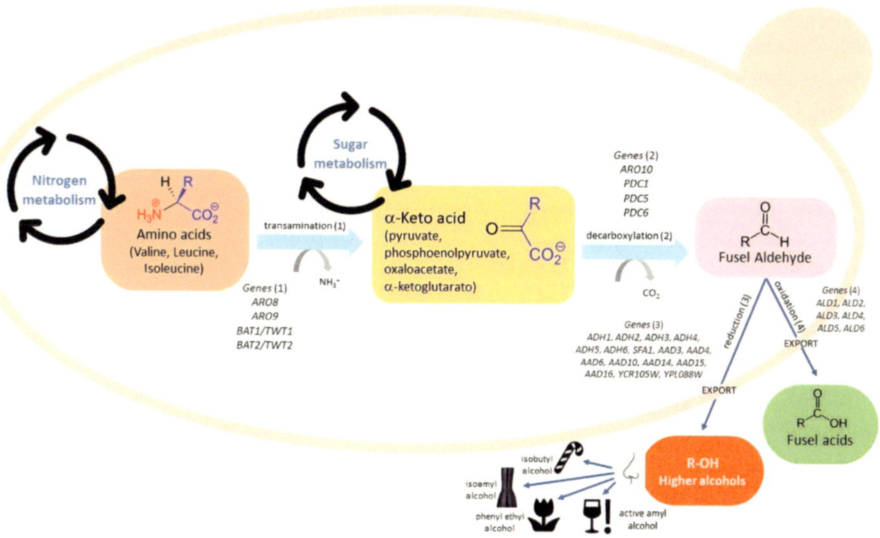

Figure 1.
Ehrlich pathway for higher alcohol production (adapted from [71]).

is heterogeneous, *S. cerevisiae* being the one with the highest level of resistance and the one in charge of leading the alcoholic fermentation. However, non-*Saccharomyces* yeast species play an important role during the early stages of spontaneous alcoholic fermentation, when the ethanol concentration is not very high [77].

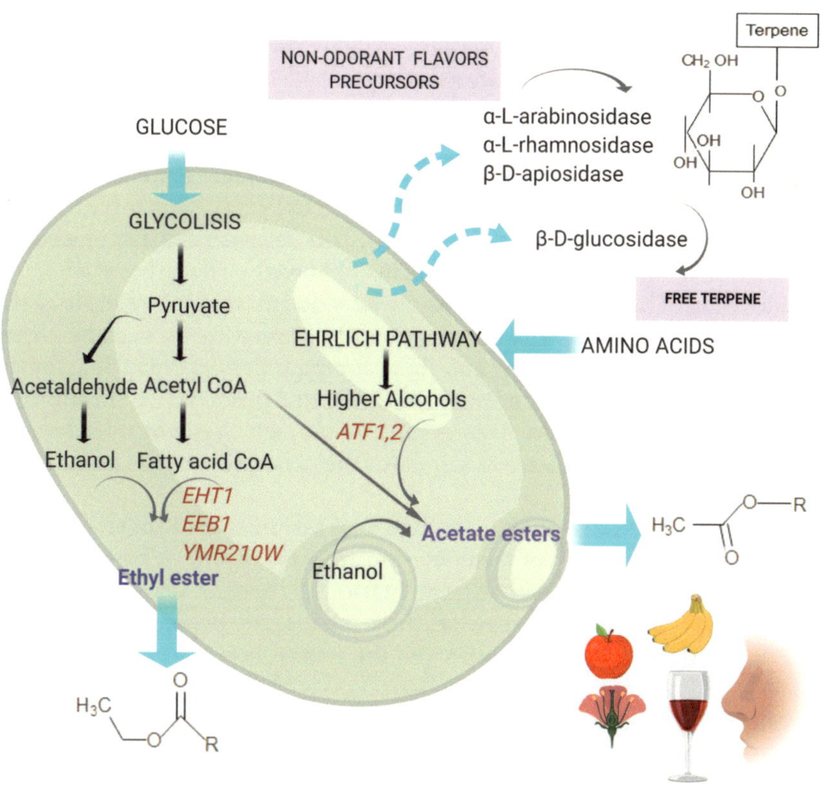

Figure 2.
Acetate ester and ethyl ester biosynthesis (adapted from [75]).

Therefore, currently, the strategy of mixed and/or sequential fermentations is used, which combines non-*Saccharomyces* yeasts with a yeast with a higher fermentative profile such as *S. cerevisiae*, which in most of the cases is necessary to properly end the industrial process of alcoholic fermentation.

T. delbrueckii has been described as capable of fermenting and tolerating up to an ethanol concentration slightly higher than 9% (v/v) [21]. On the other hand, Bely et al. [12] have reported that this value is lower, reaching only 7.4% (v/v). Nevertheless, Belda et al. [5, 6], through studies of population kinetics in sequential fermentation, observed that *T. delbrueckii* suffered a significant decrease in the cellular viability when ethanol levels exceed 8% (v/v). This suggests that the ethanol resistance of *T. delbrueckii* is limited and much lower than that of *S. cerevisiae*, which complicates its use in industrial fermentations. Nevertheless, to improve the fermentation rate of the selected nonconventional yeasts, sequential cultures are used, but this is to the detriment of the diversity of aromas that could be present in the final product. Given this context, ethanol resistance is an important factor in the selection of industrial non-*Saccharomyces* strains and particularly of *T. delbrueckii*.

Non-*Saccharomyces* yeast species can produce the aromatic volatiles that are known to be important for industrial beer and wine fermentations and that are produced by *Saccharomyces* species [78]. For the case of *T. delbrueckii*, several studies have indicated how beneficial its, from the aromatic point of view, incorporation into fermentations is [79–81]. Belda et al. [6], evaluating a sequential fermentation using Verdejo variety must, observed a higher aroma quality, intensity and fruity character. Chromatographic analysis indicated that this effect was due to an increase in the levels of the main ones, mainly 4-methyl-4-sulfanyl-pentan-2-one (4-MSP), which is represented in this grape variety. Likewise, Renault et al. [22] reported that mixed inoculations of *T. delbrueckii* and *S. cerevisiae* allowed the increase of some esters specifically produced by *T. delbrueckii*, which correlated with the maximum population reached by it in mixed cultures. Among the reported compounds were ethyl propanoate, ethyl isobutanoate and ethyl dihydrocinnamate, which are considered activity markers for this yeast.

The signaling pathways involved in the formation of aroma and flavor compounds, such as the Ehrlich pathway, or the specific enzymes responsible for the synthesis of ester, are also present in nonconventional yeast. This route has been studied extensively in *S. cerevisiae* [71, 82, 83]. This pathway consists of a step of transamination of amino acids to α-keto acids, followed by decarboxylation to "fusel aldehydes." These fusel aldehydes can be reduced or oxidized in fusel alcohols or fusel acids, respectively [71, 84]. Subsequently, aromatic esters can be formed from alcohols and fusel acids [23], and these compounds are responsible for the characteristic aroma and flavor of the final fermented product [72]. In addition, these aromatic esters have a low detection threshold, which is why minimum amounts of these compounds are required for the perception of the human olfactory senses [23].

It has been reported that the concentration of assimilable nitrogen has a significant effect on the production of fermentation aromas [85, 86]. A higher concentration of higher alcohols at the end of fermentation has been observed in media with low nitrogen content [24–26]. Likewise, there is a directly proportional relationship between the concentration of nitrogen and the synthesis of the ethyl esters, in which the initial content of nitrogen is associated with an increase in the production of esters [27, 28]. In this sense, Bloem et al. [87] observed that the nitrogen composition of the medium could influence the redox balance in the yeast cells during alcoholic fermentation and that variations in this balance could change the final concentrations of certain volatile compounds. Changes in the levels of these compounds were closely related to the effects of redox status on the availability of acetyl-CoA, an intermediate of central carbon metabolism and precursor of α-keto acids. Similar results were reported by Rollero et al. [88] who observed that a small change in the acetyl-CoA pool would affect the bioconversion of acetate esters from higher alcohols. These results suggest that it is possible to increase the aromatic potential of *T. delbrueckii* by modulating the availability of nitrogen in the medium, which would influence the redox balance of the cells directly affecting the final concentrations of certain volatile compounds.

Through next-generation sequencing, Tondini et al. [89] characterized the transcriptome of *T. delbrueckii* COFT1 observing differences in glucose fermentation pathways and the formation of aromatic and flavoring compounds, such as glycerol, esters and acetic acid with respect to *S. cerevisiae*. These differences are partly explained by the absence of paralogous genes in glycolysis and glycerol biosynthesis in *T. delbrueckii*. It has been reported that T. *delbrueckii* produces less acetic acid [29], and this phenomenon depends on increased expression of genes related to alcoholic fermentation, while acetate ester levels were influenced by the absence of esterases, *ATF1–2*. Likewise, a lower production of ethyl esters was observed in *T. delbrueckii* COFT1, which suggests a negative regulation in the fatty acid pathway biosynthesis.

2.2 *Kluyveromyces* spp.

Kluyveromyces species do not usually intervene in spontaneous fermentation processes because they have a low fermentation capacity and slow multiplication [31]. However, they are capable of producing considerable amounts of lactic acid (1.5–1.8 g/L) and low amounts of acetic acid. It has been reported that *Kluyveromyces* species produce aromatic compounds such as esters, monoterpenic alcohols, carboxylic acids, ketones, furans and isoamyl acetate in liquid phase fermentation. Of all these compounds, the production of 2-phenylethanol (2-PE) stands out [32], with the aroma of rose petals, which is commercially important, since it gives characteristics that positively influence wine quality, among others [90]. In particular, the influence of the carbon source [91, 92], the aeration rate [92], the composition of the medium [93] and growing conditions [94] on the production of aromas in *K. marxianus* has been studied.

K. marxianus produces polygalacturonases, enzymes that added in the fermentation of musts favor the release of aromatic compounds, resulting in citrus, balsamic and floral wines [33]. Other studies have demonstrated the fermentation capacity of *K. marxianus* in pure culture for the production of tequila; however, in mixed cultures with *S. cerevisiae*, the activity of *K. marxianus* is negatively affected [95].

Another group of important aromatic compounds is monoterpenoids. The common precursor of these compounds is geranyl pyrophosphate (GPP). Although plants, such as *Vitis vinifera* and *Humulus lupulus*, produce monoterpenoids [36], it has been reported that yeasts can also produce them [37], highlighting *K. lactis* [38–40].

Marcišauskas et al. [34], using the strain of *K. marxianus* iSM996, constructed the first genome-scale metabolic model for this yeast. This model contains several unique biosynthetic pathways for aromatic compounds such as 2-PE, phenethyl acetate and ethyl acetate. The *K. marxianus* iSM996 model is a solid tool to evaluate the metabolic characteristics of *K. marxianus*, allowing the integration of experimental data and strain design based on the model.

Ivanov et al. [35] studied the production potential of 2-PE by the strain of *K. marxianus* 35. The results revealed that the enzymatic activity of aminotransferase, pyruvate decarboxylase and alcohol dehydrogenase, key enzymes of the Ehrlich pathway, was almost twice as large compared to *S. cerevisiae*. In addition, the residual concentration of 2-PE was twice lower in *K. marxianus* 35 and the efficiency was found to be 73% for this strain. Additionally, the sequence variability in the genes encoding the key enzymes of the Ehrlich pathway suggests that in addition to the physiological advantages *Kluyveromyces* have probably undergone substantial evolutionary genetic alterations that result in higher enzymatic activities and a better transformation potential of 2-PE.

2.3 *Hanseniaspora*

Species of the genus *Hanseniaspora* are ubiquitous in the winemaking environment, and some of them have been proposed as wine yeast starters [96].

Fermentations of mixed cultures by wild yeasts, such as *H. guilliermondii*, together with *S. cerevisiae* have shown higher concentrations of acetate ester compared to fermentations with *S. cerevisiae* alone, without significantly affecting acetaldehyde, acetic acid, glycerol and higher total alcohols [43]. However, Lleixà et al. [44] reported that the use of the *H. vineae* species as an initiator is capable of granting aromatic complexity in wines, producing key aromatic compounds. However, the sensory evaluation of the wines produced by this apiculate yeast is still limited and the results have not been consistent. In this regard, Medina et al. [45] reported that fermentation using *H. vineae* produced up to 10 times higher levels of 2-phenylethyl acetate in wine, compared to conventional and spontaneous fermentations. However, the opposite was observed for the concentration of 2-PE, which was significantly lower. Similar results were reported by Viana et al. [46, 47], regarding the high production of 2-phenylethyl acetate by *H. vineae*.

It should be noted that the aromatic contribution of 2-PE is controversial. Fuente-Blanco [97] reported that the contribution of 2-PE in the aroma of red wine was insignificant, in addition to depending on the aromatic context.

On the other hand, Viana et al. [16] reported that *Hanseniaspora* spp. produce high levels of ethyl acetate. In this regard, it is important to highlight that ethyl acetate at low levels, below 80 mg/L, confers aromatic complexity on the wine, giving it a "fruity" aroma. However, over 150 mg/L is responsible for the typical altered sensory properties of acescence [98].

The acetic acid concentration in wines is also important, becoming a defect near its flavor threshold of 0.7–1.1 g/L. Some *H. uvarum* species have been reported to produce acetic acid levels of up to more than 3.4 g/L [42].

Other compounds have been associated with the metabolism of *H. vineae* such as β-damascenone, isoamyl acetate and phenylacetaldehyde, which have been identified in ice wine fermentations [48].

Seixas et al. [99] reported the reconstruction of the metabolic network for *H. guilliermondii* UTAD222, noting that this strain of yeast contains four genes that code for β-glucosidases, as well as the genes necessary for the synthesis of acetaldehyde, ethyl esters and higher alcohols. Surprisingly, no *S. cerevisiae* acetyl transferase-like proteins, involved in the synthesis of acetate esters, were found in the ORFeome of *H. guilliermondii* UTAD222. This is contradictory because it has been described that the synthesis of these compounds is high in this species [43]. Likewise, no sequences associated with aryl alcohol dehydrogenases were found, enzymes necessary for the synthesis of higher alcohols from aldehydes, which could contribute to the lower reported capacity of this species to produce these compounds, especially in comparison with *S. cerevisiae*.

Giorello et al. [100] recently reported genome sequencing, assembly and phylogenetic analysis of two strains of *H. vineae*. When these genomes were compared with 14 genomes of *S. cerevisiae*, specific flavor gene duplications and absences were identified in the *H. vineae* genome. In this regard, the increase observed in the formation of 2-phenylethyl acetate and phenylpropanoids, such as 2-phenylethyl and benzyl alcohol, could be explained by duplications of *ARO8, ARO9* and *ARO10* genes. Similarly, the high level of acetate esters produced by *H. vineae* compared to that of *S. cerevisiae* is related to the identification of six proteins with domains of alcohol acetyltransferase (AATase). The opposite occurs with the reduced production of higher branched chain alcohols, fatty acids and ethyl esters, which responds to the absence of branched chain amino acid transaminases (*BAT2*) and acyl coenzyme A (acyl-CoA)/ethanol O-acyltransferases (*EEB1*).

2.4 *Metschnikowia*

Metschnikowia pulcherrima is one of the non-*Saccharomyces* yeast species with the greatest capacity to express extracellular hydrolytic enzymes. In *M. pulcherrima*, the presence of enzymes with pectinase, protease, glucanase, lichenase, β-glucosidase, cellulase, xylanase, amylase, sulfite reductase, lipase and β-lyase activity [49, 101–103] has been described. Also, its high proteolytic activity makes it a candidate to be used in fermentations with *S. cerevisiae*, releasing amino acids and increasing the available nitrogen sources for the growth of *S. cerevisiae* [104, 105]. It also stands out for its glucosidase-dependent strain activity [106, 107], which increases in aerobic conditions [50], promoting the release of varietal aromas by hydrolyzing bound monoterpenes. The expression of β-D-glucosidase favors the release of free terpenes and this activity has been evaluated using the 4-methylumbelliferyl-β-D-glucoside (MUG) and p-nitrophenyl-β-D-glucoside (pNPG) substrates [108].

Terpenes are relevant in the varietal character of various white grape varieties, being the main descriptors of varieties such as Muscat, Riesling or Alvariño [51]. Their presence and relevance in certain red grape varieties are also specific. However, the composition of free terpenes in the must is scarce, with a large amount of glycosylated terpenes [52]. These can be released by enzymatic hydrolysis by glycosidase enzymes [53, 109]. Within this group, linalool, geraniol, nerol, citronellol and alpha-terpineol stand out [51, 53].

The enzymatic hydrolysis of glycosides is mainly carried out by several enzymes that act sequentially, according to two steps: first, α-L-rhamnosidase, α-L-arabinosidase or β-D-apiosidase make the cleavage from terminal sugar and rhamnose, arabinose or apiose and the corresponding β-D-glycosides are released. Subsequently, the release of terpene occurs after the action of a β-D-glucosidase [110].

Likewise, mixed fermentations between *M. pulcherrima* and *S. cerevisiae* have identified higher levels of acetate esters and β-damascenone, and lower levels of C6 alcohols in ice wines of Vidal Blanc grape variety [48]. Similarly, a higher production of higher alcohols has been reported, with a greater amount of isobutanol and phenylethanol [54].

Another aspect to highlight for *M. pulcherrima* is that it has the ability to produce biogenic amines (histamine, tyramine and putrescine); however, this phenomenon would be strain dependent [49].

To date, only the genome of one *M. pulcherrima* strain has been reported [111], and genetic studies are scarce. Reid et al. [103] identified and characterized the gene that codes for an aspartic protease of *M. pulcherrima* IWBT Y1123, called MpAPr1. The results indicated that this protein presented homology with proteases of the yeast genera. Likewise, aspartic protease activity was confirmed by heterologous expression in *S. cerevisiae* YHUM272. This gene was found in 12 other strains of *M. pulcherrima*; however, analyzes revealed that the intensity of the enzyme activity was strain dependent and was not related to the gene sequence.

2.5 *Brettanomyces* spp.

The yeast *Brettanomyces bruxellensis* is one of the main contaminant yeasts in wines, with the ability to metabolize hydroxycinnamic acids, which are naturally present in grapes, into volatile phenols [55, 56]. It has been described that this yeast can grow in various stages of wine production, for example, after alcoholic fermentation, during malolactic fermentation, during maturation in barrels or in already bottled wine. This characteristic is due to its ability to tolerate high ethanol variables [57].

Volatile phenols represent a large family of aromatic compounds where vinyl and ethyl derivatives are involved with product deterioration [55, 58]. These volatile phenols, especially 4-ethylphenol, are responsible for odors that have been described as "animal," "medicine," "leather" and "stable," which at concentrations above their perception threshold are detrimental to the aromatic profile of wines [55, 58].

The production of these compounds by *Brettanomyces* spp. is the result of the enzymatic transformation of hydroxycinnamic acids (3-methoxy-4-hydroxycinnamic acid (ferulic acid) and 4-hydroxycinnamic acid (*p*-coumaric acid)) by the action of two specific enzymes: cinnamate decarboxylase (CD) and vinylphenol reductase (VR) [112–115]. Also, *Brettanomyces* yeast species are capable of producing 2-acetyl-3,4,5,6-tetrahydropyridine, 2-acetyl-1,2,5,6-tetrahydropyridine and 2-ethyl-3,4,5,6-tetrahydropyridine. These compounds are responsible for "mousy taint" produced by microorganisms in the presence of lysine and ethanol [59].

It has been described that the ability of these yeasts to produce volatile phenols is variable [116, 117]. Factors such as the pH of the wine, the concentration of sugar and the moment in which this yeast is inoculated influence this capacity [118]. Along with this, it has been observed that the production of 4-ethylphenol in red wines is related to population growth, a phenomenon that would be strain dependent [119].

From the genetic point of view, there is a great intraspecific diversity of strains of *B. bruxellensis* [120–123], which translates into the different phenotypes of production of reported volatile phenols. The number of chromosomes in this species can vary between 4 and 9, with chromosome sizes in the range of 1 to 6 Mb, and total genome size between 20 and 30 Mb [124, 125]. Also, karyotypic studies suggest speciation due to genome rearrangements. However, available genetic studies are of a limited number of strains [121, 126–130]. In this regard, a transcriptomic analysis of the strain of *B. bruxellensis* LAMAP2480 exposed to *p*-coumaric acid indicates that this acid generates a stress condition, inducing the expression of the proton pump together with the output of toxic compounds, as well as the output of nitrogen compounds, reducing intracellular concentration and triggering the expression of nitrogen metabolism genes (**Figure 3**) [121].

Additionally, sequencing and genome analysis of the strain of *B. bruxellensis* AWRI1499 reported the presence of three homologous proteins with the isoamyl acetate hydrolysis enzymes of *S. cerevisiae* that are related to isoamyl alcohol concentrations and isoamyl acetate produced in fermentation. This strain was evaluated under fermentation conditions in model wine and produced higher levels of esters [60]. In this sense, it has been described that the formation of esters between *Brettanomyces* strains is variable.

The positive aromatic contribution of these yeasts has been studied mainly in beer. *Brettanomyces* spp. are able to esterify medium and long chain fatty acids in their respective esters, influencing the sensory profile of beers. Likewise, it has been reported that *Brettanomyces* has β-glucosidase activity, which would be responsible for breaking down the cellobiose present in the barrels, explaining its survival during the wine aging stage [131]. Crauwels et al. [120] reported that *B. bruxellensis* has

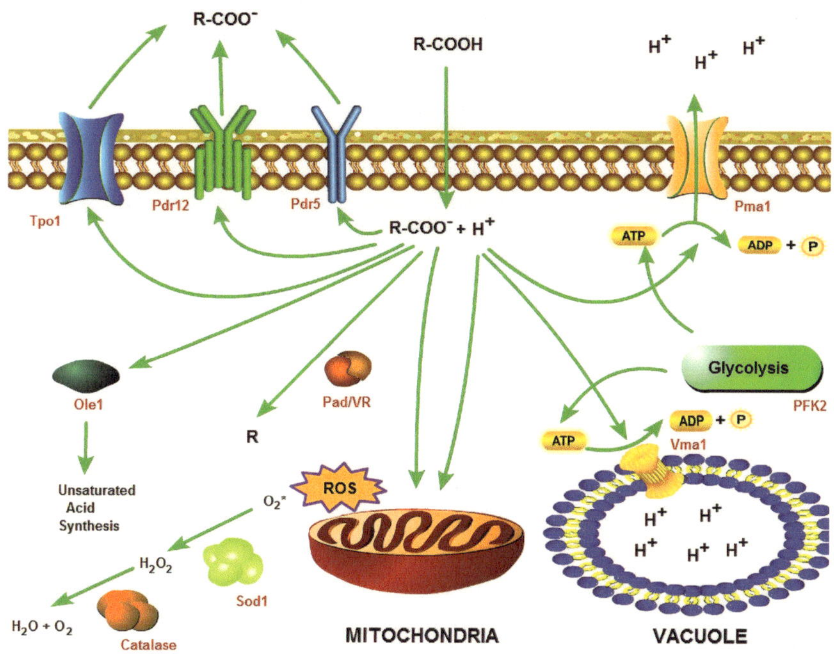

Figure 3.
Model of early response to stress by p*-coumaric acid.*

two genes that encode β-glucosidases. Most strains from beer, with the exception of strain ST05.12/22, have only one copy, while strains isolated from the wine have both ORFs.

2.6 *Schizosaccharomyces* spp.

While *Schizosaccharomyces* genus yeasts have been associated with the production of compounds such as hydrogen sulfide (H_2S) and acetaldehyde [61] that negatively impact the aromatic quality of wines, *S. pombe* stands out mainly for its ability to degrade malic acid into ethanol and deacidify musts of grapes and wines.

L-Malic acid is a compound that is present in grape must and its concentration depends on the grape varieties and climatic conditions. When malolactic fermentation (MLF) occurs, the lactic acid bacteria transform L-malic acid into lactic acid, reducing the total acidity and thereby increasing the pH of the grape must [132]. However, factors such as ethanol concentration, pH, temperature and sulfur dioxide (SO_2) level affect the successful completion of MLF [133].

An alternative to this process is the malo-ethanolic deacidification carried out by *S. pombe* [62, 134]. This yeast exhibits a high tolerance to low pH and high levels of SO_2, characteristics that make it highly compatible for use during the winemaking process [135]. Benito et al. [63] reported that the conversion of malic acid to ethanol decreases the total acidity by approximately 4 g/L and increases the final pH by approximately 0.4.

Other interesting characteristics of this yeast are associated with its ability to reduce gluconic acid concentrations [64, 65]. It has also been reported that the urease activity of *Schizosaccharomyces* strains could reduce the content of ethyl carbamate and biogenic amines in wine by reducing the concentrations of urea [66, 67]. Another application that *Schizosaccharomyces* has is aging on the lees, thanks to the strong autolytic release of the polysaccharides from the cell wall [136, 137].

The contribution from the aromatic point of view of *S. pombe* has been recently reported where it has been observed that it stands out mainly for producing fewer amounts of higher alcohols in comparison to *S. cerevisiae*, which could be attributed as a strain-dependent characteristic [62, 63, 66, 68, 69].

Benito et al. [63] reported a lower production of isobutanol, 2-methyl-butanol, 3-methyl-butanol and 2-phenyl-ethanol in white wines by *S. pombe* in comparison to *S. cerevisiae*. Similar results have been reported by Mylona et al. [66] where, fermenting red must, they observed a decrease in 2-methyl-butanol, 3-methyl-butanol and isobutanol by *S. pombe* in comparison to *S. cerevisiae*. On the other hand, Chen et al. [69] observed that *S. pombe* possesses a special ability to produce more 2, 3-butanediol, which contributes to the fruity aroma described as banana to wines.

In the case of esters, a similar phenomenon occurs, observing that *Schizosaccharomyces* shows a tendency to produce lower concentrations of esters in comparison to *S. cerevisiae.* It has been reported to produce lower concentrations of isoamyl acetate and 2-phenyl-ethyl acetate in comparison to *S. cerevisiae* [62]. Likewise, lower production of total esters was reported by Del Fresno et al. [68] in comparison to *S. cerevisiae*.

Finally, *S. pombe* fermentations have been reported to show higher levels of acetoin in comparison to *S. cerevisiae* controls. Also, they are commonly associated with high levels of acetic acid. These levels might vary from strain to strain [68–70].

3. Conclusion

There are many physiological studies on the contribution of non-*Saccharomyces* yeasts to the aromatic profile of wines. However, reports at the genetic level that explain the differences observed in these yeasts with respect to *S. cerevisiae* are scarce.

Despite the little information available, it is possible to establish that the differences in aromatic potential observed in non-*Saccharomyces* yeasts are mainly due to modifications in the Ehrlich pathway and the biosynthesis of acetate esters and ethyl esters. These changes can be summarized as follows:

a. Differences in the regulation of gene expression of these routes

b. Absence of paralogous genes

c. Gene duplications

d. Modification of enzymatic activities

The identification of most of these biological mechanisms has been possible thanks to the use of massive sequencing technology (NGS).

Given the relevance of the contribution of non-*Saccharomyces* yeasts to the quality and typicity of wines and their impact on taste, more studies with genetic approaches that explain the metabolic diversity of these yeasts are required.

Acknowledgements

The authors thank Dr. Tania Zaviezo, Paula Reyes-Bravo and Dagoberto Silva for their support in writing this manuscript. Fondo Nacional de Desarrollo Científico y Tecnológico program (FONDECYT 11180979) by CONICYT and Fondo de Inserción Académica PIA-UC-2018.

Conflict of interest

The authors declare no conflict of interest.

Notes

Figure 2 was created with BioRender.

Author details

Liliana Godoy*, Andrea Acuña-Fontecilla and Daniela Catrileo
Facultad de Agronomía e Ingeniería Forestal, Departamento de Fruticultura y Enología, Pontificia Universidad Católica de Chile, Santiago, Chile

*Address all correspondence to: liliana.godoy@uc.cl

References

[1] Fleet GH, Lafon-Lafourcade S, Ribéreau-Gayon P. Evolution of yeasts and lactic acid bacteria during fermentation and storage of Bordeaux wines. Applied and Environmental Microbiology. 1984;**48**:1034-1038

[2] Coghe S, Benoot K, Delvaux F, Vanderhaegen B, Delvaux FR. Ferulic acid release and 4-vinylguaiacol formation during brewing and fermentation: Indications for feruloyl esterase activity in *Saccharomyces cerevisiae*. Journal of Agricultural and Food Chemistry. 2004;**52**:602-608

[3] Gawel R, Van Sluyter S, Waters EJ. The effects of ethanol and glycerol on the body and other sensory characteristics of Riesling wines. Australian Journal of Grape and Wine Research. 2007;**13**:38-45

[4] Jolly NP, Augustyn OPH, Pretorius IS. The use of *Candida pulcherrima* in combination with *Saccharomyces cerevisiae* for the production of Chenin Blanc wine. South African Journal of Enology and Viticulture. 2003;**24**:63-69

[5] Belda I, Navascués E, Marquina D, Santos A, Calderon F, Benito S. Dynamic analysis of physiological properties of *Torulaspora delbrueckii* in wine fermentations and its incidence on wine quality. Applied Microbiology and Biotechnology. 2015;**99**:1911-1922

[6] Belda I, Ruiz J, Beisert B, Navascués E, Marquina D, Calderón F, et al. Influence of *Torulaspora delbrueckii* in varietal thiol (3-SH and 4-MSP) release in wine sequential fermentations. International Journal of Food Microbiology. 2017;**257**:183-191

[7] Balikci EK, Tanguler H, Jolly NP, Erten H. Influence of *Lachancea thermotoleranson* cv. Emir wine fermentation. Yeast. 2016;**33**:313-321

[8] Benito Á, Calderón F, Benito S. The combined use of *Schizosaccharomyces pombe* and *Lachancea thermotolerans*—Effect on the anthocyanin wine composition. Molecules. 2017;**22**(5):739

[9] Ciani M, Comitini F, Mannazzu I, Domizio P. Controlled mixed culture fermentation: A new perspective on the use of non-*Saccharomyces* yeasts in winemaking. FEMS Yeast Research. 2010;**10**:123-133

[10] Toro ME, Vazquez F. Fermentation behaviour of controlled mixed and sequential cultures of *Candida cantarellii* and *Saccharomyces cerevisiae* wine yeasts. World Journal of Microbiology and Biotechnology. 2002;**18**:351-358. DOI: 10.1023/A:1015242818473

[11] García V, Vásquez H, Fonseca F, Manzanares P, Viana F, Martínez C, et al. Effects of using mixed wine yeast cultures in the production of Chardonnay wines. Revista Argentina de Microbiología. 2010;**42**:226-229

[12] Bely M, Stoeckle P, Masneuf-Pomarède I, Dubourdieu D. Impact of mixed *Torulaspora delbrueckii*-*Saccharomyces cerevisiae* culture on high-sugar fermentation. International Journal of Food Microbiology. 2008;**122**:312-320

[13] Garcia A, Carcel C, Dulau L, Samson A, Aguera E, Agosin E, et al. Influence of a mixed culture with *Debaryomyces vanriji* and *Saccharomyces cerevisiae* on the volatiles of a Muscat wine. Journal of Food Science. 2002;**67**:1138-1143

[14] Anfang N, Brajkovich M, Goddard MR. Co-fermentation with *Pichia kluyveri* increases varietal thiol concentrations in Sauvignon Blanc. Australian Journal of Grape and Wine Research. 2009;**15**:1-8

[15] Clemente-Jimenez JM, Mingorance-Cazorla L, Martínez-Rodríguez S, Las Heras-Vázquez FJ, Rodríguez-Vico F. Influence of sequential yeast mixtures on wine fermentation. International Journal of Food Microbiology. 2005;**98**:301-308

[16] Viana F, Gil JV, Genovés S, Vallés S, Manzanares P. Rational selection of non-*Saccharomyces* wine yeasts for mixed starters based on ester formation and enological traits. Food Microbiology. 2008;**25**:778-785

[17] Fleet GH. Wine Microbiology and Biotechnology. Chur, Switzerland: Harwood Academic, CRC Press; 1993; ISBN: 9780415278508

[18] Azzolini M, Tosi E, Lorenzini M, Finato F, Zapparoli G. Contribution to the aroma of white wines by controlled *Torulaspora delbrueckii* cultures in association with *Saccharomyces cerevisiae*. World Journal of Microbiology and Biotechnology. 2015;**31**:277-293

[19] Renault P, Coulon J, Moine V, Thibon C, Bely M. Enhanced 3-sulfanylhexan-1-ol production in sequential mixed fermentation with *Torulaspora delbrueckii/Saccharomyces cerevisiae* reveals a situation of synergistic interaction between two industrial strains. Frontiers in Microbiology. 2016;**7**:293

[20] Contreras A, Hidalgo C, Henschke PA, Chambers PJ, Curtin C, Varela C. Evaluation of non-*Saccharomyces* yeasts for the reduction of alcohol content in wine. Applied and Environmental Microbiology. 2014;**80**:1670-1678

[21] Ciani M, Maccarelli F. Oenological properties of non-*Saccharomyces* yeasts associated with wine-making. World Journal of Microbiology and Biotechnology. 1997;**14**:199-203

[22] Renault P, Coulon J, de Revel G, Barbe J-C, Bely M. Increase of fruity aroma during mixed *T. delbrueckii/S. cerevisiae* wine fermentation is linked to specific esters enhancement. International Journal of Food Microbiology. 2015;**207**:40-48

[23] Saerens SMG, Delvaux FR, Verstrepen KJ, Thevelein JM. Production and biological function of volatile esters in *Saccharomyces cerevisiae*. Microbial Biotechnology. 2010;**3**:165-177

[24] Hernandez-Orte P, Bely M, Cacho J, Ferreira V. Impact of ammonium additions on volatile acidity, ethanol, and aromatic compound production by different *Saccharomyces cerevisiae* strains during fermentation in controlled synthetic media. Australian Journal of Grape and Wine Research. 2006;**12**:150-160

[25] Jiménez-Martí E, Aranda A, Mendes-Ferreira A, Mendes-Faia A, del Olmo ML. The nature of the nitrogen source added to nitrogen depleted vinifications conducted by a *Saccharomyces cerevisiae* strain in synthetic must affects gene expression and the levels of several volatile compounds. Antonie Van Leeuwenhoek. 2007;**92**:61-75

[26] Vilanova M, Ugliano M, Varela C, Siebert T, Pretorius IS, Henschke PA. Assimilable nitrogen utilisation and production of volatile and non-volatile compounds in chemically defined medium by *Saccharomyces cerevisiae* wine yeasts. Applied Microbiology and Biotechnology. 2007;**77**:145-157

[27] Ugliano M, Travis B, Francis IL, Henschke PA. Volatile composition and sensory properties of Shiraz wines as affected by nitrogen supplementation and yeast species: Rationalizing nitrogen modulation of wine aroma. Journal of Agricultural and Food Chemistry. 2010;**58**:12417-12425

[28] Torrea D, Varela C, Ugliano M, Ancin-Azpilicueta C, Leigh Francis I, Henschke PA. Comparison of inorganic and organic nitrogen supplementation of grape juice - Effect on volatile composition and aroma profile of a Chardonnay wine fermented with *Saccharomyces cerevisiae* yeast. Food Chemistry. 2011;**127**:1072-1083

[29] Renault P, Miot-Sertier C, Marullo P, Hernández-Orte P, Lagarrigue L, Lonvaud-Funel A, et al. Genetic characterization and phenotypic variability in *Torulaspora delbrueckii* species: Potential applications in the wine industry. International Journal of Food Microbiology. 2009;**134**:201-210

[30] Whitener MB, Beckner Whitener ME, Carlin S, Jacobson D, Weighill D, Divol B, et al. Early fermentation volatile metabolite profile of non-*Saccharomyces* yeasts in red and white grape must: A targeted approach. LWT- Food Science and Technology. 2015;**64**:412-422

[31] Fonseca GG, Heinzle E, Wittmann C, Gombert AK. The yeast *Kluyveromyces marxianus* and its biotechnological potential. Applied Microbiology and Biotechnology. 2008;**79**:339-354

[32] Wittmann C, Hans M, Bluemke W. Metabolic physiology of aroma-producing *Kluyveromyces marxianus*. Yeast. 2002;**19**:1351-1363

[33] Sieiro C, Villa TG, da Silva AF, García-Fraga B, Vilanova M. Albariño wine aroma enhancement through the use of a recombinant polygalacturonase from *Kluyveromyces marxianus*. Food Chemistry. 2014;**145**:179-185

[34] Marcišauska S, Ji B, Nielsen J. Reconstruction and analysis of a *Kluyveromyces marxianus* genome-scale metabolic model. BMC Bioinformatics. 2019;**20**:551

[35] Ivanov SL, Petrova VY, Pisareva EI, Kujumdzieva AV. Ehrlich pathway study in *Saccharomyces* and *Kluyveromyces* yeasts. Biotechnology and Biotechnological Equipment. 2013;**27**:4103-4110

[36] King A, Richard DJ. Biotransformation of monoterpene alcohols by *Saccharomyces cerevisiae*, *Torulaspora delbrueckii* and *Kluyveromyces lactis*. Yeast. 2000;**16**:499-506

[37] Larsen TO, Frisvad JC. A simple method for collection of volatile metabolites from fungi based on diffusive sampling from petri dishes. Journal of Microbiological Methods. 1994;**19**:297-305

[38] Drawert F, Barton H. Biosynthesis of flavor compounds by microorganisms. 3. Production of monoterpenes by the yeast *Kluyveromyces lactis*. Journal of Agricultural and Food Chemistry. 1978;**26**:765-766

[39] Jiang J. Volatile metabolites produced by *Kluyveromyces lactis* and their changes during fermentation. Process Biochemistry. 1995;**30**:635-640

[40] Chen D, Yap ZY, Liu S-Q. Evaluation of the performance of *Torulaspora delbrueckii*, *Williopsis saturnus*, and *Kluyveromyces lactis* in lychee wine fermentation. International Journal of Food Microbiology. 2015;**206**:45-50

[41] Strauss ML, Jolly NP, Lambrechts MG, van Rensburg P. Screening for the production of extracellular hydrolytic enzymes by non-*Saccharomyces* wine yeasts. Journal of Applied Microbiology. 2001;**91**:182-190

[42] Romano P. Function of yeast species and strains in wine flavour. Journal of Food Microbiology. 2003;**86**:169-180

[43] Rojas V, Gil JV, Piñaga F, Manzanares P. Acetate ester formation

in wine by mixed cultures in laboratory fermentations. Journal of Food Microbiology. 2003;**86**:181-188

[44] Lleixà J, Martín V, Portillo MDC, Carrau F, Beltran G, Mas A. Comparison of fermentation and wines produced by inoculation of *Hanseniaspora vineae* and *Saccharomyces cerevisiae*. Frontiers in Microbiology. 2016;**7**:338

[45] Medina K, Boido E, Fariña L, Gioia O, Gomez ME, Barquet M, et al. Increased flavour diversity of Chardonnay wines by spontaneous fermentation and co-fermentation with *Hanseniaspora vineae*. Food Chemistry. 2013;**141**:2513-2521

[46] Viana F, Gil JV, Vallés S, Manzanares P. Increasing the levels of 2-phenylethyl acetate in wine through the use of a mixed culture of *Hanseniaspora osmophila* and *Saccharomyces cerevisiae*. International Journal of Food Microbiology. 2009;**135**:68-74

[47] Viana F, Belloch C, Vallés S, Manzanares P. Monitoring a mixed starter of *Hanseniaspora vineae–Saccharomyces cerevisiae* in natural must: Impact on 2-phenylethyl acetate production. International Journal of Food Microbiology. 2011;**151**:235-240

[48] Zhang B-Q, Shen J-Y, Duan C-Q, Yan G-L. Use of indigenous *Hanseniaspora vineae* and *Metschnikowia pulcherrima* co-fermentation with *Saccharomyces cerevisiae* to improve the aroma diversity of Vidal Blanc Icewine. Frontiers in Microbiology. 2018;**9**:2303

[49] Barbosa C, Lage P, Esteves M, Chambel L, Mendes-Faia A, Mendes-Ferreira A. Molecular and phenotypic characterization of *Metschnikowia pulcherrima* strains from Douro wine region. Fermentation. 2018;**4**:8

[50] Ferreira AM, Mendes Ferreira A, Climaco MC, Mendes FA. The role of non-*Saccharomyces* species in releasing glycosidic bound fraction of grape aroma components - A preliminary study. Journal of Applied Microbiology. 2001;**91**:67-71

[51] Marais J. Terpenes in the aroma of grapes and wines: A review. South African Journal of Enology and Viticulture. 1983;**4**:49-58

[52] Maicas S, Mateo JJ. Hydrolysis of terpenyl glycosides in grape juice and other fruit juices: A review. Applied Microbiology and Biotechnology. 2005;**67**:322-335

[53] Gunata Z, Vallier MJ, Sapis JC, Baumes R, Bayonove C. Enzymatic synthesis of monoterpenyl β-d-glucosides by various β-glucosidases. Enzyme and Microbial Technology. 1994;**16**:1055-1058

[54] Prior KJ, Bauer FF, Divol B. The utilisation of nitrogenous compounds by commercial non-*Saccharomyces* yeasts associated with wine. Food Microbiology. 2019;**79**:75-84

[55] Chatonnet P, Dubourdie D, Boidron JN, Pons M. The origin of ethylphenols in wines. Journal of the Science of Food and Agriculture. 1992;**60**:165-178

[56] Fleet G. Yeast interactions and wine flavour. International Journal of Food Microbiology. 2003;**86**:11-22

[57] Conterno L, Fondazione E, Henick-Kling T. *Brettanomyces/Dekkera* off-flavours and other wine faults associated with microbial spoilage. Managing Wine Quality, Chapter 02. 2010:346-387

[58] Suárez R, Suárez-Lepe J, Morata A, Calderón F. The production of ethylphenols in wine by yeasts of the genera *Brettanomyces* and *Dekkera*: A review. Food Chemistry. 2007;**102**:10-21

[59] Heresztyn T. Formation of substituted tetrahydropyridines by species of *Brettanomyces* and *Lactobacillus* isolated from mousy wines. American Journal of Enology and Viticulture. 1986;**37**:127-132

[60] Curtin CD, Borneman AR, Chambers PJ, Pretorius IS. De-novo assembly and analysis of the heterozygous triploid genome of the wine spoilage yeast *Dekkera bruxellensis* AWRI1499. PLoS One. 2012;7:e33840

[61] Benito S, Palomero F, Morata A, Calderón F, Suárez-Lepe JA. New applications for *Schizosaccharomyces pombe* in the alcoholic fermentation of red wines. International Journal of Food Science and Technology. 2012;**47**:2101-2108

[62] Benito Á, Jeffares D, Palomero F, Calderón F, Bai FY, Bähler J, et al. Selected *Schizosaccharomyces pombe* strains have characteristics that are beneficial for winemaking. PLoS One. 2016;**11**:e0151102

[63] Benito S, Palomero F, Morata A, Calderón F, Palmero D, Suárez-Lepe JA. Physiological features of *Schizosaccharomyces pombe* of interest in making of white wines. European Food Research and Technology. 2013;**236**:29-36

[64] Peinado RA, Moreno JJ, Maestre O, Mauricio JC. Removing gluconic acid by using different treatments with a *Schizosaccharomyces pombe* mutant: Effect on fermentation byproducts. Food Chemistry. 2007;**104**:457-465

[65] Peinado A, Maestre O, Mauricio JC, Moreno JJ. Use of a *Schizosaccharomyces pombe* mutant to reduce the content in gluconic acid of must obtained from rotten grapes. Journal of Agricultural and Food Chemistry. 2009;**57**:2368-2377

[66] Mylona AE, del Fresno JM, Palomero F, Loira I, Bañuelos MA, Morata A, et al. Use of *Schizosaccharomyces* strains for wine fermentation - Effect on the wine composition and food safety. International Journal of Food Microbiology. 2016;**232**:63-72

[67] Benito Á, Calderón F, Benito S. *Schizosaccharomyces pombe* biotechnological applications in winemaking. Methods in Molecular Biology. 1721;**2018**:217-226

[68] Fresno JMD, del Fresno JM, Morata A, Loira I, Bañuelos MA, Escott C, et al. Use of non-*Saccharomyces* in single-culture, mixed and sequential fermentation to improve red wine quality. European Food Research and Technology. 2017;**243**:2175-2185

[69] Chen K, Escott C, Loira I, del Fresno JM, Morata A, Tesfaye W, et al. Use of non-*Saccharomyces* yeasts and oenological tannin in red winemaking: Influence on colour, aroma and sensorial properties of young wines. Food Microbiology. 2018;**69**:51-63

[70] Snow PG, Gallander JF. Deacidification of white table wines through partial fermentation with *Schizosaccharomyces pombe*. American Journal of Enology and Viticulture. 1979;**30**:45-48

[71] Hazelwood LA, Daran J-M, van Maris AJA, Pronk JT, Dickinson JR. The Ehrlich pathway for fusel alcohol production: A century of research on *Saccharomyces cerevisiae* metabolism. Applied and Environmental Microbiology. 2008;**74**:2259-2266

[72] Cordente AG, Curtin CD, Varela C, Pretorius IS. Flavour-active wine yeasts. Applied Microbiology and Biotechnology. 2012;**96**:601-618

[73] Bisson LF, Karpel JE. Genetics of yeast impacting wine quality. Annual Review of Food Science and Technology. 2010;**1**:139-162

[74] Selvaraju K, Gowsalya R, Vijayakumar R, Nachiappan V. MGL2/YMR210w encodes a monoacylglycerol lipase in *Saccharomyces cerevisiae*. FEBS Letters. 2016;**590**:1174-1186

[75] Belda I, Ruiz J, Esteban-Fernández A, Navascués E, Marquina D, Santos A, et al. Microbial contribution to wine aroma and its intended use for wine quality improvement. Molecules. 2017;**22**(2):189

[76] Gamero A, Belloch C, Ibáñez C, Querol A. Molecular analysis of the genes involved in aroma synthesis in the species *S. cerevisiae, S. kudriavzevii* and *S. bayanus* var. uvarum in winemaking conditions. PLoS One. 2014;**9**:e97626

[77] Taillandier P, Lai QP, Julien-Ortiz A, Brandam C. Interactions between *Torulaspora* delbrueckii and *Saccharomyces cerevisiae* in wine fermentation: Influence of inoculation and nitrogen content. World Journal of Microbiology and Biotechnology. 2014;**30**:1959-1967

[78] Gamero A, Quintilla R, Groenewald M, Alkema W, Boekhout T, Hazelwood L. High-throughput screening of a large collection of non-conventional yeasts reveals their potential for aroma formation in food fermentation. Food Microbiology. 2016;**60**:147-159

[79] Loira I, Vejarano R, Bañuelos MA, Morata A, Tesfaye W, Uthurry C, et al. Influence of sequential fermentation with *Torulaspora delbrueckii* and *Saccharomyces cerevisiae* on wine quality. LWT - Food Science and Technology. 2014;**59**:915-922

[80] Minnaar PP, Ntushelo N, Ngqumba Z, van Breda V, Jolly NP. Effect of *Torulaspora delbrueckii* yeast on the anthocyanin and flavanol concentrations of Cabernet Franc and Pinotage wines. South African Journal of Enology and Viticulture. 2016;**36**:50-58

[81] Cordero-Bueso G, Esteve-Zarzoso B, Cabellos JM, Gil-Díaz M, Arroyo T. Biotechnological potential of non-*Saccharomyces* yeasts isolated during spontaneous fermentations of Malvar (*Vitis vinifera* cv. L.). European Food Research and Technology. 2013;**236**:193-207

[82] Gamero A, Hernández-Orte P, Querol A, Ferreira V. Effect of aromatic precursor addition to wine fermentations carried out with different *Saccharomyces* species and their hybrids. International Journal of Food Microbiology. 2011;**147**:33-44

[83] Gamero A, Tronchoni J, Querol A, Belloch C. Production of aroma compounds by cryotolerant *Saccharomyces* species and hybrids at low and moderate fermentation temperatures. Journal of Applied Microbiology. 2013;**114**:1405-1414

[84] Styger G, Prior B, Bauer FF. Wine flavor and aroma. Journal of Industrial Microbiology & Biotechnology. 2011;**38**:1145-1159

[85] Chen H, Fink GR. Feedback control of morphogenesis in fungi by aromatic alcohols. Genes & Development. 2006;**20**:1150-1161

[86] Zupan J, Avbelj M, Butinar B, Kosel J, Šergan M, Raspor P. Monitoring of quorum-sensing molecules during minifermentation studies in wine yeast. Journal of Agricultural and Food Chemistry. 2013;**61**:2496-2505

[87] Bloem A, Sanchez I, Dequin S, Camarasa C. Metabolic impact of redox cofactor perturbations on the formation of aroma compounds in *Saccharomyces cerevisiae*. Applied and Environmental Microbiology. 2016;**82**:174-183

[88] Rollero S, Mouret JR, Sanchez I, Camarasa C, Ortiz-Julien A, Sablayrolles JM, et al. Key role of lipid management in nitrogen and aroma

metabolism in an evolved wine yeast strain. Microbial Cell Factories. 2016;**15**:32

[89] Tondini F, Lang T, Chen L, Herderich M, Jiranek V. Linking gene expression and oenological traits: Comparison between *Torulaspora delbrueckii* and *Saccharomyces cerevisiae* strains. International Journal of Food Microbiology. 2019;**294**:42-49

[90] van Breda V, Jolly N, van Wyk J. Characterisation of commercial and natural *Torulaspora delbrueckii* wine yeast strains. International Journal of Food Microbiology. 2013;**163**:80-88

[91] Fabre CE, Blanc PJ, Goma G. Production of 2-phenylethyl alcohol by *Kluyveromyces marxianus*. Biotechnology Progress. 1998;**14**:270-274

[92] Medeiros ABP, Pandey A, Christen P, Fontoura PSG, de Freitas RJS, Soccol CR. Aroma compounds produced by *Kluyveromyces marxianus* in solid state fermentation on a packed bed column bioreactor. World Journal of Microbiology and Biotechnology. 2001;**17**:767-771

[93] Etschmann MMW, Sell D, Schrader J. Medium optimization for the production of the aroma compound 2-phenylethanol using a genetic algorithm. Journal of Molecular Catalysis B: Enzymatic. 2004;**29**:187-193

[94] Etschmann MMW, Schrader J. An aqueous-organic two-phase bioprocess for efficient production of the natural aroma chemicals 2-phenylethanol and 2-phenylethylacetate with yeast. Applied Microbiology and Biotechnology. 2006;**71**:440-443

[95] Lopez CLF, Beaufort S, Brandam C, Taillandier P. Interactions between *Kluyveromyces marxianus* and *Saccharomyces cerevisiae* in tequila must type medium fermentation. World Journal of Microbiology and Biotechnology. 2014;**30**:2223-2229

[96] Cian M, Capece A, Comitini F, Canonico L, Siesto G, Romano P. Yeast interactions in inoculated wine fermentation. Frontiers in Microbiology. 2016;**7**:555

[97] de-la-Fuente-Blanco A, Sáenz-Navajas MP, Ferreira V. On the effects of higher alcohols on red wine aroma. Food Chemistry. 2016;**210**:107-114

[98] Plata C, Millán C, Mauricio JC, Ortega JM. Formation of ethyl acetate and isoamyl acetate by various species of wine yeasts. Food Microbiology. 2003;**20**:217-224

[99] Seixas I, Barbosa C, Mendes-Faia A, Güldener U, Tenreiro R, Mendes-Ferreira A, et al. Genome sequence of the non-conventional wine yeast *Hanseniaspora guilliermondii* UTAD222 unveils relevant traits of this species and of the *Hanseniaspora* genus in the context of wine fermentation. DNA Research. 2019;**26**:67-83

[100] Giorello F, Valera MJ, Martin V, Parada A, Salzman V, Camesasca L, et al. Genomic and transcriptomic basis of *Hanseniaspora vineae's* impact on flavor diversity and wine quality. Applied and Environmental Microbiology. 2019;**85**(1):e01959-18

[101] Jolly NP, Augustyn OPH, Pretorius IS. The role and use of non-*Saccharomyces* yeasts in wine production. South African Journal of Enology and Viticulture. 2006;**27**:15-39

[102] Ganga MA, Martinez C. Effect of wine yeast monoculture practice on the biodiversity of non-*Saccharomyces* yeasts. Journal of Applied Microbiology. 2004;**96**:76-83

[103] Reid VJ, Theron LW, du Toit M, Divol B. Identification and partial characterization of extracellular aspartic protease genes from *Metschnikowia pulcherrima* IWBT Y1123 and *Candida apicola* IWBT Y1384. Applied and

Environmental Microbiology. 2012;**78**:6838-6849

[104] Romano P, Capece A, Jespersen L. Taxonomic and ecological diversity of food and beverage yeasts. Yeasts in Food and Beverages, Chapter 2. 2006;**2**:13-53

[105] Morata A, Loira I, Escott C, del Fresno JM, Bañuelos MA, Suárez-Lepe JA. Applications of *Metschnikowia pulcherrima* in wine biotechnology. Fermentation. 2019;**5**:63

[106] Fernández M, Ubeda JF, Briones AI. Typing of non-*Saccharomyces* yeasts with enzymatic activities of interest in wine-making. International Journal of Food Microbiology. 2000;**59**:29-36

[107] Zott K, Thibon C, Bely M, Lonvaud-Funel A, Dubourdieu D, Masneuf-Pomarede I. The grape must non-*Saccharomyces* microbial community: Impact on volatile thiol release. International Journal of Food Microbiology. 2011;**151**:210-215

[108] Manzanares P, Rojas V, Genovés S, Vallés S. A preliminary search for anthocyanin-β-D-glucosidase activity in non-*Saccharomyceswine* yeasts. International Journal of Food Science and Technology. 2000;**35**:95-103

[109] Mateo JJ, Di Stefano R. Description of the β-glucosidase activity of wine yeasts. Food Microbiology. 1997;**14**:583-591

[110] Capozzi V, Garofalo C, Chiriatti MA, Grieco F, Spano G. Microbial terroir and food innovation: The case of yeast biodiversity in wine. Microbiological Research. 2015;**181**:75-83

[111] Gore-Lloyd D, Sumann I, Brachmann AO, Schneeberger K, Ortiz-Merino RA, Moreno-Beltrán M, et al. Snf2 controls pulcherriminic acid biosynthesis and antifungal activity of the biocontrol yeast *Metschnikowia pulcherrima*. Molecular Microbiology. 2019;**112**:317-332

[112] Godoy L, Garrido D, Martínez C, Saavedra J, Combina M, Ganga M. Study of the coumarate decarboxylase and vinylphenol reductase activities of *Dekkera bruxellensis* (anamorph *Brettanomyces bruxellensis*) isolates. FEMS Microbiology Letters. 2009;**48**:452-457

[113] Godoy L, García V, Peña R, Martínez C, Ganga MA. Identification of the *Dekkera bruxellensis* phenolic acid decarboxylase (PAD) gene responsible for wine spoilage. Food Control. 2014;**45**:81-86

[114] Tchobanov I, Gal L, Guilloux-Benatier M, Remize F, Nardi T, Guzzo J, et al. Partial vinylphenol reductase purification and characterization from *Brettanomyces bruxellensis*. FEMS Microbiology Letters. 2008;**284**:213-217

[115] Granato TM, Romano D, Vigentini I, Foschino RC, Monti D, Mamone G, et al. New insights on the features of the vinyl phenol reductase from the wine-spoilage yeast *Dekkera/Brettanomyces* bruxellensis. Annals of Microbiology. 2015;**65**:321-329

[116] Agnolucci M, Vigentini I, Capurso G, Merico A, Tirelli A, Compagno C, et al. Genetic diversity and physiological traits of *Brettanomyces bruxellensis* strains isolated from Tuscan Sangiovese wines. International Journal of Food Microbiology. 2009;**130**:238-244

[117] Harris V, Ford CM, Jiranek V, Grbin PR. *Dekkera and Brettanomyces* growth and utilisation of hydroxycinnamic acids in synthetic media. Journal of Applied Microbiology. 2008;**78**:997-1006

[118] Romano A, Perello MC, de Revel G, Lonvaud-Funel A. Growth and volatile

compound production by *Brettanomyces/Dekkera bruxellensis* in red wine. Journal of Applied Microbiology. 2008;**104**:1577-1585

[119] Barata A, Pagliara D, Piccininno T, Tarantino F, Ciardulli W, Malfeito-Ferreira M, et al. The effect of sugar concentration and temperature on growth and volatile phenol production by *Dekkera bruxellensis* in wine. FEMS Yeast Research. 2008;**8**:1097-1102

[120] Crauwels S, Zhu B, Steensels J, Busschaert P, De Samblanx G, Marchal K, et al. Assessing genetic diversity among *Brettanomyces* yeasts by DNA fingerprinting and whole-genome sequencing. Applied and Environmental Microbiology. 2014;**80**:4398-4413

[121] Godoy L, Vera-Wolf P, Martinez C, Ugalde JA, Ganga MA. Comparative transcriptome assembly and genome-guided profiling for *Brettanomyces bruxellensis* LAMAP2480 during *p*-coumaric acid stress. Scientific Reports. 2016;**6**:34304

[122] Avramova M, Cibrario A, Peltier E, Coton M, Coton E, Schacherer J, et al. *Brettanomyces bruxellensis* population survey reveals a diploid-triploid complex structured according to substrate of isolation and geographical distribution. Scientific Reports. 2018;**8**:4136

[123] Cibrario A, Avramova M, Dimopoulou M, Magani M, Miot-Sertier C, Mas A, et al. *Brettanomyces bruxellensis* wine isolates show high geographical dispersal and long remanence in cellars. PLoS One. 2019;**14**(12):e0222749. DOI: 10.1101/763441

[124] Woolfit M, Rozpedowska E, Piskur J, Wolfe KH. Genome survey sequencing of the wine spoilage yeast *Dekkera* (*Brettanomyces*) *bruxellensis*. Eukaryotic Cell. 2007;**6**:721-733

[125] Hellborg L, Piskur J. Complex nature of the genome in a wine spoilage yeast, *Dekkera bruxellensis*. Eukaryotic Cell. 2009;**8**:1739-1749

[126] Avramova M, Grbin P, Borneman A, Albertin W, Masneuf-Pomarède I, Varela C. Competition experiments between *Brettanomyces bruxellensis* strains reveal specific adaptation to sulfur dioxide and complex interactions at intraspecies level. FEMS Yeast Research. 2019;**19**

[127] Crauwels S, Van Assche A, de Jonge R, Borneman AR, Verreth C, Troels P, et al. Comparative phenomics and targeted use of genomics reveals variation in carbon and nitrogen assimilation among different *Brettanomyces bruxellensis* strains. Applied Microbiology and Biotechnology. 2015;**99**:9123-9134

[128] Tiukova I, Petterson ME, Tellgren-Roth C, Bunikis I, Eberhard T, Pettersson OV, et al. Transcriptome of the alternative ethanol production strain *Dekkera bruxellensis* CBS 11270 in sugar limited, low oxygen cultivation. PLoS One. 2013;**8**:e58455

[129] Valdes J, Tapia P, Cepeda V, Varela J, Godoy L, Cubillos FA, et al. Draft genome sequence and transcriptome analysis of the wine spoilage yeast *Dekkera bruxellensis* LAMAP2480 provides insights into genetic diversity, metabolism and survival. FEMS Microbiology Letters. 2014;**361**:104-106

[130] Curtin CD, Pretorius IS. Genomic insights into the evolution of industrial yeast species *Brettanomyces bruxellensis*. FEMS Yeast Research. 2014;**14**:997-1005

[131] Serra Colomer M, Funch B, Forster J. The raise of *Brettanomyces* yeast species for beer production. Current Opinion in Biotechnology. 2019;**56**:30-35

[132] Lonvaud-Funel A. Lactic acid bacteria and malolactic fermentation in wine. Biotechnology of Lactic Acid Bacteria, Chapter 15. 2015:231-247

[133] Wibowo D, Fleet GH, Lee TH, Eschenbruch RE. Factors affecting the induction of malolactic fermentation in red wines with *Leuconostoc oenos*. The Journal of Applied Bacteriology. 1988;**64**:421-428

[134] Ciani M. Continuous deacidification of wine by immobilized *Schizosaccharomyces pombe* cells: Evaluation of malic acid degradation rate and analytical profiles. The Journal of Applied Bacteriology. 1995;**79**:631-634

[135] Benito Á, Calderón F, Benito S. New trends in *Schizosaccharomyces* use for winemaking. In: Grape and Wine Biotechnology. IntechOpen; 2016

[136] Palomero F, Morata A, Benito S, Calderón F, Suárez-Lepe JA. New genera of yeasts for over-lees aging of red wine. Food Chemistry. 2009;**112**:432-441

[137] Domizio P, Liu Y, Bisson LF, Barile D. Cell wall polysaccharides released during the alcoholic fermentation by *Schizosaccharomyces pombe* and *S. japonicus*: Quantification and characterization. Food Microbiology. 2017;**61**:136-149

Chapter 5

Chemistry and Technology of Wine Aging with Oak Chips

Maurizio Petrozziello, Tiziana Nardi, Andriani Asproudi, Maria Carla Cravero and Federica Bonello

Abstract

The use of wood chips is a common winemaking practice that has been permitted in Europe since the early 2000s. The use of oak chips, or other wood alternative products, has not always been favorably viewed by both producers and wine consumers. Beyond possible misuse, however, wood chips are a useful tool for the optimal achievement of numerous oenological objectives, including the extraction of certain volatile odor compounds from oak wood chips as well as compounds that will improve wine quality. This chapter deals with the main oenological uses of oak wood chips, the chemical transformations that underlie this practice and the effect of their utilization on wine quality. A final aspect concerns the main compositional and sensory differences between wines aged in barrel and those aged with alternative products, as well as the discriminative analytical methods used for this purpose.

Keywords: oak wood chips, alternative products, volatile compounds, tannins, wine aging

1. Introduction

The use of wood shavings in wine production has been documented in France since the nineteenth century. It was an unusual, but well-known, practice to improve the sensory characteristics of some wines using untoasted chips of oak wood or service trees [1]. The modern use of wood fragments in winemaking began in new winemaking countries in the early 1960s [2]. In 1993, the United States regulated the use of such products in oenology. In Europe, their use was opposed until the early 2000s, but finally in 2006, Regulation (EC) No. 1507/2006 of the Commission authorized the use of pieces of oak wood in oenology, enabling European producers to compete in a rapidly evolving world market. The new EU Delegated Regulation 2019/934 [3] currently regulates the use of oak wood chips in oenology, which are used for several technological purposes. The main objective is the release of desirable compounds, mainly aromas and polyphenolics, from the wood fragments into the wine during aging.

The main reason supporting the increased use of wood chips in wine production was essentially economic. The cost of wood chips is considerably lower than that of wooden casks. The production of wood fragments involves industrial-type technologies and therefore, is much cheaper than the artisanal processing of barrels. In addition, the wood, although from the same botanical species, in the case of barrels, is obtained from the most valuable part of the trunk, whereas woody fragments are usually recovered from less valuable parts or remnants of *barrique* production.

Nowadays, wood fragments represent an opportunity for wine producers to diversify their product in order to satisfy different market needs. From a technological point of view, the oenologist can choose the size of the wood pieces, the duration of the contact with the wine and the moment of application.

On the other hand, the same reasons that supported the spread of wood chips in wine production also prevented its acceptance from a regulatory point of view. Indeed, as mentioned above, there is a cost-reduction associated with the use of oak wood chips, obtained by giving a woody touch to the wine without the need to use barrels; however, without proper regulation, this could lead to fraud. If such wine is offered as barrel-aged wine [4, 5], the false use of quality indications on its label represents counterfeiting, which is detrimental to consumers and legitimate producers. Therefore, the challenge to distinguish between wines obtained using one refinement technique or another is particularly crucial, even if complicated by the multiplicity of variables involved. However, it is equally clear that, due to the need of control for both consumer protection and quality assessment, the continuous improvement of investigation methods is essential, either analytical or sensory, that allow distinguishing wines aged in barrels from those treated with chips or alternative woods.

2. Alternative products in oenology

2.1 The choice of wood

The current European legislation stipulates that the wood used in winemaking must come exclusively from the *Quercus* genus, so the use of alternative products, obtained from woods of other botanical origin is not permitted. From a technological point of view, oak wood is certainly the most suitable for the storage and aging of wine due to several positive technological characteristics such as natural durability, porosity, resistance, a high degree of impermeability, and low water content, as well as a good amount of extractable compounds useful for the positive evolution of wine. There are approximately 600 species of oak, but only a few of these are used for oenological purposes [6]. The most commonly used European species in oenology are *Q. sessilis* (or *petraea*) and *Q. robur* (or *pedunculate*), which usually grow mixed in the same area. More rarely, *Q. pyrenaica* belonging to the same genus and mainly spread between the Iberian Peninsula and western France [7] is used for winemaking purposes. American wood is mainly from *Q. alba* (White Oak) but also several other species namely, *Q. macrocarpa* and *Q. lylata* can be useful for winemaking purposes [6]. The differences among species essentially pertain to the content of extractable compounds from wood, the ellagitannins (generally higher in European oaks) and volatile aromatic compounds, namely whiskey lactone (present in a greater amount in American species). The compositional differences between individual trees, due to forestry conditions and anatomical localization of the wood or exposure, can have a greater impact than differences due to the botanical or geographical origin of the oak.

2.2 Types of alternative products and applications

The oak wood chips employed for oenological use are small pieces of wood with dimensions ranging from a minimum of 2 mm (sometimes called granulates, tabacs or wood rice) up to about 20 mm (sometimes called shavings, *copeaux*, or fragments) and are sold in single packs of a few kg. Following the Commission Delegated Regulation -EU- 2019/934 of 12 March 2019 - (Appendix 7) [3], manufacturers distinguish products based on the heating degree, suggesting their best use in different process phases. The doses vary, on average, between 0.5 and 4 g/L for white winemaking and between 1 and 6 g/L for red winemaking, being 2 g/L a balanced and orientated usual dosage. The small dimensions guarantee a high exchange surface and consequently the refinement time is limited to a range of 4–6 weeks to a few months, depending on the size [8].

Currently, the market offers numerous alternatives to classic wood fragments. They are frequently larger products than typical chips, able to simulate aging in barrels more effectively. Commercially, they are called cubes, beans, blocks, dominoes, segments and so on. Xoakers are wooden spheres of small dimensions, measuring a few centimeters in diameter. The doses are variable, generally between 2 and 4 g/L. The infusion time varies from 1 to 6 months.

Oak staves, ministaves, and sticks are wooden slats or cylinders of variable size from a few centimeters to a meter long, width range of 25–75 mm and a thickness variable between 7 and 18/22 mm. Commercially, they are denominated in a variety of ways, but all these products share the same oenological objective, which is to optimally simulate aging in barrel. The recommended doses for the smaller ones are 1–5 g/L for about 6–12 months of contact time. Sometimes doses are reported as wood area/wine volume.

Products with dimensions less than 2 mm are not permitted by European legislation [3] but can be used by New World manufacturers and take the commercial name of dust or flour. The small size of the wood considerably increases the exchange surface with the wine and consequently, the extraction processes of aromatic compounds and tannins are very rapid (15 days to 4 months). The cost of the treatment is therefore extremely low due to the low cost of the product and the limited quantities that are used. Oak powders have the advantage of being able to be pumped together with the wine during racking operations; however, it is much more difficult to remove from wine than chips or staves. The doses vary between 0.5 and 2 g/L of wine.

All these products can be used at different stages of winemaking and for various purposes. Generally, untoasted fresh chips or low toasted wood chips are used in the early stages of winemaking in order to allow color stabilization improving the anthocyanin extraction in young wines and their color characteristics during wine stabilization [9]. Toasted products can be used both during the alcoholic fermentation and after the end of malolactic fermentation.

The use of alternative products assumes that a huge array of variables must be considered including the type of wood, the size and shape of fragments, the mode of wood seasoning, the grade of toasting, the moment of use during winemaking, duration and the timing of contact as well as the interaction with yeasts and bacteria involved in winemaking. The main technological variables are examined below.

2.3 Technological factors driving product quality

Commonly, the wood used for alternative products is obtained from remnants of the barrel-making process, especially in the production of granulates or chips. This is not a negligible part of oak wood, but almost 50–75% of the total production, depending on the method used for barrel stave production, that is, traditional via splitting or by sawdust. Oak wood is otherwise obtained from trees with small diameters or presenting some physical defects [10]. Sometimes, for example, for high-quality alternative staves, the wood is the same as that used to produce barrels. In all cases, to obtain a high-quality product, it is necessary to pay particular attention to the seasoning phases that should occur in the best possible conditions [1]. Seasoning is a fundamental process useful in eliminating excess water present in the wood from 70% to about 14–18% [11]; it can be carried out naturally, alternatively, by forced drying.

During natural seasoning, the wooden planks are stacked outdoors in the open air for a variable period, which depends on the thickness, and ranges from about 2–3 years. Slats are periodically moistened to remove, via leaching, the excess astringent and bitter compounds, such as tannins and coumarins, present in the wood [12, 13]. Furthermore, the presence of unpleasant compounds is attenuated, primarily *trans*-2-nonenal which gives the wood a hint of fresh wet wood. From a microbiological point of view, this process allows the development of a varied micro-flora on the surface of the wood which promotes the formation of fungicidal substances, the transformation of phenolics of the wood and eventually the evolution of some aromatic wood precursors [14, 15]. In this regard, eugenol decreases significantly during this process, whereas other aromas such as vanillin or oak lactone (see Section 3.1.1) are subject to contrasting phenomena of neosynthesis from aromatic precursors and degradation or leaching during seasoning [16].

Artificial seasoning allows cost containment and a considerable reduction in processing time. However, natural seasoning leads to a greater accumulation of odorous compounds in the wood, in particular, volatile phenols, phenolic aldehydes, furanic compounds, and *cis*- and *trans*-β-oak lactones compared with artificial seasoning, it also appears to be more effective in reducing the excess tannins present in the wood [17]. The loss of some important compounds during artificial seasoning as polyphenolics and some aromatic compounds (lactones, phenols, fatty acids, and norisoprenoids), as well as the formation of furanic compounds deriving from the degradation of hemicellulose, is proportional to the initial moisture content of the wood along with the drying temperature [18]. Although these differences are certainly relevant, the influence of the wood piece size and the toasting intensity on the volatile composition of alternative products is higher than the method of seasoning, that is, natural or artificial [10].

After seasoning, it is necessary to eliminate residual sapwood and bark that have a very different composition to heartwood, which is the most precious part of the wood. Oak wood is then processed to reduce it to the most appropriate size and is eventually toasted. During toasting, numerous transformations take place such as, the partial degradation of the wood polyosides that leads, in turn, to the formation of numerous odorous compounds. At the same time, a large portion of the tannins undergoes degradation, with the extent depending upon the degree of toasting. Unlike the production of barrels, the toasting of alternative products is, generally, an easy, automatic process. The technological solutions for their toasting are varied and include: direct contact of the pieces of wood with a suitably heated surface; by means of a suitably heated air jet; by irradiation with IR rays, which does not allow deep toasting of the pieces; by direct contact with a flame, used almost exclusively for the production of alternative staves. Two main benefits must be considered: the first is reducing production costs; the second is standardization in terms of quality. The toasting degree of the alternative products follows that for wooden barrels; therefore, they can be distinguished as untoasted or with a light, medium or high (heavy) toasting level. However, this classification does not represent an absolute reference as the technologies used by individual companies may differ considerably [19].

3. Oak wood chips chemistry

3.1 The general composition of oak wood

From an oenological point of view, the oak heartwood is the wood of major interest. Its chemical composition includes three large groups of compounds.

The first one consists of certain polymers constituting the cell wall and the median lamella of the vegetal cells with supporting functions. The second group is composed of several extractable substances accumulated during the natural transformation from sapwood to heartwood, as well as tannin deposits which protect against plant parasites. The third group includes, in smaller quantities, several compound residues of cellular metabolism: amino acids, fatty acids, terpene compounds, carotenoids, and various minerals [20, 21].

From a quantitative point of view, the main components of oak heartwood are cellulose (40–45% of dry weight), hemicellulose (20–25%) and lignin (25–30%) and, overall, they represent by weight, the predominant portion of the wood. These polymers form a three-dimensional structure, trapping cellulose in an insoluble and rigid matrix of lignin and hemicellulose, which gives the wood its typical technological characteristics.

Chemically, cellulose is a crystalline homopolymer consisting of units of glucose 1,4-β-bonded, which has an average molecular weight of 10^6 Da corresponding to 10,000–15,000 monosaccharide units. Hemicellulose is a complex polymer that may contain pentoses (β-d-xylose, α-l-arabinose), hexoses (β-d-mannose, β-d-glucose, α-d-galactose), and uronic acids [22]. Structurally, it has two roles: binding cellulose microfibers and strengthening the cell wall. It is worth noting that both the concentration and structure of this polymer differ between sapwood and wood. Finally, lignin, which from a structural point of view is a *p*-coumaryl alcohol polymer, is responsible for the typical mechanical properties of wood, making it more resistant to both chemical and biochemical degradation. As for the mechanical properties of wood, similarly to hemicellulose, structural differences have been found between sapwood lignin and heartwood lignin [22]. The average composition of structural polymers can vary significantly according to different factors. In this regard, it is worth mentioning that the wood composition within each tree is very different and depends on the location and anatomic position of tissues. In particular, the concentration of tannins is higher in the trunk and near the base of the tree [19]. The distance from the central part of the trunk influences the chemical composition and therefore, the technological characteristics of the wood [23]; the resistance to heat treatment for instance depends on whether the wood portions are central or radial. These aspects are very important from a technological point of view because the wood used for alternative products is generally obtained from different parts to those used for barrel production.

Finally, both the rate of growth and botanical origin affect strongly the chemical-physical characteristics of the wood. Slow growth leads to fine grains, less dense wood, which is of greater resilience and, most of all, richer in extractable compounds, whereas rapid growth leads to the formation of medium or coarse grain woods. As regards botanical origin the main differences concern the tannin content, generally higher in the heartwood of European origin and the volatile compounds (lactones, norisoprenoids, sesquiterpenes, and fatty acids), usually more abundant in *Q. alba* wood [19].

3.2 Volatile compounds

3.2.1 Oak-derived volatile compounds

β-methyl-γ-octalactone is included among the minority aromatic compounds, but it has a crucial role from a sensory point of view. This is a natural lactone present in fresh oak wood, and it is characterized by intense notes of coconut, celery, and pastry with a very low perception threshold. Also called whiskey lactone or oak lactone, because it was discovered in the 1970s for the first time in whiskey and shortly thereafter in the oak wood used for their processing [24, 25], oak lactone is formed from the cyclization of 3-methyl-4-hydroxyoctanoic acid. This compound is present in oak wood as a glycoconjugate precursor namely, galloylglucoside, glucoside, and a rutinoside derivative [26, 27] and can undergo hydrolysis during both seasoning and toasting operations or during the maturation of the wine in contact with the oak wood.

From a structural point of view, oak lactone is a molecule with two chiral carbons, therefore, four different optical configurations are possible. Only two stereoisomers of oak lactone are present in oak wood [28], the "*cis*" isomer with configuration (3S, 4S) and the "*trans*" isomer with configuration (3S, 4R). From a sensory point of view, the "*cis/trans*" ratio strongly influences the wine aroma, because the *cis* isomer is several times more odorous than the *trans* isomer [29]. The *cis* form is more abundant in *Q. alba* than in the European oaks, and among the latter, the *cis* isomer is more abundant in *Q. petraea* than in *Q. robur* [30]. The *cis/trans* lactone ratio was about 3.5 and 1 in American or French oak chip-treated wines, respectively [31]. It is interesting to note that the aging method (chips or barrels) did not influence this ratio which is related to the oak's origin [32].

Generally, this compound is more abundant in American oak wood. A recent work showed that wines macerated with *Q. pyrenaica* chips presented levels of oak lactone and other wood-related compounds, more similar to those macerated with French oak wood chips and lower than American chips [33].

Among the aromatic compounds from fresh oak wood that have a significant impact on the aroma of wine can be included eugenol, which is characterized by typical clove notes and is present in small quantities, especially in sapwood [34]. Another important group of naturally occurring compounds is that of norisoprenoids [35]. These compounds originate from the degradation of carotenoids and xanthophylls present in the wood. Some of them have very low perception thresholds and perfumes that vary from floral to balsamic (see Section 3.4).

3.2.2 Compounds derived from the degradation of polyosides

The degradation of wood polymers under an inert atmosphere proceeds gradually following the increase of temperature [36]. The decomposition of hemicellulose and cellulose takes place at 200–380°C and 250–380°C, respectively, while lignin decomposition occurs over the range of 180–900°C. Moreover in normal cooperage conditions, the degradation of cellulose occurs with great difficulty and the sensory impact of its derivatives remains negligible [22]. On the contrary, hemicellulose is more susceptible to hydrolysis, which can occur during both toasting and wine aging, and leads to an increase in the total content of galactose, fructose or xylose [37], up to a few hundred mg/L. From a microbiological point of view, this aspect should be considered carefully for the development of undesired microflora in wines, taking into account the ability of some spoilage microorganism, namely *Brettanomyces* spp. [38], to consume these sugars. The formation of monosaccharides during wood toasting leads to their thermodegradation and of odorous volatile compounds neogenesis. These compounds, named furanic aldehydes (mainly furfural and 5-hydroxymethylfurfural), are contained in negligible quantities in seasoned oak wood, used to produce barrels and chips, but their content increases dramatically passing from light to medium toasting level and tends to decrease with strong toasting. From a sensory point of view their importance is modest and linked primarily to a general increase in "overall oak" perceived intensity rating and decreased "fruity" aroma [39].

Other compounds, namely maltol and cyclotene, originating from hemicellulose degradation [22], are characterized by specific caramel notes. Like furanic aldehydes, they have a limited impact on empyreumatic notes of wine aged using oak wood, because of their high perception threshold. Sugar condensation products such as DDMP (2,3-dihydro-3,5-dihydroxy-6-methyl-4(H)-pyran-4-one), HDMF (4-hydroxy-2,5- dimethylfuran-3(2H)-one) and DHM (dihydromaltol), derived from glucose and proline condensation, seem to have a greater impact on the toasty/caramel aroma [40].

Finally, the thermodegradation of lignin leads to the formation of several key aromatic compounds including aromatic hydrocarbons, phenols (mainly mono-methoxylated and dimethoxylated derivatives), aromatic aldehydes (benzoic aldehydes), and syringyl-derived compounds. Various factors such as the moisture of the wood, the intensity of the heat applied, the presence of other polymers such as cellulose can influence the chemical yield of these reactions [41]. Specially, the toasting intensity is the factor that most influences the final composition of these compounds in wine. Low/medium levels of heating lead to the formation of cinnamic and benzoic aldehydes as synapic aldehyde, coniferyl aldehyde, vanillin, and siringaldehyde. Among these, vanillin has an important aromatic impact thanks to its recognizable scent and low perception threshold [39]. More intense toasting levels lead to the formation of volatile phenols such as phenol, cresol isomers, guaiacol, 4-methylguaiacol, eugenol, isoeugenol and propiovanillone, some of them characterized by a low perception threshold and clear spiced or smoky notes.

Finally, some heterocyclic compounds present in small quantities, such as pyrazine, pyrrole pyridine, and triazole derivatives have been identified in toasted wood extracts. These compounds could be generated by Maillard's reaction during the toasting operations [42].

3.2.3 Factors affecting xylovolatiles compounds during winemaking

The type of alternative wood has an important influence on the diffusion kinetics of aromatic compounds. Generally volatile compound accumulation is faster using wood chips than staves, on the other hand, staves lead to a greater accumulation of aromas, in all cases, the extraction seems to be complete after 3–12 months of aging [10, 43].

The botanical origin of wood has great importance in defining the transfer of aromatic compounds namely, oak lactone to wine. Wines aged in contact with American oak chips showed a significant increase of *cis*-oak lactone and guaiacol [31]. On the other hand, wines aged with French oak chips exhibited a major increase of furfural, 5-methylfurfural, 4-vinylguaiacol and *trans*-oak lactone.

The aging time was related to a higher content of esters [44]; the type of wood pieces was correlated to *cis*-oak lactone levels, octanal and 5-methyl furfural, and *cis*-oak lactone with the toasting degree.

The accumulation of phenols depends on the degree of toasting but, in general, a higher accumulation of these compounds occurs in wines aged with staves compared with those aged with chips [45]. For guaiacol, 4-methylguaiacol and eugenol, the maximum accumulation has been registered between 6 and 12 months.

The main variations during wine aging involve furan aldehydes; these changes are certainly decisive for wine quality. During the first months of storage, a high accumulation of furan aldehydes is observed [45], more remarkable in aging with staves than with chips; then their content decreases sharply, similar to that occurring in wines aged in *barrique* [46, 47]. This reduction is likely due to microbiological rather than chemical reactions. The notable reductase activity of yeasts and bacteria leads to the formation of furanic alcohols from their respective aldehydes. The observed decrease of furan aldehydes in wine during aging is also due to their involvement in reactions with polyphenols and, in particular, to the formation of condensation compounds with polyphenols, mainly flavanols [48]. Vanillin and syringaldehyde exhibit accumulation and degradation curves during wine aging similar to those described for furan aldehydes [45].

Yeasts can also transform furfural to 2-furanmethanethiol (2-FMT), through the addition of hydrogen sulfide present during fermentation to furfural [49]. 2-FMT, with a very low perception threshold (0.4 ng/L) and its distinguishable odor of coffee [50], is the key aroma compound of the *boisée* aroma of wines. A similar biosynthetic mechanism has been hypothesized for the formation of benzenemethanethiol, characterized by subtle mineral notes, starting from benzaldehyde [51] and for vanillylthiol, a chemical compound reminiscent of cloves, and smoke originating from vanillin [52].

3.3 Tannins and micro-oxygenation

The "extractable fraction" of wood represents up to 10–15% of dry heartwood. Certainly, ellagitannins are the most abundant components in this fraction and, together with other compounds, are the source of many of the interesting sensory characteristics found in aged wines [53]. Eight ellagitannins have been identified in traditional oak species: castalagin, vescalagin, granidin, and roburins (A, B, C, D, and E), with the two most abundant compounds being the stereoisomers, vescalagin and castalagin [53–55]. Ellagitannins are transferred to the wine during aging, contributing to sensations of bitterness and astringency and behaving as antioxidants due to their capacity to consume oxygen [15, 53–55]. Moreover, ellagitannins directly affect wine color via reactions with anthocyanins forming red orange anthocyanin-ellagitannin complexes that are much more stable over time than free anthocyanins [55]. They often also occur in association with flavonoids to form flavano-ellagitannin derivatives (such as acutissimin A and acutissimin B) detected in aged wine and are also involved in tannin condensation [54]. Variation in ellagitannin concentration in the same wood, due to the different cooperage processes has been reported in various papers [53]; focusing on ellagitannin and alternative products, a recent study [56] performed with model wine showed that French oak chips released significantly higher amounts of ellagitannins than American oak chips at any toasting level. Their release by oak chips decreased as the toasting level increased in the French oak but this trend was not so clear in American oak.

Oxidation, condensation and polymerization reactions, in which phenolic compounds are involved, are oxygen dependent. During the aging of the wine in barrels, oxygen intake may promote, disappearance of reduction off-flavors and reduction of vegetal characteristics, but also color intensity and stabilization. This process, in barrels, is not a controlled process, but depends on wood characteristics. On the contrary, a monitored oxygenation process in stainless steel tanks can control the changes in the phenolic structure and aroma of the wine by managing oxygen-requiring reactions [57, 58].

Oxygenation of wine, which is defined as the diffusion of air or oxygen into the wine, is an authorized oenological practice in the International Code of Oenological Practices of the OIV [59]. Micro-oxygenation (MOX), consisting of dispensing micro quantities of oxygen in a controlled way, was developed in France by Patrick Ducournau at the beginning of the 1990s, then Ducournau and Laplace registered a patent for the MOX method [60, 61]. Several researches have experimented MOX combined with the application of oak alternative products with the purpose to simulate the evolution and stabilization of the phenolic compounds that spontaneously takes place in barrel.

The formation of acetaldehyde from ethanol oxidation during wine MOX favors the creation of ethyl bridges between flavanols and between flavanols and anthocyanins leading to an increase of both color intensity and color stability [62]. Ellagitannins contributes to the formation of ethylidene-bridged anthocyanin-tannin adducts through the formation of hydroperoxy radicals [63]. Moreover, MOX generally favors the formation of pyranoanthocyanins rising from the reaction of anthocyanins with small molecules, namely acetaldehyde. These compounds are likely to contribute to the red/orange hues observed in red wines during aging [64]. Acetaldehyde can also form bridges between tannin molecules, creating macromolecular structures that precipitate, leading to a decrease in astringency [65].

In general, MOX in combination with ellagitannins increased color intensity, even after 5 months of bottle aging due to increase of polymeric pigments including ethylidene-bridged compounds. These compounds contributed to the red and violet color range, but reduced hue levels, due to larger contributions to the 520 nm range [66] as shown for the treatment MOX with added tannin.

The oxygen consumption rate was clearly related to the level of released ellagitannins. Therefore, oak chips should be chosen considering their potential to release ellagitannins, not only because they can have a direct impact on the flavor and body of the wine, but also as they can protect against oxidation. Moreover, the origin and size of the oak chips seem to influence results when their addition is combined with MOX technique. The effect of American, French and Spanish oak chips or staves on in combination with MOX during red wine aging was researched; wine treated with staves (larger pieces of wood) and also aged with French oak products consumed more oxygen [67]. Finally, grape variety and especially MOX had more influence on phenolic composition and wine color than the type of oak chips which did not modify the chromatic characteristics of the red wines [68].

3.4 Secondary compounds found in oak wood

Various isoprenoid compounds and derivatives have been isolated and described in oak wood. Among them terpenes are compounds with a very low perception threshold and remarkable olfactory pleasantness; however their contribution of wood to wine is rather limited [69]. On the contrary, the presence of carotenoids is an important factor to consider for barrel production. The content of carotenoids in oak wood is found to be generally low and considerably variable between samples, depending mainly on the color of the piece of wood. Pinkish woods are mostly considered for barrel making being significantly richer in carotenoids than other colored woods. In particular, the molecules responsible for the pinkish hue of woods are found to be principally β-carotene and lutein [70]. Carotenoid compounds are highly sensitive to oxygen, light and temperature.

Pyrolysis/Gas Chromatography/Mass Spectrometry (PY/GC/MS) used on samples of French oak, to simulate the heating of barrels demonstrated that the thermal degradation products obtained after pyrolysis of β-carotene and lutein respectively were essentially norisoprenoids and sesquiterpenes [71]. During the natural seasoning the wood barrels are exposed to light and oxidation, then to heat during toasting and new aromatic compounds could be produced. With regard the norisoprenoids, over 30 different highly odorous compounds derived from the degradation of carotenoids, have been highlighted in oak wood, among which the main ones are 3-oxo-a-ionol, and dehydrovomifoliol [35]. American oak wood seems to be richer in norisoprenoids than that of European origin, while numerous compounds are common to both oak and grapes.

Finally, several pyrazines and pyridine derivatives have been detected in toasted oak wood [72]. Among them 2,5-disubstituted pyrazines seem to be responsible for rancid butter off-flavor [73]. Moreover 2-methoxy-3,5-dimethylpyrazine is linked to "corky," potato, green hazelnut, and dusty odor [74]. However this compound, synthesized by some proteobacteria, degrades at temperatures above 220° C, consequently the wood toasting reduces significantly its content.

Other extractable compounds present in smaller quantities in oak wood are amino acids, fatty acids and minerals.

4. Sensory profile of wines refined with alternative products

Several research studies have highlighted the effects of the application of wood fragments on the sensory profile of wines. Some findings on the topic are reported below.

Generally, the oak-chip treatment favors the polymerization of anthocyanins and tannins, leading to a reduction of monomers in wines. In a research involving two Italian red wines, Aglianico and Montepulciano, after 1 year of aging, the content of polymeric phenols in both red wines was about 40% higher in oak-treated samples compared with the control wines. This effect, however, strongly depends on grape variety and on the polyphenolic profile of each cultivar. The same oak chip-treated samples showed attenuation of floral and fruity descriptors and the introduction of oak notes (woody, vanilla, spicy notes, and black pepper), accompanied by a higher astringency. After 1 year of aging, the flavor complexity decreased, especially the spicy notes and astringency, which were even more reduced in the oak-treated samples than in the control wines [75].

The sensory profile of wine also depends on the phase of the winemaking process at which the oak chips are added, as well as on the oak chip dose. A study on the red wine Bobal [76] showed that the wines with oak wood chips added during alcoholic fermentation had a similar sensory profile to control wines with olfactory attributes of red fruits, liquorice, pepper, leather, tobacco, and cassis, but with some woody notes. On the other hand, wines with oak chips added during malolactic fermentation showed higher intensities for the oak descriptors than wines with the oak chips added when the malolactic fermentation had finished. The intensity of woody attributes was higher when the chips were added in higher doses (6 g/L).

Another experiment with white wines, Verdejo, showed the different effects of medium-toasted oak chips added during the alcoholic fermentation or aging stage [75]. With respect to the control (no chips were added), oak-treated wines showed a decrease of the descriptors fresh, green apple, fruity, tropical fruit, and citric and higher intensities for "ripe fruit" and "sweet" and for new attributes like coconut, sweet spices, woody (oak), and toasty in the oak chip-treated wines. Specifically, the wines with oak chips added during the alcoholic fermentation presented a lower content of volatile oak-extractable compounds, thus intensity of wood-related sensory attributes, but higher concentrations of fermentative volatile substances than the new wines aged with oak chip. The toasting status of oak chips (toasted or not toasted) seemed to be more relevant than the origin of oak in a study carried out on Chardonnay wine [77]. The differences highlighted among wines with added oak chips of different origin, did not influence the preference of the panel. On the other hand, the level of toasting was more important than the origin of the oak chips (German or French) [78] . Finally, the quantity of oak chips can have greater impact than the origin of the oak [79]. The sensory profile of the white wines *Listan blanco,* indeed, was mainly influenced by the amount of oak chips, less by the geographical origin of the oak, which was more relevant only in the case of comparing oak chips and barrels.

Nevertheless, other research results have shown sensory differences attributed to the origin of the oak chips [31]. The wines (Romanian red wines Fetească neagră) aged in contact with American oak chips had a higher intensity of vanilla, toasty and cacao aromas compared with the wines aged using French oak chips, which had a higher intensity of smoky, licorice, and toasty aromas. The degree of toasting of the wood chips or staves also influenced aromas in the same Romanian red wine [44]. A woody attribute was more evident in wines aged with oak pieces with a low degree of toasting, whereas medium plus toasted wood increased the fruity aroma descriptors.

With the aim of simulating the effect of an oak barrel, the wood pieces can be added to wine in combination with micro-oxygenation [80]. This technique (oak chips or staves combined with micro-oxygenation) can produce wines with sensory characteristics very similar to those of products aged in new American and French oak barrels for 6 months [57].

5. Oak wood chips vs. barrel

5.1 Sensory differences and consumer preferences

As previously described, the quantity of oak chips, their dimensions and shape, their degree of toasting, contact time, and the stage of the winemaking process of their application can influence the sensory characteristics of the wine in numerous ways. Moreover, traditional refinement in wooden containers can lead to different results based on factors such as the time of contact, the use of new woods or used barrels, the period and frequency of an eventual *batonnage*, together with the possibility of using containers of different sizes. Therefore, the challenge to distinguish between wines obtained using barrel or chip-aged wines is particularly complicated due to the multiplicity of variables involved.

From a sensory point of view, wines refined with chips in steel containers are not clearly different from those long preserved in new barrels [81]. On the other hand, young wines made with chips are almost sensorially indistinguishable from those stored in new barrels for short periods (about 3 months); both are characterized by light *boisé* olfactory notes. Moreover, considering the effects of oak wood chip size, wood contact time (3, 6, and 9 months) and type of container (tank or used wood barrel) in wines analyzed after 6 months of bottling [82], some evident sensory differences were observed. The sensory effect of the addition of oak wood chips was more evident for the wines in tanks than for those in used barrels. Moreover, wines in used wood barrels with wood chips were not easily distinguishable from wines with no added wood chips. Considering wines treated with alternative products, the more appreciated were those aged in contact with cubes for 3 or 6 months. The authors suggested to use oak wood chips for short-aged wines and to reuse wood barrels still in good condition, pointing out that new wood barrels give wines of higher quality [82].

The sensory differences between wines aged in wood barrels or with oak wood fragments do not always reflect significant differences regarding consumer preference [83]. In a survey on consumer preferences, a large disparity of preferences was found among the participants that traditionally consume quality wines. A large proportion of respondents (55%) said that they would not buy wines produced using oak wood chips, whereas others declared that they would buy them only if, after tasting they had perceived the same quality as the wines in wood barrels. However, young people seemed to be less traditional and more open to buy a wine produced using oak wood fragments. Another online survey [84] involving Australian wine consumers reached similar conclusions.

Finally, the use of oak wood fragments could be an interesting alternative to barrels in emerging wine countries, such as Mexico [85] or Brazil [86], due to their lower costs. Nevertheless, before using oak wood in winemaking, local producers should investigate the preferences of their potential consumers.

5.2 Main analytical techniques that can be used for the discrimination and evaluation of product quality

The challenge of explaining the sensory differences described previously from a chemical point of view, is not easy. Considerable works have addressed the chemical characterization of wines aged with alternative products. However, only some of these studies have directly compared the composition of a barrel-aged wine and that of a wine in contact with wood oak fragments [45, 87–91]. Nevertheless, it is equally clear that, because of the need of control for consumer protection and quality assessment, it is essential to develop methods of investigation that allow distinguishing wines arising from barrel aging from those that used alternative woods. To our knowledge, the main techniques that have been used so far for this purpose are: (i) methods examining differences in phenolic composition (based on High Performance Liquid Chromatography (HPLC) or on classical phenolics determination) and color analyses; (ii) chromatographic techniques investigating.

Variances in xylovolatiles and other Volatile Organic Compounds (VOCs), mostly Gas chromatography (GC) and GC–Mass Spectrometry-based (GC-MS); (iii) infrared spectroscopy (IR) for discriminating wines through an overall chemometric approach, mainly through NIR (near infrared) and MIR (Mid infrared) spectroscopy; and (iv) Nuclear Magnetic Resonance (NMR)-based methods. Most of the following research work is based on studies carried out by coupling 2 or more of the abovementioned techniques, and treating the composite data sets through multivariate statistical analyses [5].

5.2.1 Discrimination methods based on phenolics and color profiles

A primary work in 2004 analyzed changes in phenolic compounds, phenolic acids, aldehydes, and the color of wine aged for 5 months in order to determine the influence of the type of aging and oak wood origin used for storage. A discriminant analysis was performed with the aging system as the discriminant factor, using over 66 samples. Wines aged in wood barrels differed considerably in their characteristics with respect to wines aged with wood chips. Based on the obtained model, 94 and 83% of the samples aged in barrel and chips, respectively, were correctly assigned [88]. A further, comprehensive study was carried out to discriminate wine aged in wood barrels from those aged with alternative products, both during the wood contact period and bottling stage [89]. During the first 6 months of aging, wines treated with wood staves obtained characteristics that were halfway between wines with chips and those aged in wood barrels [89]. However, as the wood contact period increased so did the differences between wines stored in traditional and alternative systems (staves or chips): after a 2-year bottling period, the wines from the three systems became unique enough to tell them apart. Discriminant analysis revealed the most meaningful variables: the yellow color component, anthocyanins (cyanidin-3-glucoside, vitisin A and sum of *p*-coumaryl derivates), vanillic acid, protocatechuic aldehyde, and epicatechin.

The factorial analysis performed during the investigation of both phenolic and sensory profile of red wines aged in wood barrels (French and American) and wines aged with oak wood chips, [81] highlighted several differences; wines aged with wood chips showed characteristics closer to the wines aged in wood barrels for 3 months. Moreover, total and polymeric anthocyanins, together with acetylated and glucoside anthocyanins and pigments from the direct condensation of anthocyanin flavonol, seem to be the main variables able to differentiate wines [81]. Authors also pointed out the strong effect of grape variety on the discriminant variables associated with "alternative" or "traditional" wines.

A recent study [91], aimed to characterize the flavonoid and non-flavonoid phenolic composition of wines in contact with wood barrels, chips and staves during a 12-month aging period has hypothesized that the effect of wood on the phenolic composition was mostly associated with the original and intrinsic characteristics of each grape variety. Therefore, this work concluded that the extraction of phenolic compounds from oak wood during wine aging is closely related to the wood format, to grape variety and aging time. Consistent with papers previously presented, the study confirmed that the final effect of wood on wine is not related only to the transference of polyphenols from wood, but also to structural modifications of grape polyphenols.

5.2.2 Discrimination methods based on volatile compounds

As a matter of fact, GC-MS is successfully used for the characterization and quantitative determination of volatile and semi-volatile compounds directly issued from oak wood [92, 93].

Among the many research studies using GC-MS in order to distinguish wines aged with alternative products or barrels, the work presented by Triacca et al. [94] stands out for the number of samples analyzed. A database made up of 352 new barrel wines, 665 used barrel wines and 600 chip wines, was created in order to verify compliance with laws and regulations prohibiting the use of chips in Switzerland. Wood-related volatiles (xylovolatiles) were elaborated using chemometrics techniques (logistic regression analysis). The authors were able to assign new unknown samples, with good certainty, to the chips or barrel group [94].

A research study carried out in 2008 [95] reported the influence on aroma compounds of adding oak wood chips either in stainless steel tanks or in used barrels, comparing these wines with those aged in new barrels. Both the size of the oak chips and the contact time were considered. To separate the samples according to wood format, three discriminant functions were obtained (81.5% correct classification). Wines in new wood barrels were separated from wines with wood chips, and lactones and 5-methylfurfural were the variables with the highest discriminant power. In a recent comprehensive study [90], 75 volatile compounds were determined by applying GC-MS and flame ionization detection (GC-FID) to a wide set of wines with differing aging processes. The authors found that compounds directly related to wood have greater discriminative power for separating wines aged in barrels from those macerated with oak fragments, but no single compound permits flawless classification. Therefore, they studied the overall effect of the addition of oak fragments on a set of 231 samples and compared them to those same wines aged in oak barrels. Thus, they developed a set of criteria that enables distinguishing, with a high degree of accuracy. The application of these criteria allowed the correct classification in over 90% of cases. It was found that out of the 75 analyzed compounds, those which best enable discrimination are the following: oak lactone isomers, vanillin, acetovanillone, syringaldehyde, furfural, furfuryl alcohol, 5-methylfurfural, 5-hydroxymethylfurfural, eugenol, methyl vanillate, and ethyl vanillate. Vanillin, acetovanillone, and syringaldehyde are the compounds that were present in higher concentrations in wines fermented or macerated with wood fragments than in wines aged in barrels. Eugenol (significantly higher) and oak lactone isomers are the compounds that explain the variance in wines aged in barrels. The authors also point out that the extraction of wood-derived compounds is affected by many factors such as the age of the barrel, the application during fermentation or maceration and the dose. Nevertheless, the vanillin + acetovanillone/eugenol ratio was an essential marker for discrimination.

A more recent study [96] focused on the characterization of xylovolatile aromatic compounds using GC-MS of wines aged in barrels and those produced using oak chips. Approximately 200 Italian wines aged using oak chips or wood barrels were analyzed and 60 xylovolatile compounds were identified. Wines aged in barrels had a higher concentration of ethylvanillate, 4-ethylphenols, eugenol and whiskey-lactones than wines aged with chips, which were characterized by a generally higher concentration of furanic compounds and hydroxybenzaldehyde derivatives. The presence of 4-ethylphenols at higher concentrations in barrel-aged wines indicated that there is still, in general, a higher risk of contamination from *Brettanomyces bruxellensis* compared with chip refinement. Overall, promising perspectives arose from applying DA (discriminant analysis) to classify wines depending on the aging method (barrels vs. chips) with >96.5% success.

On the other hand, other work has demonstrated that discrimination based on VOCs is not always easy and generalized patterns are hard to establish [97]. The analytical profile of the wood-related volatiles would be expected to exhibit large variations, precluding the detection of a generalized pattern in several cases. Moreover, the aging of red and white wines would certainly follow quite different evolution routes, in terms of the enrichment in wood-related volatiles.

An innovative approach recently proposed in some papers [32, 45] couples VOC analyses with other analytic techniques to overcome the problems described above. Indeed, most previous studies investigating the influence of the type and length of the aging process have dealt with a particular subgroup of compounds (volatile or phenolic), which has recently been considered to be a limiting condition to obtain a more comprehensive view of the subject [45]. Therefore, a study carried out in 2008 on Spanish wines [45] followed the evolution of both aromatic and phenolic composition of wine during the contact time with oak wood: chips and staves, with and without micro-oxygenation. These aging procedures were compared with the traditional oak barrel aging method. Two canonical discriminant analyses were carried out to classify the different treatments based on the volatile compounds from oak wood and low-molecular weight phenols. A good separation was achieved between the three treatments (chips, staves and barrels) for both groups of compounds. In the case of VOCs, 100% of cases were correctly classified. Wines aged in barrels were correlated with a high concentration of 5-hydroxymethylfurfural and 5-methylfurfural; wines in contact with staves were correlated with a high content of vanillin and *trans*-isoeugenol. As for low-molecular weight polyphenols, 100% of the cases were also correctly classified. The variables with the highest discriminant power were syringic acid and quercetin (related to staves) and caftaric acid (associated with barrels). On the other hand, the discriminant function 2 correlates caftaric acid with the wines aged in barrels and with fragments and gallic acid with the wines treated with chips.

The main aim of another research was to evaluate the same wine after 10 years in bottles when aged in barrels and when treated with alternative products and oxygen, by means of volatile compound quantitation and color analyses [32]. Overall, the traditionally aged wines suffered a smaller decrease in color intensity, hue, significance of reds and blues, followed by wines treated with staves + MOX and chips + MOX. Furthermore, the highest concentration of *cis*-oak lactone was found in wines aged in barrels. While, generally, oak-related compounds were found at lower concentrations in barrel-wines than in alternative-products wines, except for eugenol.

It is worth noting that the GC-MS technique and VOCs analyses have been used in several studies on wood alternatives, although without direct comparison with barrel aging, investigating various technological implications: the impact of time of the addition of chips during winemaking [98, 99]; the effect of wood toasting degree and contact time with oak fragments [31, 44]; and the geographical origin of oak wood from which alternative products are issued [77, 100].

5.2.3 Discrimination based on spectroscopy methods

In the last decade, vibrational spectroscopy, namely infrared-based techniques (IR), supported by chemometric methods, was found to be a powerful technique because of its widespread use in analytical laboratories, its versatility and low economic impact. Moreover, IR technology requires minimal sample processing prior to analysis [5].

Certainly, IR spectroscopy coupled with multivariate data analysis has been used for the determination of oak volatile compounds [101–103] and for classifying barrels [104]. Subsequently, in recent years, this method has been proposed for discriminating wines aged in different types of wood containers and for different time periods [96, 105, 106], taking advantage on PLS-based calibration between GC-MS and near infrared spectroscopy (NIR).

The first publication appeared in 2012 [107], which investigated two different levels of information on fusion of NIR spectra and midinfrared (MIR) spectra from red wines aged in different ways. A total of 96 red wines, including wines aged in oak barrels, wines aged in stainless steel tanks with oak chips and without, were analyzed. Discriminant models of the three different aged wines were established, and the FDA method was applied to build the classification models of three different aged wines using the NIR, MIR and the merged spectra, reaching up to 98% correct classification with the latter. The results suggest that the spectral fusion of NIR and MIR is a promising technology for discriminating different aged wines.

A recent paper aimed to identify if spectroscopic techniques allow discriminating wines aged with alternative oak products (chips and staves) from different oak woods (American, French and Spanish) and floating micro-oxygenation (20 μg/L), compared with those aged in barrels, after 10 years of bottling [108]. The spectral information and analysis were performed in an FTIR-ATR (Fourier-transform IR-Attenuated Total Reflection). The results indicated that with this technique it is possible to clearly separate the wines aged by the three systems (chips, staves, and barrels) in the case of American oak. In the case of French oak, wines aged with chips were clearly differentiated for wines from a single grape variety and with similar oenological features.

The most recent paper on this topic [96] analyzed approximately 90 red wines issued from the same wine appellation in Italy, including commercial barrel-aged wines and wines aged using different types of commercial oak chips. Wines were analyzed in transmittance using NIR. In order to test if combined explanatory variables made it possible to discriminate treatments, an orthogonal partial least squares discriminant analysis (OPLS-DA) was carried out. Several factors were considered, including the aging process, the type of oak used for aging (wood *barriques*, big barrels or chips) and the wine typologies (differing for some oenological parameters). OPLS-DA application reached >96.5% success in classifying wines depending on the aging process (no wood, barrel/*barrique* and chips) in the internal validation check and > 90% in an external test. Certainly, further studies are needed to validate the potential of the technique in order to improve it for wine authentication, enlarging both the array of wine types (e.g., testing different appellations and grape varieties), and the range of oak alternatives (staves, cubes, blocks, and other sizes of product) to be tested, possibly in the presence of micro-oxygenation.

It is worth noting that IR spectroscopy has been used in other studies on wine aging, although without direct comparison between wood alternatives and barrel aging. For instance, FT-IR and UV-visible (UV-Vis) spectroscopic techniques combined with multivariate analysis were used to obtain regression models to study the aging level of high-quality Sherry wines [106]. Moreover, [105] wines aged in barrels made from different wood species and in stainless steel tanks, were analyzed. A complete differentiation of the samples was achieved according to grape variety, the container type and the aging time based on two spectral regions of their FT-IR spectra. The overall significance of these studies is still provisional because of the restricted data sets and the inhomogeneity of spectral ranges analyzed in the different papers. However, the results show the potential of IR spectroscopy and chemometric analysis for discriminating wines issued from different aging processes. In this regard, identifying discriminating algorithms and creating robust databases is a necessary condition to obtain a method that can be used routinely in a laboratory for quickly acquiring specific information about the type of wood used for wine aging.

Finally, nuclear magnetic resonance (NMR) is a powerful tool for analysis, quality control and authentication of wines. The main advantage of non-targeted wine analysis by ^{1}H NMR spectroscopy is the ability to collect, relatively simply and fastly, a huge amount of compositional information relating to a single sample [109]. Unlike other analytical techniques, however, it requires complex instrumentation and specialized personnel to be realized, but thanks to its versatility and reproducibility, this analytical technique seems to be very promising for the purposes described in this chapter.

6. Conclusions

In modern winemaking, oak products alternative to barrels are useful and flexible tools for wineries, enabling them to meet the needs of an increasingly wide and varied market. The compounds from oak wood impart a typical aromatic profile to wine and contribute to the increased polyphenolic patrimony of the final wine. Recently, the use of wood fragments has expanded, and new formats of alternative products have been added to those already known, significantly improving wine quality and dramatically increasing the opportunities of choice. Major advantages associated with the use of these products concern the reduction of both costs and production times. Furthermore, it enables the possibility of obtaining standardized wines whose compositional and sensory characteristics are very close to those defined in the phase of product design; in addition, the reduction of microbiological risks associated with the use of barrels is not negligible. Among the factors that most influence the sensory quality of the final wine, the dose of oak chips, the level of toasting and the moment of application must be considered. The different geographical and botanical origin of the chips seems to secondarily affect the result.

From a sensory point of view, it is not easy to distinguish products obtained by aging with chips from those obtained traditionally. Considering the consumer preference between oak chips and barrel wines, the variability among tasters is very high depending on the age and experience (e.g., young tasters and experienced tasters) and winemakers could take in consideration the specific preferences of different consumer groups. The distinction of the two different products is therefore mainly possible through laboratory investigation techniques: analytical methods allow expanding the compositional differences between the two products both for commercial purposes, but also for the control of fraudulent activities if necessary. Among the analytical techniques used, the spectroscopic ones are certainly those that, combined with multivariate statistics techniques, allow to obtain the best results in this field.

The possibility of refining wines with chips, staves or other alternative products, does not therefore oppose the use of barrels, but represents an additional option and a further opportunity for winemakers to obtain high-quality products. The two types of wines can today coexist in the wine global market, appreciated by different segments of consumers.

Author details

Maurizio Petrozziello*, Tiziana Nardi, Andriani Asproudi, Maria Carla Cravero and Federica Bonello
CREA Research Centre for Viticulture and Enology, Conegliano, Italy

*Address all correspondence to: maurizio.petrozziello@crea.gov.it

References

[1] Chatonnet P. Situation et évolution de l'utilisation des alternatives dans le monde. Revue des Oenologues. 2008;**2**:45-48

[2] Singleton VL, Draper DE. Wood chips and wine treatment; the nature of aqueous alcohol extracts. American Journal of Enology and Viticulture. 1961;**12**:152

[3] Commission Delegated Regulation (EU) 2019/934 of 12 March 2019 supplementing Regulation (EU) No 1308/2013 of the European Parliament and of the Council as regards wine-growing areas where the alcoholic strength may be increased, authorised oenological practices and restrictions applicable to the production and conservation of grapevine products, the minimum percentage of alcohol for by-products and their disposal, and publication of OIV files. Vol. 149; 2019

[4] Ebeler SE, Takeoka GR, Winterhalter P. Progress in Authentication of Food and Wine. Washington, DC: ACS Publications; 2011

[5] Versari A, Laurie VF, Ricci A, Laghi L, Parpinello GP. Progress in authentication, typification and traceability of grapes and wines by chemometric approaches. Food Research International. 2014;**60**:2-18. DOI: 10.1016/j.foodres.2014.02.007

[6] Chatonnet P, Dubourdieu D. Comparative study of the characteristics of American white oak (Quercus alba) and European oak (Quercus petraea and Q. robur) for production of barrels used in barrel aging of wines. American Journal of Enology and Viticulture. 1998;**49**:79-85

[7] Fernández de Simón B, Muiño I, Cadahía E. Characterization of volatile constituents in commercial oak wood chips. Journal of Agricultural and Food Chemistry. 2010;**58**:9587-9596. DOI: 10.1021/jf101301a

[8] Campbell JI, Pollnitz AP, Sefton MA, Herderich MJ, Pretorius IS. Factors affecting the influence of oak chips on wine flavour. Australian and New Zealand Wine Industry Journal. 2006;**21**:38-42

[9] Gordillo B, Baca-Bocanegra B, Rodriguez-Pulído FJ, González-Miret ML, García Estévez I, Quijada-Morín N, et al. Optimisation of an oak chips-grape mix maceration process. Influence of chip dose and maceration time. Food Chemistry. 2016;**206**:249-259. DOI: 10.1016/j.foodchem.2016.03.041

[10] Fernández de Simón B, Cadahía E, del Álamo M, Nevares I. Effect of size, seasoning and toasting in the volatile compounds in toasted oak wood and in a red wine treated with them. Analytica Chimica Acta. 2010;**660**:211-220. DOI: 10.1016/j.aca.2009.09.031

[11] Auer J, Rawyler A, Dumont-Beboux N. Impact of Oak Staves Seasoning Time on Barrels and Wines Quality. Switz: Rev Suisse Vitic Arboric Hortic; 2008

[12] Chatonnet P. Origin and processing of oak used in cooperage. Influence of the origin and the seasoning on the composition and on the quality of the oak. "Le bois et la qualité des vins et des eaux-de-vie", Special issue of Journal International des Sciences de la Vigne et du Vin. 1992:39-49

[13] Sefton MA, Francis IL, Pocock KF, Williams PJ. The influence of natural seasoning on the concentrations of eugenol, vanillin, and cis- and trans-β-methyl-γ-octalactone extracted from French and American Oakwood. Sciences des aliments. 1993;**13**:629-643

[14] Vivas N, Amrani-Joutei K, Glories Y, Doneche B, Brechenmacher C.

Développement de microorganismes dans le bois de cøeur de chêne (Quercus petraea Liebl) au cours du séchage naturel à l'air libre. Annales des Sciences Forestières. 1997;**54**:563-571

[15] Vivas N, Glories Y. Étude et optimisation des phénomènes impliqués dans le séchage naturel du bois de chêne. Revue Francaise d'Oenologie. 1996;**158**:28-35

[16] Spillman PJ, Sefton MA, Gawel R. The effect of oak wood source, location of seasoning and coopering on the composition of volatile compounds in oak-matured wines. Australian Journal of Grape and Wine Research. 2004;**10**:216-226

[17] Martínez J, Cadahía E, Fernández de Simón B, Ojeda S, Rubio P. Effect of the seasoning method on the chemical composition of oak heartwood to cooperage. Journal of Agricultural and Food Chemistry. 2008;**56**:3089-3096

[18] Masson E, Puech J-L. Incidence de différents modes de séchage en étuve sur les teneurs en ellagitanins et en composés volatils de merrains de chêne sessile (Quercus petraea (Matt.) Liebl.). Journal of Science & Technology. 2000;**6**:39-78

[19] Jackson RS. Wine Science: Principles and Applications. California, USA: Academic Press; 2008

[20] Moutounet M, Feuillat F, Puech JL. Caractéristiques du bois de chêne en relation avec la qualité des vins. Quad Vitic Enol Univ Torino. 2003;**26**:101-120

[21] Moutounet M, Puech J-L, Keller R, Feuillat F. Les caractéristiques du bois de chêne en relation avec son utilisation en œenologie: Le phénomène de duramisation et ses conséquences: De l'utilisation du bois sous toutes ses formes. Revue Francaise d'Oenologie. 1999;**174**:12-17

[22] Le Floch A, Jourdes M, Teissedre P-L. Polysaccharides and lignin from oak wood used in cooperage: Composition, interest, assays: A review. Carbohydrate Research. 2015;**417**:94-102. DOI: 10.1016/j.carres.2015.07.003

[23] Hamada J, Pétrissans A, Mothe F, Ruelle J, Pétrissans M, Gérardin P. Intraspecific variation of European oak wood thermal stability according to radial position. Wood Science and Technology. 2017;**51**:785-794. DOI: 10.1007/s00226-017-0910-0

[24] Suomalainen H, Nykänen L. Investigations on the aroma of alcoholic beverages. Naeringsmiddelindustrien. 1970;**23**:15-30

[25] Masuda M, Nishimura K. Branched nonalactones from some Quercus species. Phytochemistry. 1971;**10**:1401-1402. DOI: 10.1016/S0031-9422(00)84355-1

[26] Wilkinson KL, Prida A, Hayasaka Y. Role of glycoconjugates of 3-methyl-4-hydroxyoctanoic acid in the evolution of oak lactone in wine during oak maturation. Journal of Agricultural and Food Chemistry. 2013;**61**:4411-4416. DOI: 10.1021/jf400175h

[27] Masson E, Baumes R, Le Guernevé C, Puech J-L. Identification of a precursor of β-methyl-γ-octalactone in the wood of sessile oak (Quercus petraea (Matt.) Liebl.). Journal of Agricultural and Food Chemistry. 2000;**48**:4306-4309. DOI: 10.1021/jf0002950

[28] Guichard E, Fournier N, Masson G, Puech J-L. Stereoisomers of ß-methyl-γ-Octalactone. I. Quantification in brandies as a function of wood origin and treatment of the barrels. American Journal of Enology and Viticulture. 1995;**46**:419-423

[29] Abbott N, Puech J-L, Bayonove C, Baumes R. Determination of the aroma threshold of the cis and trans racemic forms of β-methyl-γ-octalactone by gas chromatography-sniffing analysis. American Journal of Enology and Viticulture. 1995;**46**:292-294

[30] Masson G, Guichard E, Fournier N, Puech J-L. Stereoisomers of ß-methyl-γ-octalactone. II. Contents in the wood of French (Quercus robur and Quercus petraea) and American (Quercus alba) oaks. American Journal of Enology and Viticulture. 1995;**46**:424-428

[31] Dumitriu G-D, Teodosiu C, Gabur I, Cotea VV, Peinado RA, López de Lerma N. Evaluation of aroma compounds in the process of wine ageing with oak chips. Food. 2019;**8**:662. DOI: 10.3390/foods8120662

[32] Del Alamo-Sanza M, Nevares I, Martínez-Gil A, Rubio-Bretón P, Garde-Cerdán T. Impact of long bottle aging (10 years) on volatile composition of red wines micro-oxygenated with oak alternatives. LWT. 2019;**101**:395-403. DOI: 10.1016/j.lwt.2018.11.049

[33] Rodríguez-Bencomo JJ, Ortega-Heras M, Pérez-Magariño S, González-Huerta C. Volatile compounds of red wines macerated with Spanish, American, and French oak chips. Journal of Agricultural and Food Chemistry. 2009;**57**:6383-6391. DOI: 10.1021/jf900739k

[34] Castino M. Ruolo delle cessioni del legno nel vino. Quad Della Scuola Spec Vitic Ed Enol. 2002;**26**:83-100

[35] Sefton MA, Francis IL, Williams PJ. Volatile norisoprenoid compounds as constituents of oak woods used in wine and spirit maturation. Journal of Agricultural and Food Chemistry. 1990;**38**:2045-2049. DOI: 10.1021/jf00101a009

[36] Gašparovič L, Koreňová Z, Jelemenský Ľ. Kinetic study of wood chips decomposition by TGA. Chemical Papers. 2010;**64**:174-181. DOI: 10.2478/s11696-009-0109-4

[37] del Alamo M, Bernal JL, del Nozal MJ, Gómez-Cordovés C. Red wine aging in oak barrels: Evolution of the monosaccharides content. Food Chemistry. 2000;**71**:189-193. DOI: 10.1016/S0308-8146(00)00145-X

[38] Galafassi S, Merico A, Pizza F, Hellborg L, Molinari F, Piškur J, et al. Dekkera/Brettanomyces yeasts for ethanol production from renewable sources under oxygen-limited and low-pH conditions. Journal of Industrial Microbiology & Biotechnology. 2011;**38**:1079-1088. DOI: 10.1007/s10295-010-0885-4

[39] Prida A, Chatonnet P. Impact of oak-derived compounds on the olfactory perception of barrel-aged wines. American Journal of Enology and Viticulture. 2010;**61**:408-413

[40] Cutzach I, Chatonnet P, Henry R, Dubourdieu D. Identification of volatile compounds with a "toasty" aroma in heated oak used in Barrelmaking. Journal of Agricultural and Food Chemistry. 1997;**45**:2217-2224. DOI: 10.1021/jf960947d

[41] Brebu M, Vasile C. Thermal degradation of lignin–a review. Cellulose Chemistry & Technology. 2009;**44**:353-363

[42] Chatonnet P. Les composés aromatiques du bois de chêne cédés aux vins, influence des opérations de chauffe en tonnellerie. Conn. Vigne Vin. 1989:81-91. (Special issue)

[43] Garde-Cerdán T, Ancín-Azpilicueta C. Review of quality factors on wine ageing in oak barrels. Trends in Food Science and Technology. 2006;**17**:438-447. DOI: 10.1016/j.tifs.2006.01.008

[44] Dumitriu G-D, Peinado RA, Cotea VV, de Lerma NL. Volatilome fingerprint of red wines aged with chips or staves: Influence of the aging time and toasting degree. Food Chemistry. 2020;**310**:125801

[45] Rubio-Bretón P, Garde-Cerdán T, Martínez J. Use of oak fragments during the aging of red wines. Effect on the phenolic, aromatic, and sensory composition of wines as a function of the contact time with the wood. Beverages. 2018;**4**:102

[46] Towey JP, Waterhouse AL. The extraction of volatile compounds from French and American oak barrels in chardonnay during three successive vintages. American Journal of Enology and Viticulture. 1996;**47**:163-172

[47] Cerdán TG, Ancín-Azpilicueta C. Effect of oak barrel type on the volatile composition of wine: Storage time optimization. LWT- Food Science and Technology. 2006;**39**:199-205. DOI: 10.1016/j.lwt.2005.01.009

[48] Nonier Bourden MF, Vivas N, Absalon C, Vitry C, Fouquet E, Vivas de Gaulejac N. Structural diversity of nucleophilic adducts from flavanols and oak wood aldehydes. Food Chemistry. 2008;**107**:1494-1505. DOI: 10.1016/j.foodchem.2007.10.012

[49] Blanchard L, Tominaga T, Dubourdieu D. Formation of Furfurylthiol exhibiting a strong coffee aroma during oak barrel fermentation from furfural released by toasted staves. Journal of Agricultural and Food Chemistry. 2001;**49**:4833-4835. DOI: 10.1021/jf010539w

[50] Tominaga T, Blanchard L, Darriet P, Dubourdieu D. A powerful aromatic volatile Thiol, 2-Furanmethanethiol, exhibiting roast coffee aroma in wines made from several Vitis vinifera grape varieties. Journal of Agricultural and Food Chemistry. 2000;**48**:1799-1802. DOI: 10.1021/jf990660r

[51] Roland A, Schneider R, Razungles A, Cavelier F. Varietal Thiols in wine: Discovery, analysis and applications. Chemical Reviews. 2011;**111**:7355-7376. DOI: 10.1021/cr100205b

[52] Floch M, Shinkaruk S, Darriet P, Pons A. Identification and organoleptic contribution of vanillylthiol in wines. Journal of Agricultural and Food Chemistry. 2016;**64**:1318-1325

[53] Martínez-Gil A, del Alamo-Sanza M, Sánchez-Gómez R, Nevares I. Alternative woods in enology: Characterization of tannin and low molecular weight phenol compounds with respect to traditional oak woods. A review. Molecules. 2020;**25**:1474. DOI: 10.3390/molecules25061474

[54] Vivas N, Glories Y. Role of oak wood Ellagitannins in the oxidation process of red wines during aging. American Journal of Enology and Viticulture. 1996;**47**:103-107

[55] Puech JL, Feuillat F, Mosedale JR, Puech C. Extraction of ellagitannins from oak wood of model casks. Vitis. 1996;**35**:211-214

[56] Navarro M, Kontoudakis N, Giordanengo T, Gómez-Alonso S, García-Romero E, Fort F, et al. Oxygen consumption by oak chips in a model wine solution; influence of the botanical origin, toast level and ellagitannin content. Food Chemistry. 2016;**199**:822-827. DOI: 10.1016/j.foodchem.2015.12.081

[57] Oberholster A, Elmendorf BL, Lerno LA, King ES, Heymann H, Brenneman CE, et al. Barrel maturation, oak alternatives and micro-oxygenation: Influence on red wine aging and quality. Food Chemistry. 2015;**173**:1250-1258

[58] Paul R. Micro-oxygenation-where now. In: Proceeding Australian Society of Viticolture and Oenology. Seminar on Use of gases in winemaking. Allen M, et al. editor. Adelaide: ASVO; October 2002:18-30

[59] OIV. International Code of Oenological Practices. Paris: International Organization of Vine and Wine; 2012, 1990

[60] Blaauw DA. Micro-Oxygenation in Contemporary Winemaking [Thesis]. Stellenbosch, South Africa: Cape Wine Academy; 2009

[61] Lemaire T, Gilis JF, Fort JP, Ducournau P. Micro-oxygenation in extended maceration and early stages of red wine maturationimmediate effects and long term consequences. In: Proc. Elev. Aust. Wine Ind. Tech. Conf. Adelaide, Australia: Australian Wine Industry Technical Conference; 2002. pp. 33-43

[62] Es-Safi N-E, Fulcrand H, Cheynier V, Moutounet M. Studies on the acetaldehyde-induced condensation of (−)-epicatechin and malvidin 3-O-glucoside in a model solution system. Journal of Agricultural and Food Chemistry. 1999;**47**:2096-2102

[63] Vivas N, Bourgeois G, Vitry C, Glories Y, de Freitas V. Determination of the composition of commercial tannin extracts by liquid secondary ion mass spectrometry (LSIMS). Journal of the Science of Food and Agriculture. 1996;**72**:309-317

[64] Oliveira J, de Freitas V, Mateus N. Chapter 14 - polymeric pigments in red wines. In: Morata A, editor. Red Wine Technol. California, USA: Academic Press; 2019. pp. 207-218. DOI: 10.1016/B978-0-12-814399-5.00014-1

[65] del Carmen LM, Canals R, González-Manzano S, Canals JM, Santos-Buelga C, Zamora F. Influence of micro-oxygenation treatment before oak aging on phenolic compounds composition, astringency, and color of red wine. Journal of Agricultural and Food Chemistry. 2006;**54**:4246-4252

[66] Sartini E, Arfelli G, Fabiani A, Piva A. Influence of chips, lees and micro-oxygenation during aging on the phenolic composition of a red Sangiovese wine. Food Chemistry. 2007;**104**:1599-1604

[67] Del Álamo M, Nevares I, Gallego L, de Simón BF, Cadahía E. Micro-oxygenation strategy depends on origin and size of oak chips or staves during accelerated red wine aging. Analytica Chimica Acta. 2010;**660**:92-101

[68] Pérez-Magariño S, Ortega-Heras M, Cano-Mozo E, Gonzalez-Sanjose ML. The influence of oak wood chips, micro-oxygenation treatment, and grape variety on colour, and anthocyanin and phenolic composition of red wines. Journal of Food Composition and Analysis. 2009;**22**:204-211

[69] Di Stefano R. Composti volatili di vini rossi invecchiati in legno. Vini Ltalia L; 1988;**30**:9-18

[70] Masson G, Baumes R, Puech J-L, Razungles A. Demonstration of the presence of carotenoids in wood: Quantitative study of cooperage oak. Journal of Agricultural and Food Chemistry. 1997;**45**:1649-1652

[71] Nonier M-F, De Gaulejac NV, Vivas N, Vitry C. Characterization of carotenoids and their degradation products in oak wood. Incidence on the flavour of wood. Comptes Rendus Chimie. 2004;**7**:689-698

[72] Maga JA, Chen Z. Pyrazine composition of wood smoke as influenced by wood source and smoke generation variables. Flavour and Fragrance Journal. 1985;**1**:37-42

[73] Shinkaruk S, Floch M, Prida A, Darriet P, Pons A. Identification of Dialkylpyrazines off-flavors in oak wood. Journal of Agricultural and Food Chemistry. 2019;**67**:10137-10144. DOI: 10.1021/acs.jafc.9b03185

[74] Chatonnet P, Fleury A, Boutou S. Origin and incidence of 2-methoxy-3, 5-dimethylpyrazine, a compound with a "fungal" and "corky" aroma found in cork stoppers and oak chips in contact with wines. Journal of Agricultural and Food Chemistry. 2010;**58**:12481-12490

[75] Baiano A, De Gianni A, Mentana A, Quinto M. Effects of the treatment with oak chips on color-related phenolics, volatile composition, and sensory profile of red wines: The case of Aglianico and Montepulciano. European Food Research and Technology 2016;**242**:745-767

[76] García-Carpintero EG, Gallego MAG, Sánchez-Palomo E, Viñas MAG. Sensory descriptive analysis of Bobal red wines treated with oak chips at different stages of winemaking. Australian Journal of Grape and Wine Research. 2011;**17**:368-377. DOI: 10.1111/j.1755-0238.2011.00161.x

[77] Guchu E, Díaz-Maroto MC, Pérez-Coello MS, González-Viñas MA, Ibáñez MDC. Volatile composition and sensory characteristics of chardonnay wines treated with American and Hungarian oak chips. Food Chemistry. 2006;**99**:350-359. DOI: 10.1016/j.foodchem.2005.07.050

[78] Návojská J, Brandes W, Nauer S, Eder R, Frančáková H. Influence of different oak chips on aroma compounds in wine. Journal of Microbiology, Biotechnology and Food Sciences. 2020;**9**:957-971

[79] Gutiérrez Afonso VL. Sensory descriptive analysis between white wines fermented with oak chips and in barrels. Journal of Food Science. 2002;**67**:2415-2419

[80] Gómez-Plaza E, Bautista-Ortín AB. Chapter 10 - emerging technologies for aging wines: Use of chips and micro-oxygenation. In: Morata A, editor. Red Wine Technol. UK, USA: Academic Press; 2019. pp. 149-162. DOI: 10.1016/B978-0-12-814399-5.00010-4

[81] Ortega-Heras M, Pérez-Magariño S, Cano-Mozo E, González-San José ML. Differences in the phenolic composition and sensory profile between red wines aged in oak barrels and wines aged with oak chips. LWT- Food Science and Technology. 2010;**43**:1533-1541. DOI: 10.1016/j.lwt.2010.05.026

[82] Cano-Lopez M, Bautista-Ortin AB, Pardo-Minguez F, Lopez-Roca JM, Gomez-Plaza E. Sensory descriptive analysis of a red wine aged with oak chips in stainless steel tanks or used barrels: Effect of the contact time and size of the oak chips. Journal of Food Quality. 2008;**31**:645-660. DOI: 10.1111/j.1745-4557.2008.00226.x

[83] Pérez-Magariño S, Ortega-Heras M, González-Sanjosé ML. Wine consumption habits and consumer preferences between wines aged in barrels or with chips. Journal of the Science of Food and Agriculture. 2011;**91**:943-949. DOI: 10.1002/jsfa.4269

[84] Crump A, Johnson T, Bastian S, Bruwer J, Wilkinson K. Consumers' knowledge of and attitudes toward the role of oak in winemaking. International Journal of Wine Research. 2014;**6**:21-30. Available from: https://doi.org/10.2147/IJWR.S70458

[85] Espitia-López J, Escalona-Buendía HB, Luna H, Verde-Calvo JR. Multivariate study of the evolution of phenolic composition and sensory profile on mouth of Mexican red merlot wine aged in barrels vs wood chips. CyTA Journal of Food. 2015;**13**:26-31. DOI: 10.1080/19476337.2014.907826

[86] Alencar NMM, Ribeiro TG, Barone B, Barros APA, Marques ATB, Behrens JH. Sensory profile and check-all-that-apply (cata) as tools for evaluating and characterizing syrah wines aged with oak chips. Food Research International. 2019;**124**:156-164

[87] Arapitsas P, Antonopoulos A, Stefanou E, Dourtoglou VG. Artificial aging of wines using oak chips. Food Chemistry. 2004;**86**:563-570. DOI: 10.1016/j.foodchem.2003.10.003

[88] Del Alamo SM, Escudero JAF, De Castro Torío R. Changes in phenolic compounds and colour parameters of red wine aged with oak chips and in oak barrels. Food Science and Technology International. 2004;**10**:233-241. DOI: 10.1177/1082013204046095

[89] Del Álamo M, Nevares I, Gallego L, Martin C, Merino S. Aging markers from bottled red wine aged with chips, staves and barrels. Analytica Chimica Acta. 2008;**621**:86-99. DOI: 10.1016/j.aca.2008.05.014

[90] Hernández-Orte P, Franco E, Huerta CG, García JM, Cabellos M, Suberviola J, et al. Criteria to discriminate between wines aged in oak barrels and macerated with oak fragments. Food Research International. 2014;**57**:234-241. DOI: 10.1016/j.foodres.2014.01.044

[91] Laqui-Estaña J, López-Solís R, Peña-Neira Á, Medel-Marabolí M, Obreque-Slier E. Wines in contact with oak wood: The impact of the variety (Carménère and cabernet sauvignon), format (barrels, chips and staves), and aging time on the phenolic composition. Journal of the Science of Food and Agriculture. 2019;**99**:436-448. DOI: 10.1002/jsfa.9205

[92] Vichi S, Santini C, Natali N, Riponi C, López-Tamames E, Buxaderas S. Volatile and semi-volatile components of oak wood chips analysed by accelerated solvent extraction (ASE) coupled to gas chromatography–mass spectrometry (GC–MS). Food Chemistry. 2007;**102**:1260-1269. DOI: 10.1016/j.foodchem.2006.07.023

[93] De Rosso M, Cancian D, Panighel A, Dalla Vedova A, Flamini R. Chemical compounds released from five different woods used to make barrels for aging wines and spirits: Volatile compounds and polyphenols. Wood Science and Technology. 2009;**43**:375-385. DOI: 10.1007/s00226-008-0211-8

[94] Triacca M, Boldi M-O, Rawyler A. Différenciation des vins de barrique et de copeaux. Revue Suisse de Viticulture, d'Arboriculture et d'Horticulture. 2013;**45**:240-247

[95] Bautista-Ortín AB, Lencina AG, Cano-López M, Pardo-Mínguez F, López-Roca JM, Gómez-Plaza E. The use of oak chips during the ageing of a red wine in stainless steel tanks or used barrels: Effect of the contact time and size of the oak chips on aroma compounds. Australian Journal of Grape and Wine Research. 2008;**14**:63-70. DOI: 10.1111/j.1755-0238.2008.00008.x

[96] Petrozziello M, Bonello F, Asproudi A, Nardi T, Tsolakis C, Bosso A, et al. Differences in xylovolatiles composition between chips or barrel aged wines. OENO One. 2020;**54**(3):513-522. DOI: 10.20870/oeno-one.2020.54.3.2923

[97] Loupassaki S, Abouzer M, Basalekou M, Fyssarakis I, Makris DP. Evolution pattern of wood-related volatiles during traditional and artificial ageing of commercial red and white wines: Association with sensory analysis. International Food Research Journal. 2016;**23**(4):1459-1465

[98] Gómez García-Carpintero E, Gómez Gallego MA, Sánchez-Palomo E, González Viñas MA. Impact of alternative technique to ageing using oak chips in alcoholic or in malolactic fermentation on volatile and sensory composition of red wines. Food Chemistry. 2012;**134**:851-863. DOI: 10.1016/j.foodchem.2012.02.194

[99] Gómez García-Carpintero E, Sánchez-Palomo E, González Viñas MA. Volatile composition of Bobal red wines subjected to alcoholic/malolactic fermentation with oak chips. LWT- Food Science and Technology. 2014;**55**:586-594. DOI: 10.1016/j.lwt.2013.10.024

[100] de Simón BF, Cadahía E, Muiño I, Del Álamo M, Nevares I. Volatile composition of toasted oak chips and staves and of red wine aged with them. American Journal of Enology and Viticulture. 2010;**61**:157-165

[101] Garde-Cerdán T, Lorenzo C, Alonso GL, Salinas MR. Employment of near infrared spectroscopy to determine oak volatile compounds and ethylphenols in aged red wines. Food Chemistry. 2010;**119**:823-828

[102] Garde-Cerdán T, Lorenzo C, Zalacain A, Alonso GL, Salinas MR. Using near infrared spectroscopy to determine haloanisoles and halophenols in barrel aged red wines. LWT- Food Science and Technology. 2012;**46**:401-405

[103] Zhang Y, Chen J, Lei Y, Zhou Q, Sun S, Noda I. Discrimination of different red wine by Fourier-transform infrared and two-dimensional infrared correlation spectroscopy. Journal of Molecular Structure. 2010;**974**:144-150. DOI: 10.1016/j.molstruc.2010.03.021

[104] Michel J, Jourdes M, Le Floch A, Giordanengo T, Mourey N, Teissedre P-L. Influence of wood barrels classified by NIRS on the Ellagitannin content/composition and on the organoleptic properties of wine. Journal of Agricultural and Food Chemistry. 2013;**61**:11109-11118. DOI: 10.1021/jf403192y

[105] Basalekou M, Pappas C, Tarantilis P, Kotseridis Y, Kallithraka S. Wine authentication with Fourier transform infrared spectroscopy: A feasibility study on variety, type of barrel wood and ageing time classification. International Journal of Food Science and Technology. 2017;**52**:1307-1313. DOI: 10.1111/ijfs.13424

[106] Ferreiro-González M, Ruiz-Rodríguez A, Barbero GF, Ayuso J, Álvarez JA, Palma M, et al. FT-IR, Vis spectroscopy, color and multivariate analysis for the control of ageing processes in distinctive Spanish wines. Food Chemistry. 2019;**277**:6-11. DOI: 10.1016/j.foodchem.2018.10.087

[107] Tao S, Li J, Li J, Tang J, Mi J, Zhao L, et al. Discriminant analysis of red wines from different aging ways by information fusion of NIR and MIR spectra. In: Comput. Comput. Technol. Agric. V. Berlin Heidelberg: Springer; 2012. pp. 478-483

[108] Sanchez-Gomez R, Anjos O, Nevares I, Delgado T, Del Alamo-Sanza M. Discrimination of aging wines with alternative oak products and micro-oxygenation by FTIR-ATR. Vitis. 2019;**58**:77-82

[109] Amargianitaki M, Spyros A. NMR-based metabolomics in wine quality control and authentication. Chemical and Biological Technologies in Agriculture. 2017;**4**:9. DOI: 10.1186/s40538-017-0092-x

Chapter 6

Port Wine: Production and Ageing

Juliana Milheiro, Fernanda Cosme, Luís Filipe-Ribeiro and Fernando M. Nunes

Abstract

Port wine is a traditional and worldwide renowned fortified wine produced in the Douro Demarcated Region (DDR) Northeast of Portugal by specific and traditional winemaking practices. The final quality and uniqueness of the different Port wine styles are highly dependent on the ageing process, namely time, temperature and oxygen levels that will ultimately dictate the type and extension of the chemical changes that occur during this process. These chemical changes occurring during the Port wine ageing process results in significant changes in colour and aroma according to the different ageing conditions employed for the different Port wine styles. This chapter intends to give a broad and generic overview of the known and potential chemical changes occurring during ageing of Port wine that are responsible for the changes in the sensory profile observed during the ageing process. Also, the known chromatic and aromatic characteristics of the different Port wine styles and the specific ageing processes, reductive or oxidative, are reviewed.

Keywords: Port wine, Douro Demarcated Region, ageing process, colour, aroma

1. Introduction

Port wine is a traditional fortified wine produced in the Douro Demarcated Region (Northeast of Portugal in the Douro Valley, **Figure 1**) under very specific conditions. There are several Port wine styles being related to the winemaking and ageing process and also to the ageing time, which enhances uniqueness to the wines and recognition throughout the world. The Douro Demarcated Region is located within the Douro River basin, surrounded by mountains, having a total area of approximately 250,000 hectares. This region has singular climate and soil characteristics for the production of selected grape varieties for the Port winemaking that contributes to the distinctive characteristics of Port wines and guarantee that these wines are irreproducible elsewhere. This area is divided into three naturally distinct sub-regions (**Figure 1**) concerning the climatic as well as socio-economic factors, "Baixo Corgo" (Lower Corgo), "Cima Corgo" (Upper Corgo), and "Douro Superior" (Upper Douro) [1].

Different grape varieties are used in the production of Port wine, being usually produced by their blend. There is evidence that the grape varieties determines the wine character, even after the extended ageing process. The grape varieties that may be cultivated in the Douro Demarcated Region are regulated by Decree-Law n°104/85, 10th of April, 1985 [2]. Accordingly, Port wine is produced only from the authorised grape varieties, being the principal red grape varieties recommended for Port wine production "Touriga Nacional", "Mourisco Tinto", "Bastardo", "Tinta

Figure 1.
Location of the Douro Demarcated Region in the northeast of Portugal and of the three sub-regions: "Baixo Corgo", "Cima Corgo" and "Douro Superior" according to [1].

Roriz", "Tinta Cão", "Tinta Amarela", "Tinta Barroca", "Touriga Franca" and "Tinta Francisca", since these grape varieties produce wines with stable colour, fruity aroma and sugar content, characteristics required to produce good quality Port wines. The white grape varieties used for White Port wine production are "Malvasia Fina", "Viosinho", "Donzelinho Branco", "Gouveio", "Rabigato", and "Códega" [3].

In the Port wine vinification process, the alcoholic fermentation is stopped according to the desired residual sugar content by the addition of a wine spirit known as "aguardente vínica" (with an alcohol content of about 77% (v/v)), to an alcohol content up to 18–22% (v/v) of the final product. Therefore, Port wine is a naturally sweet fortified wine since the natural sugar from the grapes is not completely transformed into alcohol. After the vinification process, Port wine is usually stored and aged in wood barrels of different sizes, from 2 years to many decades in accordance with the intended Port wine style. The Port wine ageing process can take place either in the Douro Valley or in Vila Nova de Gaia (Porto), in order to qualify for a Certificate of Origin from the "Instituto dos Vinhos do Douro e Porto" (IVDP).

2. Port wine production and styles

Port wine is subjected to an extensive set of legislation and regulations. According to the Decree-Law n°173/2009 of 3rd of August [4], the IVDP, located in Oporto city, has the responsibility of promoting and perform the quality control of Port wine, as well as the amount of Port wine that can be produced annually, regulating all the production process, and the protection of the denominations of origin Douro and Port and the geographical indication of the Douro Region. The panel of expert tasters of IVDP is responsible for the certification and approval of wines and wine spirits, as well as the granting of the guarantee seal [5].

In the Port wine vinification process, the alcoholic fermentation is stopped, between 6% and 9% (v/v) alcohol content, according to the Port wine style and sweetness desired. Therefore, the wine is runoff from the skins, and it is fortified with a wine spirit containing 77% (v/v) of ethanol to raise the alcohol concentration to 18–22% (v/v). The average proportions of wine spirit added are 115 L for each 435 L of fermenting wine. The wine spirit allowed to be used in Port wine production required rigorous quality standards regulated by the laboratories and panel of tasters of IVDP. The sensory characteristics evaluated by the panel are turbidity, colour, aroma, and taste. The analytical parameters (ethyl carbamate, total higher alcohols, acetaldehyde (ethanal), ethyl acetate, methanol, 2-butanol, 1-butanol, allylic acid, cyanidric acid, calcium, copper, iron, alcohol content, total acidity, and density) must be below of the allowed limits described in Regulation n° 84/2010 [6]. Contrasting with most other fortifying spirits, the wine spirit used in Port wine production is not highly rectified; therefore, it contains many flavourants, especially higher alcohols, and aldehydes and this fortification process results in a high concentration of acetaldehyde in these initial wines.

Port wine can be extra dry, dry, semi-dry, sweet or very sweet, according to the levels of unfermented sugars remaining (**Table 1**) that is dependent on the time of wine spirit addition to stop the alcoholic fermentation [1].

There are four main different styles of Port wine, designated as Tawny, Ruby, White and Rosé (**Figure 2**). Port wine intended for Tawny style are obtained from different wines in different stages of ageing. During the ageing in the wood barrels, the red colour of the wines gradually develops into tawny, medium tawny or light tawny, with an aroma of dried fruits and wood [1]. In this style, there are some special categories like Tawny Reserve, Tawny with Indication of Age (10, 20, 30 and 40 years) and "Colheita". This last category is an exception, as these wines are from a single vintage [7]. All these wines are ready to drink when they are bottled [1]. Port wines that belong to Ruby style are wines that the evolution of their deep red colour is limited and the fruity character is maintained [1]. Within this Port style, special categories

Sweetness	Specific gravity (g/cm^3, 20°C)	°Baumé (°Bé)	Sugar content (g/L)
Extra dry	<0.9980	0.0	<40
Dry	0.9980–1.0079	0.0–1.3	40–65
Semi-dry	1.0080–1.0179	1.4–2.7	65–85
Sweet	1.0180–1.0339	2.8–5.0	85–130
Very sweet	>1.0340	> 5.0	>130

Table 1.
Port wine classification according to their sugar content.

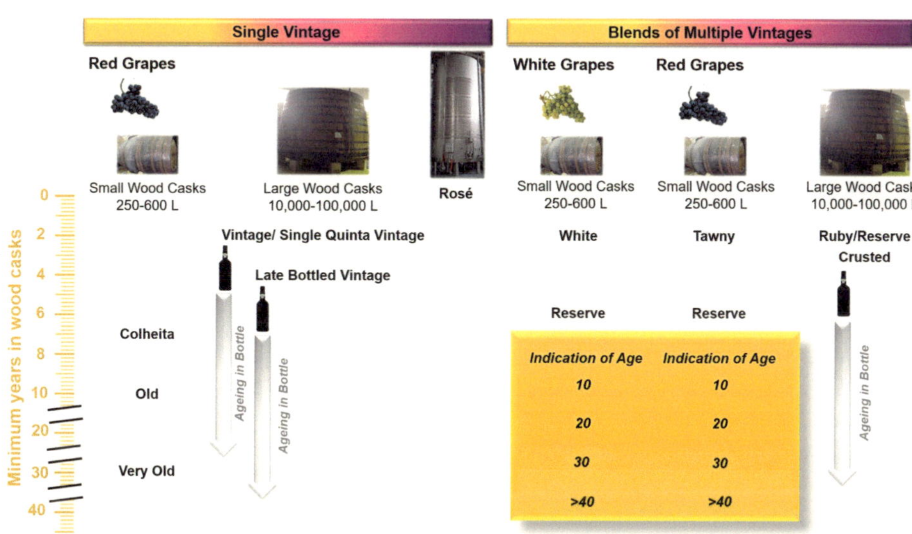

Figure 2.
Port wines styles and categories according to [1, 7].

can be found like Crusted, Reserve, Late Bottled Vintage (LBV) and Vintage [7]. Crusted Ports are high quality, very full-bodied, deep coloured wines, obtained by blending wines from numerous vintages, aged for a minimum of 2 years in wood barrels and then bottled and aged during further 3 years. The year in which the wine was bottled must be indicated on the bottle label [1, 7]. LBV is a special single harvest and it is bottled after 4 to 6 years from the harvest, with the previous ageing in vats (wood or stainless steel) and has a deep ruby red colour, extremely full body and rich in the mouth. To be recognised as Vintage, the wines must present an outstanding quality and with a very full bodied and have a deep colour. Vintage and LBV are good for storing since they age well in the bottle [1]. The types of Port wines made from red grape varieties vary in colour from deep purple to light gold, with transitional hues like tawny, golden tawny, golden and light gold. White Port wines differ according to shorter or longer periods of ageing and different degrees of sweetness. The colour of White Port wines varies from pale yellow, straw to golden white. However, when aged in wood barrels for many years, white Port wines develop, through natural oxidation, a golden hue that is very similar to that of a very old Tawny Port wine made from red grapes [1]. The special categories inside this style are similar to those of Tawny Port wines [7]. White Port wine is a Port wine style with increasing market expression. Nowadays it represents 33% of the total Port wine sales with a higher market value (+6.2%) when compared to Tawny Port wine, and nearly 80% of the production is exported [1]. Rosé Port wine is a recent Port wine style, first released in the market in 2008 by Croft, part of the Taylor Fladgate Partnership. It is technically a Ruby Port but fermented in a similar way to a Rosé wine, with limited grape skin maceration, thus producing the pink colour. Croft came up with Rosé Port as a way to introduce the pleasures of Port wine to a younger market. The colour of this type of Port wine may go from light ruby to pale salmon. This style is commonly described as a light and fresh style of Port wine that is very fruity and enjoyable [1]. Different ageing processes leads to numerous Port wines quality categories, presenting different colours (from white to deep purple), sweetness (sweet to dry) and a wide range of flavours. The Port wines styles and categories are summarised in **Figure 2**.

3. Port wine ageing process

The ageing process is an important period for this wine and includes storage, ageing in wood barrels or vat tanks and/or bottle ageing. When aged in old wood barrels their size depends on the Port wine style. Wine intended for Ruby and Vintage Port wine production will be aged in large wood barrels and that intended for Tawnies will be aged in small ones. The Tawny Port wine undergoes an oxidative ageing process, while Ruby and Vintage Port have a much less oxidative ageing termed reductive ageing process [1, 7].

Ruby, Reserve and LBV Port wines usually age in large wood barrels for two, three years or even six years (LBV) and have a deep red youthful colour and intense fruity flavours, evocative of cherry and blackberry. Tawny Port wines (10, 20, 30 and 40 years old) age for longer periods in small wood barrels and show nuttiness and aroma of butterscotch. White Port wines usually age for two or three years in large wood barrels. Traditionally, White Port wines are fermented with skin contact like Red Port wines; in this case, the wines are aged in conditions that results in its oxidation. Nevertheless, the trend is for a shorter maceration period, to obtain White Port wines with a pale colour and fresh aromas [1]. Like Red Port wines, most White Port wines are fortified when half of the grape sugar concentration has been fermented. Semi-dry and dry White Port wines are fortified later, or when alcoholic fermentation is finished [8].

The fortification process gives a high concentration of acetaldehyde to the wines. Acetaldehyde is probably responsible for the colour stability by favouring the production of anthocyanin-tannin polymers (discussed below) [9, 10]. The high sugar concentration retained tends to mask the bitterness of small flavanols, but not their astringency [8, 11]. Young Port wines are generally sweet, intensely red with a high concentration in tannins and with a fruity aroma. The colour, aroma and flavour of young Port wines are due to compounds from the grape, from the alcoholic fermentation and from the wine spirit used for the fortification. These wines need to age to develop the complex sensory attributes typically associated with the several Port wine styles. Port wines can be aged for a minimum of three years to a decade or more in old wood barrels to develop their character; normally it is aged in an old wood barrel, ranging from 525 or 600 L capacity up to 200,000 L. The type and length of the ageing process, as well as the capacity of the ageing barrel, and the oxygenation during racking, influenced the Port wine style that will be developed depending mainly on the wine style planned. Therefore, wines destined for Ruby and Vintage Port wines will be usually aged in large wood barrels, while those intended for the production of Tawny Port wines will be aged in small wood barrels (**Figure 2**). Racking is a very important operation during Port wine ageing and may be performed periodically. Slight fortification after each racking operation to adjust the alcoholic content up to 22% (v/v), compensating the volume lost via evaporation from the wood barrels [8].

The value of aged Tawny Port wine is linked to the characteristic aroma compounds developed during the ageing process in small wood barrels that allow the admission of oxygen. This oxidative ageing is influenced by factors such as oxygen levels, temperature, and pH. The high quality Tawny Port wines generally have an 'indication of age (10, 20, 30, or over 40 years old)' on the bottle, and are a blend of wines aged in wood barrels from different years. The age indicated on the label corresponds to a wine that has the sensory characteristics recognised by the IVDP of a wine aged in wood with 10, 20, 30, or over 40 years, obtained by blending wines with different ages. Tawny Port wines produced from a single vintage are referred to as 'Colheita' Port wines, aged in wood barrels for a minimum of 7 years [1].

Ruby Port wines have red colour, full-bodied structure and often still quite fruity in character when the wines are ready to drink. Ruby Port wines are aged between 3 to 5 years before blending and bottling in old large wood barrels known as "balseiros" of larger capacity, between 10.000 and 100.000 L, and do not usually have any wood-aged characteristics. The flavour modifies from an intensely fruity, even spirity character when the wines are very young to a rich fruity ruby wine after 3 to 5 years ageing in wood. They are used to age full-bodies and fruity wines such as Ruby, LBV and Vintage Port wine. These wines age more slowly than those aged in smaller wood barrels, retaining their structure and fresh fruity aromas that are the main characteristics of these wines. Some special Ruby Port wines (the so called Vintage Port wines) have a considerable bottle ageing process, giving lighter red wines, with often a very fruity character, despite having aged for two decades or more. Vintage Port wines are aged in wood barrels for two or three years, followed by a considerable ageing time in a bottle in the so called reductive ageing (10 to 50 years or more before consumption), and so it develops a different character from those wines aged exclusively in wood barrels. These wines remain fruity and with a red colour. Consequently, Vintage Port wines develop much of its distinctive bouquet from a long process of reductive ageing in bottle.

After the initial period in wood, LBV wines are aged in dark glass bottles in cool dark cellars with controlled temperature, ventilation and relative humidity. The vintage year is always indicated on the label [1].

White Port wine is made in the same way as red Port wines. However, there is a tendency to reduce the skin contact time, and even to ferment clarified grape juice at a lower temperature (18–20°C), to obtain wines with fruity aromas. The wines are aged in small size old wood barrels for a minimum of three years before its commercialisation depending on the desired White Port wine colour type.

3.1 Chemical changes and sensory development during Port wine ageing

3.1.1 Colour development

The colour of red and white Port wines is one of the main quality parameters of the different Port wines styles. For Port wines made from red grape varieties, the initial wine colour is mainly due to the anthocyanins extracted from grape skins during vinification. Nevertheless, in a young Port wine, the percentage of colour due to the so called polymeric pigments is already 23 to 30% indicating that changes in the compounds responsible for the colour have already started during the short alcoholic fermentation and wine spirit addition to stop the alcoholic fermentation. The red Port wine colour increases up to 80% during the first months of ageing depending on the concentration of free acetaldehyde present in the young wine. After 46 weeks of ageing, the polymeric pigments can make up 78 and 98% of the wine colour [12]. The colour evolution during ageing is explained by the involvement of anthocyanins in different equilibria in solution and their simultaneous transformation through various concurring chemical reactions to a range of other pigments, many of them still unknown (**Figure 3**).

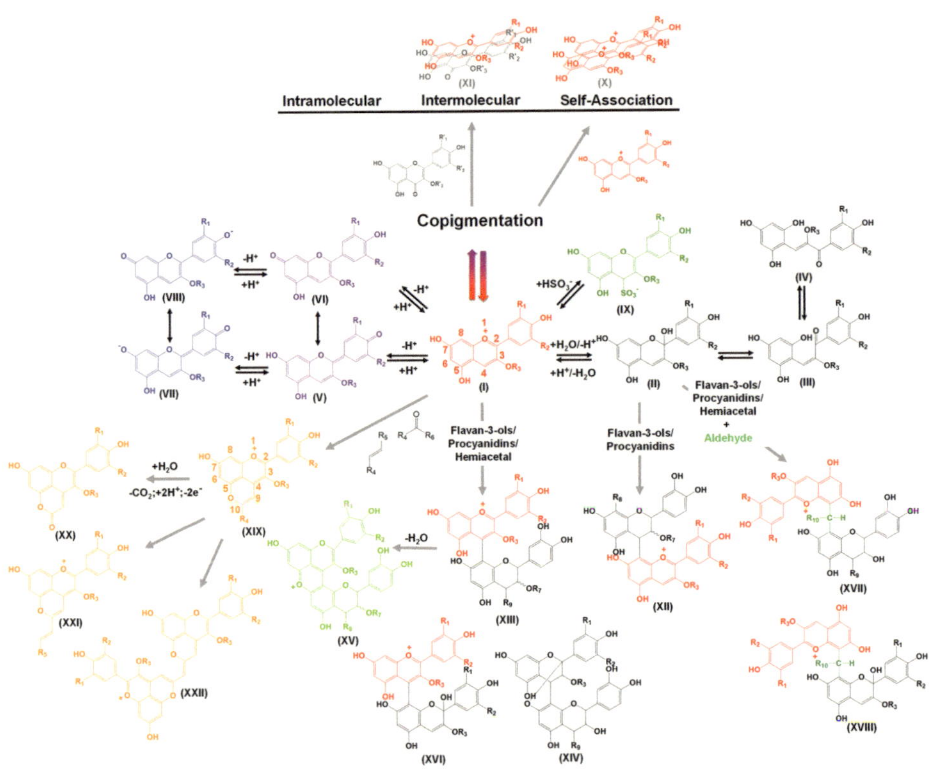

Figure 3.
*Colour evolution during ageing by the involvement of anthocyanins in different equilibria and their simultaneous transformation through various concurring chemical reactions to a range of other pigments (references are listed in **Table 2**).*

These changes are dependent on the wine composition like anthocyanin, flavonol and tannin concentrations, different processing parameters like temperature, oxygen level, pH and the presence of other compounds either produced during alcoholic fermentation, added during processing or formed during the ageing process. On the other hand, no studies have been reported about the colour changes occurring during White Port wine ageing.

Anthocyanins in aqueous solution, depending on the pH, occur in different forms present in equilibria [13–15]. At pH < 2, the red flavylium cation is the main structure present (I in **Figure 3**). With increasing pH, for values between 3 and 6, after hydration of the flavylium cation, the colourless hemiketal (II) structure is formed, this last being in equilibrium with the pale yellow *cis*-chalcone (III) through tautomerisation. This chalcone isomer is also in equilibrium with the *trans*-chalcone isomer (IV). With the pH increase, the flavylium cation is deprotonated to the corresponding violet neutral quinoidal bases (V and VI) that at higher pH yields the blue anionic quinoidal bases after further deprotonation (VII and VIII, **Figure 3**) [15]. When sulphur dioxide (SO_2) is present, there is observed reversible bleaching of anthocyanins that occurs due to the formation of the colourless anthocyanin-4-bisulphite adducts [16] (IX).

Considering all these equilibriums, at wine pH (3–4) these pigments would be expected to be present mainly in their non-coloured hemiketal form (II). However, the flavylium cation (I) is the main form present in young red wines. This is the result of its stabilisation by different copigmentation mechanisms such as self-association and interaction with other wine components [17–20]. In the copigmentation process, anthocyanins and other colourless organic compounds, such as flavonoids, amino acids, organic acids, polysaccharides, anthocyanins, or metallic ions, form molecular or complex associations [21]. The copigmentation is based in two effects [22]: (1) the formation of the π–π complex which causes changes in the spectral properties of the molecules in the flavylium ion, increasing the absorption intensity (hyperchromic effect) and its wavelength (bathochromic shift); and (2) the stabilisation of the flavylium form by the π complex displaces the equilibrium in such way that the red colour increases. This association also gives protection for the water nucleophilic attack in the 2 position of the flavylium cation [23] and for other species such as peroxides and sulphur dioxide in the 4 position [24, 25], so that the balance is displaced from hydrated forms towards the red flavylium cations. If the copigment is other anthocyanin, a self-association is formed (X); in the case of copigments with free electron pairs, an intermolecular copigmentation takes place (XI) finally, in the most complex case, the copigmentation can be carried out by a part of the structure itself (usually one of the aromatic acyl group substituents) (**Figure 3**).

During wine ageing, the concentration of monomeric anthocyanins starts to decrease leading to the formation of new anthocyanin derived pigments with different colour features and greater colour expression at high pH, important for the long-term colour stability of aged red wines [21]. The formation of most of the anthocyanin-derived pigments occurs in the first months of ageing, as the oxidative conditions in oak barrels favour their formation [26, 27]. Copigmentation has been hypothesised as the first mechanism involved in the formation of polymeric anthocyanin-derived pigments in red wines during ageing [19]. Numerous pigments have been characterised in wines and wine-like model solutions, and can be classified into three groups with respect to their formation pathways: 1) Direct condensation between anthocyanins and flavonols; 2) Condensation between anthocyanins and flavonols mediated by aldehydes, mainly acetaldehyde; and 3) Pyroanthocyanins (**Figure 3**). Although some of these pigments have only been detected in very small quantities in red wines, they have unique spectroscopic features that may, in some way, contribute together to the overall colour of aged red wines. In the first case, free anthocyanins can condense directly with flavan-3-ols and oligomeric proanthocyanins generating tannin-anthocyanins condensation products (T-A$^+$, XII) or anthocyanin-tannin condensation products (A$^+$-T, XIII) [8, 28–33]. The T-A$^+$ formation begins with the acid cleavage of the interflavanic bond of a procyanidin, giving a carbocation T$^+$ which reacts with the hydrated form of the anthocyanin (II). This mechanism leads to a colourless compound (T-AOH) which easily dehydrates to the coloured flavylium form T-A$^+$ [34]. In the formation of A$^+$-T pigments, nucleophilic addition of the flavanol takes place onto the flavylium form of the anthocyanin, yielding a colourless compound with the anthocyanin in flavene form. This flavene can be oxidised, resulting in a coloured flavylium A$^+$-T pigment (XIII) or in a colourless compound A(-O-)T with a type-A bond (XIV) [31] (**Figure 3**). As described for the monomeric anthocyanins, these pigments can also occur in a dynamic equilibrium among some molecular forms, mainly the quinoidal base, the flavylium cation and the hemiketal or carbinol pseudobase [30]. Both T-A$^+$ pigments and colourless A(-O-)T have been detected in wines [35]. Dimeric anthocyanins (XVI) consisting of one unit under flavylium cation and the other one under hydrated hemiketal form (A$^+$-AOH) were also characterised by mass spectrometry in wine like solutions [36] (**Figure 3**).

The A$^+$-T adducts can generate yellow-orange xanthylium pigments (XV) by further structural rearrangements. After the dehydration, a new heterocyclic pyran ring is formed and the xanthylium structure is generated [17, 37–40] (**Figure 3**). However, xanthylium pigments are also proposed to be formed directly from oligomeric flavan-3-ols [41, 42].

On the other hand, the acetaldehyde-mediated polymerisation between either only flavanols or with anthocyanins is the most well documented reaction in the literature [31, 37, 43–51]. Acetaldehyde is the main aldehyde (90%) present in wines as a result of yeast metabolism during the first stages of alcoholic fermentation, being also produced throughout the wine ageing process from ethanol oxidation [52]. In fortified wines like Port wines, this compound and other aldehydes (propionaldehyde, 2-methylbutyraldehyde, isovaleraldehyde, methylglyoxal, benzaldehyde) are present in higher amounts due to the addition of wine spirit (40–260 mg/L of acetaldehyde) to stop the alcoholic fermentation [53]. Ethyl-linked products, including ethyl-linked flavanols [54, 55] and ethyl-linked anthocyanin-flavanol pigments (XVII) [55] have been detected in wines (**Figure 2**). The formation of ethyl-linked anthocyanin oligomers (A$^+$-Et-AOH, XVIII) was also shown to occur both in model solution and in wine [56]. The ethyl-linked 8,8-malvidin-3-glucoside dimer was characterised by NMR under biflavylium cation forms [56]. However, physicochemical studies carried out on this pigment showed that the dimer under monoflavylium cation is the most abundant form at wine pH [57].

Another important group of anthocyanin derived pigments formed during ageing, also found in red Port wines, are the pyranoanthocyanins (XIX) (**Figure 3**). Pyranoanthocyanins are a group of anthocyanin-derived pigments [58, 59], which were first discovered in red wine by Cameira-dos-Santos et al. [60]. Pyranoanthocyanins are structurally characterised by the presence of a fourth ring between C-4 and the 5-hydroxyl group of an anthocyanin moiety, differing from each other on the type of group or molecule linked to the C-10 of the new ring [58, 61, 62]. The pyranic ring in pyranoanthocyanins provides protection against the nucleophilic attack from water or bisulphite, increasing their stability [63], making these compounds exceptionally stable pigments towards sulphite bleaching and pH variations. Both anthocyanin-flavanol derived pigments, direct ones and ethyl-linked ones, show less stability during ageing than pyranoanthocyanins. Through the reaction of anthocyanins with acetaldehyde [61, 63], pyruvic acid [58, 64], cinnamic acids [65, 66], acetoacetic acid [64], and procyanidins in the presence of acetaldehyde [67], several different classes of these pigments have been identified in the past decade such as vitisins [58, 59, 61, 68, 69], hydroxyphenyl-pyranoanthocyanins (pinotins) [59, 64, 70–72], methylpyranoanthocyanins [59, 73], vinylflavanol-pyranoanthocyanins [59], portisins [58, 59, 61, 66, 67, 74, 75], and more recently a new family of pyranoanthocyanin dimers [28, 73, 76, 77] (**Table 2**).

Pyruvic acid leads to the major pyranoanthocyanins determined in wines, i.e. carboxy-pyranoanthocyanins (R = COOH), sometimes referred to as vitisin A [58, 59, 61, 68, 69]. In red Port wines, it is the main pigment found during ageing. Due to its particular vinification process, the concentration of vitisin A is very high: 51.2 mg/L for Touriga Nacional Port wines, for example [78]. Indeed, wine fortification after alcoholic fermentation allows greater availability of pyruvic acid [79], which leads to reaching the highest contents shortly after fermentation and during the first year of ageing, followed by a slow decline [80]. After one year of ageing in barrels, the contents decrease by about 15–25% and about 70% after two years, whereas it is not so much important during bottle ageing (9–18%). Romero and Bakker [81] have demonstrated that the addition of pyruvic acid to finished Port wines from four different grape varieties resulted in an increase of malvidin-pyruvic acid adducts. It was also found that the concentration of anthocyanin-pyruvic acid adducts in wines was directly related to the original grape anthocyanin profile; the higher the initial anthocyanin precursor forms, the higher the concentration of corresponding adducts [81–83]. Morata et al. [84] have reported that the yeast strain used in the alcoholic fermentation (inoculated or not) also affects the production of malvidin-3-glucoside-pyruvate, existing a direct relation between the concentration of the pigment and the production of pyruvic acid by the yeast.

Moreover, the content of SO_2 in must can also influence the production of malvidin-3-glucoside-pyruvate, since SO_2 regulates the concentration of pyruvic acid through the formation of a weak bisulphite addition compound [85].

Romero and Bakker [68] have reported that malvidin-derived pyruvic acids adduct in model solutions provided approximately 11-fold (at pH 3) and 14-fold (at pH 2) more colour than grape anthocyanins.

Flavanol pyranoanthocyanins are formed by the cycloaddition between anthocyanins and 8-vinylflavanol adducts initially derived from the cleavage of ethyl-linked flavanol oligomers [46] or pigments [86, 87]. In red Port wines, pyranoanthocyanin-procyanidin dimers were identified in higher concentrations than the corresponding pyranoanthocyanin-catechins, representing up to 80% of the total pyranomalvidin-flavanols. This postulate is concordant with the fact that procyanidin dimers are more abundant than catechin monomers in grapes and wines from the Douro region [88, 89]. Furthermore, their concentrations decreased in older wines for both malvidin-3-glucoside derived-pigments (10.59 mg/L in 3 year aged wines, 9.16 mg/L in 4 year aged wines and 7.86 mg/L in 6 year aged wines) and associated coumaroyl pigments (6.62, 5.51 and 3.33 mg/L in 3, 4 and 6 year aged wines, respectively).

A second generation of pyranoanthocyanins can be formed by the reaction between a vitisin A and other metabolites. For example, oxovitisins (XX) are neutral yellowish pyranone structures involving the nucleophilic attack of water at the C-10 position of vitisin A [90].

Pyroanthocyanins	Precursors	References
R1 and R2 = OCH3; R3 = Glucose		
(I) In **Figure 3**		
Malvidin-3-glucoside (Oenin)		[13–15]
(XIX) In **Figure 3**		
R4 = H		
Non-substituted pyroanthocyanins (Vitisin B)	Oenin+acetaldehyde	[58–61, 68, 69]
R4 = COOH		
Carboxypyroanthocyanins (Vitisin A)	Oenin+pyruvic acid	[58–61, 68, 69]
R4 = CH_3		
Methylpyroanthocyanins	Oenin+acetoacetic acid or acetone	[59, 73]
R4 = $COCH_3$		
Acetylpyroanthocyanins	Oenin+diacetyl	[8, 28–33]
R4 = hydroxyphenyl		
Hidroxyphenylpyroanthocyanins	Oenin+*p*-coumaric acid or vinylphenol	[59, 64, 70–72]
R4 = dihydroxyphenyl		
Pinotin A	Oenin + caffeic acid or vinylcatechol	[59, 64, 70–72]
R4 = flavanol		
Flavanol-pyroanthocyanins	Oenin+vinylflavanols or flavan-3-ols + acetaldehyde	[59]
(XX) In **Figure 3**		
Pyranone-anthocyanins (oxovitisins)	Carboxypyroanthocyanins + water	[90]
(XXI) In **Figure 3**		
R6 = hydroxyphenyl		
Vinylphenyl-pyroanthocyanins (Portisin B)	Carboxypyroanthocyanins+hydroxyci nnamic acids or vinylphenols	[58, 59, 61, 66, 67, 74, 75]
R6 = flavanol		
Vinylflavanol-pyroanthocyanins (Portisin A)	Carboxypyroanthocyanins + vinylflavanols or flavan-3-ols and acetaldehyde	[58, 59, 61, 66, 67, 74, 75]
(XXII) in **Figure 3**		
Pyroanthocyanins dimers	Carboxypyroanthocyanins + methylpyroanthocyanins	[28, 73, 76, 77]

Table 2.
Pyranoanthocyanins identified in wines and precursors.

In 2003 Mateus et al. [74] reported a new group of pyranoanthocyanins-vinylpyranoanthocyanins-which were named portisins (XXI), because of their occurrence in aged red Port wine [61, 67, 74, 75], **Figure 3**. The structure of these compounds consists of a pyranoanthocyanin moiety linked through a vinyl bridge to a flavanol or phenol unit. Their pathway of formation involves the carboxypyranoanthocyanins and vinylphenolic compounds. The first of these compounds reported in the literature arise from reaction of 8-vinylflavanol with carbon C-10 of the carboxypyranoanthocyanins, followed by loss of a formic acid group yielding the vinyl bridge. Portisins have been shown to have very high colouring capacity, much higher than that of their anthocyanin or pyruvic acid adduct counterparts [91–93]. Later, other portisins (B type) were detected in aged Port wines. In these, the flavanol moiety is replaced by a phenolic moiety with different hydroxylation and methoxylation patterns [61, 67, 74, 75]. These compounds were reported to result from the reaction of carboxypyranoanthocyanins with vinylphenols and cinnamic acids, following a mechanism similar to that of vinylflavanols and involving a further decarboxylation. However, the colour features of these portisins are different from those of the portisins discussed above because they have a λmax hypsochromically shifted from that of vinylpyranoanthocyanin-catechins, and are only slightly affected by the substitution pattern of the new phenolic ring (between 533 and 540 nm at aqueous pH 1) [91].

The condensation between A-type vitisins and methylpyroanthocyanins results in the formation of pyranoanthocyanin dimers (XXII), **Figure 3**. These turquoise blue pigments were found in a 9 year aged Port wine [73].

3.1.2 Aroma composition and sensory development

The volatile compounds present in Port wines have their origin on the grapes used, are produced during the alcoholic fermentation and being also added as part of the wine spirit used for Port wine production that contains trace volatile compounds such as esters (ethyl hexanoate, ethyl octanoate, ethyl decanoate) and terpenes (α-terpineol, linalool) that can affect the quality of the Port wines, contributing to a fruity, balsamic and spicy aroma [94]. In addition, wine spirits are rich in aldehydes such as acetaldehyde, propionaldehyde, isovaleraldehyde, isobutyraldehyde, and benzaldehyde [94]. The volatile profile of young Port wines is significantly different from that of aged Tawny Port wines or bottle-aged Port wines. Producers blend wines from several vintages and vineyards to produce wines with a consistent character. The final aroma character of the Port wine is to a considerable extent determined by the processes that take place during the oxidative ageing process of these wines, such as oxidation, carbohydrate degradation, formation and hydrolysis of esters, formation of acetals and to a lesser extent extraction of components from wood [11]. More than 200 volatile components have been detected in Port wines, 141 of which have been entirely or partially identified, however, the sensory importance of the various groups of volatile compounds does not entirely explain the sensory properties of Ruby or Tawny Port wines [95]. For the Ruby Port wine sensory profile, the attributes are 'Ruby', 'Persistence', 'Red fruits', 'Fruity flavour', 'Astringency' and 'Floral' were dominant, whereas in the White Port wine attributes like 'Honey', 'Sweet taste', 'Alcoholic sensation', 'Balance', 'Acid taste' and 'Moscatel' are the ones that better characterise these wines, Tawny Port wines are characterised by the attributes 'Dried fruits flavour', 'Dried fruits', 'Spices', 'Wood' and 'Sweet/Honey' [96, 97]. The Pink Port wines sensory attributes are characterised by the attributes 'Red fruit aroma', 'Body', 'Fruit aroma', 'Fruity flavour', 'Spicy sensation' and 'Persistence' [98].

Norisoprenoids have been found to contribute significantly to the aroma of young and aged Port wines [76, 99–101]. In a one year aged Port wine produced from Touriga Franca and Touriga Nacional grape varieties the norisoprenoid, 2,6,6-trimethylcyclohex-2-ene-1,4-dione, described as having sweet honey aroma, was identified by Rogerson et al. [102]. In a young Port wine produced from Tinto Cão and Tinta Barroca grape varieties, Rogerson et al. [103] identified the 1,3-dimethoxybenzene and 2-ene-1,4-dione. Falqué-Lopez et al. [104] characterised a one year aged Touriga Nacional monovarietal Port wine as having 'plum brandy', 'mulberry', 'cherry', 'wild fruits' and 'dry raisin' aromas and Guedes de Pinho et al. [105] identified linalool and linalyl acetate as being the responsible for the bergamot descriptor.

Ferreira et al. [100] have studied the influence of several factors on the levels of norisoprenoid in Port wines such as dissolved oxygen levels, free sulphur dioxide concentration, pH, and time/temperature of ageing. These authors observed that temperature and pH had a major influence on norisoprenoids levels and oxygen saturation reduced these compounds.

The concentration of several norisoprenoids increases during ageing, as for example β-ionone and β-damascenone in Vintage Port wines, and vitispirane, 2,2,6-trimethylcyclohexanone (TCH) and 1,1,6-trimethyl-1,2-dihydronaphthalene (TDN) in Tawny Port wines [99]. Ferreira and Pinho [99] showed that the occurrence of β-damascenone, β-ionone, TCH, TDN, and vitispirane was distinct in young or aged Port wines. It was observed that in wood barrel ageing TDN, vitispirane, and TCH increased, however, the concentration of β-ionone and β-damascenone decreased. Freitas et al. [106] described that TCH was responsible for the "rock-rose-like" aroma. According to several authors, in Port wines due to the short fermentation time precursors of norisoprenoids such as carotenoids, β-carotene, lutein, neoxanthin and violaxanthin can be present [104, 107, 108]. Carotenoids are the precursor of norisoprenoids and in Port wine carotenoids persist after the vinification process [108]. Grape varieties used for Port wine production are rich in certain carotenoids and viticultural practices, such as bunch shading [108, 109] and grapevine water status [110] can influence the concentration of carotenoids in the grape berries.

Acetals, derived from glycerol and acetaldehyde, also appear to be involved in the flavour of aged Tawny Port wines [111]. The levels of aldehydes and methyl ketones increase during the oxidative ageing of Port wines. The major aliphatic aldehyde is acetaldehyde with a clear trend of increasing with the time of storage in wood barrels. Glycerol is present in wines in large amounts, in particular with concentration from 4 to 8 g/L in Port wines and therefore the formation of acetal can be high. At wine pH, four isomers are formed by condensation of glycerol and acetaldehyde: *cis*- and *trans*-5-hydroxy-2-methyl-1,3-dioxane and *cis*- and *trans*-4-hydroxymethyl-2-methyl-1,3-dioxolane. These four acetals have been studied in more detail in order to understand their impact on wine aroma and if these substances can be used as indicators of Port wines age with oxidative ageing conditions. These four isomers are found in Port wine at high concentrations. Nevertheless, this reaction is strongly dependent on free sulphur dioxide levels. When there is no free sulphur dioxide, the level of four isomers increases with the extent of ageing. On the other hand, when sulphur dioxide combines with acetaldehyde, the acetals cannot be formed because of the formation of the acetaldehyde-bisulphite adduct. The concentrations of the four acetals increases consistently with age due to the constant increase of acetaldehyde content and the nonexistence of free sulphur dioxide during Port wine storage. The acetal with the highest intensity aroma described as sweet and Port-like is *trans*-5-hydroxy-2- methyl-1,3-dioxane and the aroma threshold limit of the total concentration of the four acetals was determined as 100 mg/L [111]. Many acetals have been isolated from Tawny Port wines, but their contribution to the oxidised character of the wine is unclear [8].

Port wines with extensive wood-ageing have higher concentrations of diethyl and other succinate esters that contribute to the Port wine bouquet. Oak lactones (β-methyl-γ-octalactone isomers) and other oxygen-containing heterocycles have also been isolated. Some of the latter are furan derivatives, such as dihydro-2-(3H)-furanone and may contribute to a sugary oxidised bouquet [8]. Esters of 2-phenylethanol may contribute to the fruity, sweet bouquet of Port wines, and diacetyl can contribute to its caramel odour [99]. Some aldehydes and ketones are associated with the oxidative aged Port wines, conferring "rancio" odour to wines [112, 113].

However, in wood barrel aged Port wine the 3-hydroxy-4,5-dimethyl-2(5H)-furanone (sotolon) seems to be the most significant volatile compound [112, 113]. Some works suggested that sotolon contributes to the characteristic barrel aged aroma of Port wines [114], being the fundamental molecule to understand the "perceived age" of Port wines. The levels of sotolon were measured in "Colheita" and Tawny categories and were shown that it increases with ageing time being present in a range of concentration of some dozen μg/L in a young wine, to about 100 μg/L in wines with 10 years ageing, and to about 1000 μg/L in Port wines older than 50 years [112–114]. Albeit being a compound with an apparently important role in Port wine aged aroma, the mechanism of sotolon formation in wine is not yet fully understood. However, different pathways have been proposed, such as aldol condensation between α-ketobutyric acid and acetaldehyde [115–117] (Pathway 1 – **Figure 4**) and the reaction between ethanol and ascorbic acid [118] (Pathway 2 – **Figure 4**). The formation of sotolon in Port wines is dependent on the temperature and oxygen levels [112, 113], which are crucial parameters during the oxidative ageing of Tawny Port wines. The sensory threshold limit of sotolon was determined as 19 μg/L, which is above the amount present in Port wines older than 10 years [112, 113].

Young Port wines show higher levels of volatile sulphur compounds than aged Port wines [119]. Sulphur compounds, such as 2-mercaptoethanol, 2-(methylthio) ethanol, ethyl 3-(methylthio) propionate, 3-(methylthio)-1-propanol, *cis*-(odourless), and *trans*-2-methyltetrahydrothiophen-3-ol, 3-(ethylthio)-1-propanol, 4-(methylthio)-1-butanol, dimethyl sulfone, benzothiazole, 3-(methylthio)-1-propionic acid and N-3-(methylthiopropyl) acetamide are not present or are present in lower concentrations in aged Tawny Port wines when compared to young Tawny Port wines [119].

The changes observed in the volatile composition of Port wines during the oxidative ageing results in a complex, oxidised character defined as nutty, nuts,

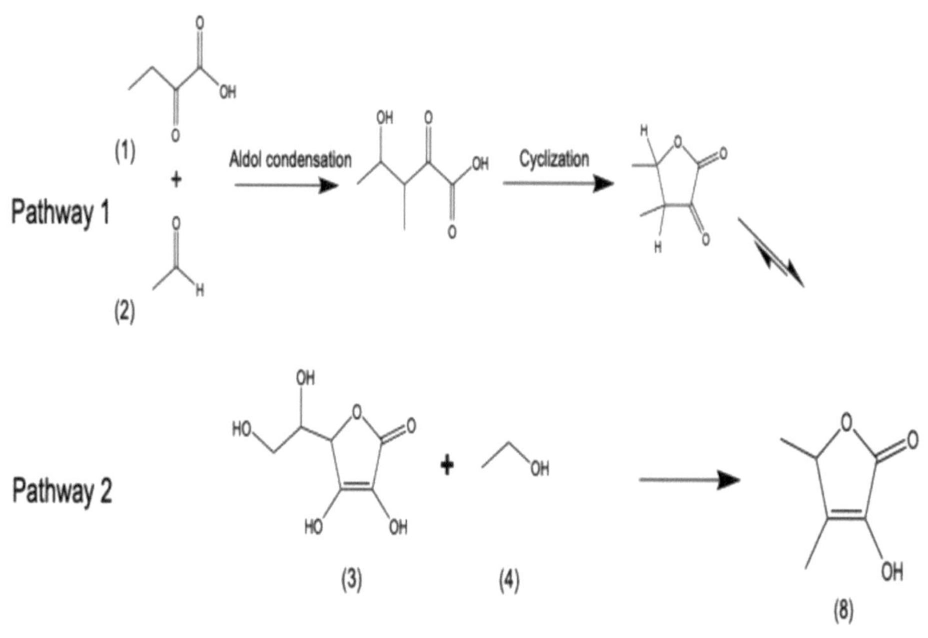

Figure 4.
Pathway 1: Reaction between α-ketobutyric acid (1) and acetaldehyde (2); pathway 2: Reaction between ascorbic acid (3) and ethanol (3). These reactions can lead to the formation of sotolon (8) according to [115–118].

raisins and crisp apples with a slightly oaky note' giving an impression of dryness. According to Falqué et al. [104] and Ferreira et al. [120] floral, bergamot-like, violet or jasmine notes are present in young Port wines that are changed during the ageing process in wood barrels. Freitas et al. [106] and Ferreira et al. [121] have described the flavours developed during ageing of Port wines as woody, burnt, dry fruit, nutty, and spicy.

4. Conclusions

Port wine is one of the most famous and old fortified wine in the world, produced in the Douro Demarcated Region (Portugal), a region with a singular "terroir". It is a traditional product with more than 250 years old and commercialised all around the world for many centuries. However, more knowledge is still needed in order to understand and control its composition and its evolution during the wine ageing process. Port wine presents a complex physicochemical matrix that results from complex and concurrent chemical reactions that occur during the ageing process significantly changing its sensory profile. Different Port wine styles are obtained using different ageing processes, related to different oxygen levels, temperature and sulphur dioxide variations. The two main Port wine styles are the Ruby and Tawny Port wines, with the first style being obtained by a reductive ageing process resulting in a ruby colour and fruity aroma, while Tawnies Port wine styles shows a brown colour and an intense dry fruit aroma. The colour of the different Port wines styles is one of their main quality parameters. For the Port wines produced with red grapes, colour changes are related to the changes in the anthocyanins composition during the ageing process. The colour attributes of the Port wines made with white grapes is still largely unknown. Sotolon plays an important role in aged Port wines aroma obtained by oxidative ageing, due to its low olfactory threshold and pleasant and potent aroma. Sotolon levels increase during Port wine oxidative ageing. The intense government legislation and specific production rules protect this important product produced in a world protected region (Unesco) in order to reduce adulterations or even imitations. However, more studies are still needed to deepen our knowledge in order to understand and control the reactions involved in Port wine ageing process that contribute to its uniqueness.

Acknowledgements

We appreciate the financial support provided to the Research Unit in Vila Real (UIDB/00616/2020 and UIDP/00616/2020) by FCT and COMPETE and JM acknowledges the financial support provided by the FCT grant PD/BD/135331/2017.

Conflict of interest

The authors declare no conflict of interest.

Author details

Juliana Milheiro, Fernanda Cosme*, Luís Filipe-Ribeiro and Fernando M. Nunes*
Chemistry Research Centre - Vila Real, (CQ-VR), Food and Wine Chemistry Laboratory, University of Trás-os-Montes and Alto Douro, Vila Real, Portugal

*Address all correspondence to: fcosme@utad.pt and fnunes@utad.pt

References

[1] IVDP, Instituto dos Vinhos do Douro e Porto [Internet]. 2020. Available from: http://www.ivdp.pt/.

[2] Decree-Law n°104/85, 10th of April. Ministério da Agricultura. Diário da República, 1ª série, no. 83.

[3] Portaria n° 195/1985 of 10th April. Ministério da Agricultura, do desenvolvimento rural e das pescas. Diário da República, 1° série-B, no. 91.

[4] Decree-Law n°173/2009 of 3rd of August. Ministério da Agricultura, do desenvolvimento rural e das pescas. Diário da República, 1ª série, no. 148.

[5] Moreira N, Guedes de Pinho P. Port Wine. In: Jackson RS, editor. Advances in Food and Nutrition Research. Vol. 63. Academic Press, 2011. p. 119-146. DOI: 10.1016/B978-0-12-384927-4.00005-1.

[6] Regulation n° 84/2010 of 8th of February. Diário da República, 2ª série, no. 26.

[7] Regulation n° 242/2010 of 15th March. Instituto dos Vinhos do Douro e do Porto, I. P. Diário da República, 2ª série, no. 51.

[8] Jackson RS. Wine Science, Principles, Practice, Perception. 4th ed. Academic Press; 2014. 776 p. DOI: 10.1016/B978-0-12-381468-5.00011-7.

[9] Saucier C, Lopes P, Mirabel M, Guerra C, Glories Y. Tannin-anthocyanin interactions: influence on wine colour. In: Waterhouse AL, Kennedy JA, editors. Red Wine Colour: Revealing the Mysteries. ACS Symposium Series 886; 2004. p. 265-273. DOI: 10.1021/bk-2004-0886.ch016

[10] Timberlake CF, Bridle P, Interactions between anthocyanins, phenolic compounds and acetaldehyde and their significance in red wines. American Journal of Enology and Viticulture. 1976; 27: 97-105.

[11] Clarke RJ, Bakker J. Wine Flavour Chemistry. 1st ed. Blackwell Publishers; 2004. 336 p.

[12] Bakker J, Timberlake CF, The mechanism of colour changes in ageing port wine. American Journal of Enology and Viticulture. 1986; 37: 288-292.

[13] Brouillard R, Lang J, The hemiacetal-*cis*-chalcone equilibrium of malvin, a natural anthocyanin. Canadian Journal of Chemistry.1990; 68: 755-761.

[14] Santos H, Turner DL, Lima JC, Figueiredo P, Pina FS, Macanita AL, Elucidation of the multiple equilibria of malvin in aqueous-solution by onedimensional and 2-dimensional NMR. Phytochemistry. 1993; 33:1227-1232.

[15] Brouillard R, Delaporte B, Chemistry of Anthocyanin Pigments. 2. Kinetic and thermodynamic study of proton transfer, hydration, and tautomeric reactions of malvidin 3-glucoside. Journal of the American Chemical Society.1977; 99: 8461-8468.

[16] Berke B, Chèze C, Vercauteren J, Deffieux G, Bisulfite addition to anthocyanins: Revisited structures of colorless adducts. Tetrahedron Letters. 1998; 39:5771-5774. doi: 10.1016/S0040-4039(98)01205-2.

[17] Liao H, Cai Y, Haslam E, Polyphenol Interactions. Anthocyanins: Co-Pigmentation and Colour Changes in Red Wines. Journal of the Science of Food and Agriculture.1992; 59 (3): 299-305.

[18] Trouillas P, Sancho-García JC, De Freitas V, Gierschner J, Otyepka M, Dangles O, Stabilizing and modulating

color by copigmentation: insights from theory and experiment. Chemical Reviews. 2016; 116: 4937-4982.

[19] Brouillard R, Dangles O, Anthocyanin molecular interactions: The first step in the formation of new pigments during wine aging?. Food Chemistry. 1994; 51 (4): 365-371.

[20] Gonzalez-Manzano S, Duenas M, Rivas-Gonzalo J C, Escribano-Bailon MT, Santos-Buelga C, Studies on the copigmentation between anthocyanins and flavan-3-ols and their influence in the colour expression of red wine. Food Chemistry. 2009; 114 (2): 649-656.

[21] Boulton R, The copigmentation of anthocyanins and its role in the color of red wine: A critical review. American Journal of Enology and Viticulture. 2001; 52 (2): 67-87.

[22] Dangles O, Brouillard R, Polyphenol interactions. The copigmentation case: Thermodynamic data from temperature variation and relaxation kinetics. Medium effect. Canadian Journal of Chemistry. 1992; 70 (8): 2174-2189.

[23] Matsufuji H, Otsuki T, Takeda T, Chino M, Takeda M, Identification of reaction products of acylated anthocyanins from red radish with peroxylradicals. Journal of Agricultural and Food Chemistry. 2003; 51(10): 3157-3161.

[24] García-Viguera C, Bridle P, Influence of structure on colour stability of anthocyanins and flavylium salts with ascorbic acid. Food Chemistry. 1999; 64(1): 21-26.

[25] Mazza G, Brouillard R, Recent developments in the stabilization of anthocyanins in food products. Food Chemistry. 1987; 25(3): 207-225.

[26] Atanasova V, Fulcrand H, Cheynier V, Moutounet M, Effect of oxygenation on polyphenol changes occuring in the course of wine-making. Analytica Chimica Acta. 2002, 458:15-27.

[27] Alcalde-Eon C, EscribanoT, Santos-Buelga C, Rivas-Gonzalo JC, Changes in the detailed pigment composition of red wine during maturity and ageing: A comprehensive study. Analytica Chimica Acta. 2006, 563(1):238-254.

[28] Nave F, Teixeira N, Mateus N, De Freitas V, The fate of flavanol–anthocyanin adducts in wines: Study of their putative reaction patterns in the presence of acetaldehyde. Food Chemistry. 2010, 121: 1129-1138.

[29] Salas E, Le Guernevé C, Fulcrand H, Poncet-Legrand C, Cheynier V, Structure determination and colour properties of a new directly linked flavanol-anthocyanin dimer. Tetrahedron Letters. 2004, 45: 8725-8729.

[30] Remy S, Fulcrand H, Labarbe B, Cheynier V, Moutounet M, First confirmation in red wine of products resulting from direct anthocyanin-tannin reactions. Journal of the Science of Food and Agriculture, 2000; 80: 745-751.

[31] Salas E, Fulcrand H, Meudec E, Cheynier V, Reactions of anthocyanins and tannins in model solutions. Journal of Agricultural and Food Chemistry. 2003, 51: 7951-7961.

[32] Salas E, Atanasova V, Poncet-LegrandC,MeudecE,MazauricJP, Cheynier V, Demonstration of the occurrence of flavanol-anthocyanin adducts in wine and in model solutions. Analytica Chimica Acta. 2004; 513(1): 325-332. DOI: 10.1016/j.aca.2003.11.084.

[33] He F, Pan Q-P, Shi Y, Duan C-Q, Chemical synthesis of proanthocyanidins in vitro and their reactions in aging wines. Molecules. 2008; 13: 3007-3032.

[34] Macz-Pop GA, Gonzalez-Paramas AM, Perez-Alonso JJ, Rivas-Gonzalo JC, New flavanol-anthocyanin condensed pigments and anthocyanin composition in guatemalan beans (Phaseolus spp.). Journal of Agricultural and Food Chemistry, 2006; 54: 536-542.

[35] Morata A, Gómez-Cordovés MC, Calderón F, Suárez JA, Effects of pH, temperature and SO_2 on the formation of pyranoanthocyanins during red wine fermentation with two species of *Saccharomyces*. International Journal of Food Microbiology, 2006; 106: 123-129.

[36] Lu Y, Foo LY, Unusual anthocyanin reaction with acetone leading to pyranoanthocyanin formation. Tetrahedron Letters, 2001; 42: 1371-1373.

[37] Santos-Buelga C, Francia-Aricha EM, De Pascual-Teresa S, Rivas-Gonzalo JC, Contribution to the identification of the pigments responsible for the browning of anthocyaninflavanol solutions. European Food Research and Technology, 1999; 209: 411-415.

[38] Baranowski ES, Nagel CW, Kinetics of malvidin-3-glucoside condensation in wine model systems. Journal of Food Science, 1983; 48: 419-429.

[39] Somers TC, The phenolic nature of wine pigments. Phytochemistry, 1971; 10: 2175-2186.

[40] Dueñas M, Fulcrand H, Cheynier V, Formation of anthocyanin-flavanol adducts in model solutions. Analytica Chimica Acta, 2006; 563:15-25.

[41] Fulcrand H, Cheynier V, Oszmiansky J, Moutounet M, An oxidised tartaric acid residue as a new bridge potentially competing with acetaldehyde in flavan-3-ol condensation. Phytochemistry, 1997; 46: 223-227.

[42] Jurd L, Somers TC, The formation of xanthylium salts from proanthocyanidins. Phytochemistry, 1970; 9: 419-427.

[43] Rivas-Gonzalo J C, Bravo-Haro S, Santosbuelga C, Detection of compounds formed through the reaction of malvidin-3-monoglucoside and catechin in the presence of acetaldehyde. Journal of Agricultural and Food Chemistry, 1995; 43: 1444-1449.

[44] Francia-Aricha E M, Guerra M T, Rivas-Gonzalo J C, Santos-Buelga C, New anthocyanin pigments formed after condensation with flavanols. Journal of Agricultural and Food Chemistry, 1997; 45: 2262-2266.

[45] Es-Safi N, Fulcrand H, Cheynier V, Moutounet M, Competition between (+)-catechin and (−)-epicatechin in acetaldehyde-induced polymerization of flavanols. Journal of Agricultural and Food Chemistry, 1999; 47:2088-2095.

[46] Es-Safi N, Fulcrand H, Cheynier V, Moutounet M, Studies on the acetaldehyde induced condensation of (−)-epicatechin and malvidin 3-O-glucoside in a model solution system. Journal of Agricultural and Food Chemistry, 1999; 47:2096-2102.

[47] Escribano-Bailon T, Alvarez-Garcia M, Rivas-Gonzalo J.C, Heredia FJ, Santos-Buelga C, Color and stability of pigments derived from the acetaldehyde-mediated condensation between malvidin-3-*O*-glucoside and (+)-catechin. Journal of Agricultural and Food Chemistry, 2001; 49: 1213-1217.

[48] Remy-Tanneau S, Guerneve C L, Meudec E, Cheynier V, Characterization of a colorless anthocyanin-flavan-3-ol dimer containing both carbon-carbon and ether interflavanoid linkages by NMR and mass spectrometries. Journal

of Agricultural and Food Chemistry, 2003; 51: 3592-3597

[49] Pissarra J, Mateus N, Rivas-Gonzalo J C, Santos-Buelga C, De Freitas V, Reaction between malvidin 3-glucoside and (+)-catechin in model solutions containing different aldehydes. Journal of Food Science, 2003; 68: 476-481

[50] Pissarra J, Lourenço S, González-Paramás AM, Mateus N, Santos-Buelga C, Silva AS, De Freitas V. Structural characterization of new malvidin-3-glucoside-catechin aryl/alkyl pigments. Journal of Agricultural and Food Chemistry, 2004; 52:5519-5526.

[51] Saucier C, Guerra C, Pianet I, Laguerres M, Glories Y, (+)-Catechin-acetaldehyde condensation products in relation to wine-ageing. Phytochemistry, 1997; 46:229-234.

[52] Liu SQ, Pilone GJ, An overview of formation and roles of acetaldehyde in winemaking with emphasis on microbiological implications. International Journal of Food Science & Technology, 2000; 35(1):49-61.

[53] Pissarra JI, Lourenço S, Machado JM, Mateus N, Guimaraens D, de FreitasV, Contribution and importance of wine spirit to the port wine final quality - initial approach. Journal of the Science of Food and Agriculture, 2005; 85: 1091-1097.

[54] Saucier C, Little D, Glories Y, First evidence of acetaldehyde-flavanol condensation products in red wine. American Journal of Enology and Viticulture, 1997; 48: 370-373

[55] Cheynier V, Fulcrand H, Sarni P, Moutounet M, Application des techniques analytiques à l'étude des composés phénoliques et de leurs réactions au cours de la vinification. Analusis, 1997; 25: M14-M21.

[56] Atanasova V, Fulcrand H, Le Guerneve C, Cheynier V, Moutounet M, Structure of a new dimeric acetaldehyde malvidin 3-glucoside condensation product. Tetrahedron Letters, 2002; 43: 6151-6153.

[57] Atanosova V, Fulcrand H, Le Guernevé C, Dangles O, Cheynier V, PolyphenolCommunications.Marrakech, 2002; Vol. 2, p 417.

[58] Fulcrand H, Benabdeljalil C, Rigaud J, Cheynier V, Moutounet M, A new class of wine pigments generated by reaction between pyruvic acid and grape anthocyanins. Phytochemistry, 1998; 47: 1401-1407

[59] de Freitas V, Mateus N, Formation of pyranoanthocyanins in red wines: a new and diverse class of anthocyanin derivatives. Analytical and Bioanalytical Chemistry, 2011; 401:1463-1473

[60] Cameira-dos-Santos PJ, Brillouet JM, Cheynier V, Moutounet M, Detection and partial characterisation of new anthocyanin derived pigments in wine. Journal of the Science of Food and Agriculture, 1996; 70:204-208.

[61] Bakker J, Timbarlake CF, The isolation and characterisation of new colour stable anthocyanins occurring in some red wines. Journal of Agricultural and Food Chemistry. 1997; 45: 35-43. DOI: doi.org/10.1021/jf960252c.

[62] Benabdeljalil C, Cheynier V, Fulcrand H, Hakiki A, Mosaddak M, Moutounet M, Evidence of new pigments resulting from reaction between anthocyanins and yeast metabolites. Sciences des Aliments, 2000; 20: 203-220

[63] He J, Santos-Buelga C, Silva AMS, Mateus N, Freitas V, Isolation and structural characterization of new anthocyanin-derived yellow pigments in aged red wines. Journal of Agricultural

and Food Chemistry, 2006; 54(25): 9598-9603. DOI: 10.1021/jf062325q

[64] Fulcrand H, Santos P-JC, Sarni-Manchado P, Cheynier V, Favre-Bonvin J, Structure of new anthocyanin-derived wine pigments. Journal of the Chemical Society, Perkin Transactions 1. 1996; 7: 735-739.

[65] Schwarz M, Wabnitz TC, Winterhalter P, Pathway leading to the formation of anthocyanin vinylphenol adducts and related pigments in red wines. Journal of Agricultural and Food Chemistry. 2003; 51(12): 3682-3687. DOI: 10.1021/jf0340963

[66] Rentzsch M, Schwarz M, Winterhalter P, Hermosin-Gutierrez I, Formation of hydroxyphenyl-pyranoanthocyanins in Grenache wines: Precursor levels and evolution during aging. Journal of Agricultural and Food Chemistry. 2007; 55(12): 4883-4888. DOI: 10.1021/jf0702491

[67] Mateus N, Oliveira J, Santos-Buelga C, Silva AMS, Freitas, V, NMR structure characterization of a new vinylpyranoanthocyanin-catechin pigment (a portisin). Tetrahedron Letters. 2004; 45(17): 3455-3457. DOI: doi.org/10.1016/j.tetlet.2004.03.007

[68] Romero C, Bakker J, Interactions between grape anthocyanins and pyruvic acid, with effect of pH and acid concentration on anthocyanin composition and color in model solutions. Journal of Agricultural and Food Chemistry, 1999; 47: 3130-3139.

[69] Bakker J, Bridle P, Honda T, Kuwano H, Saito N, Terahara N, Timberlake CF, Identification of an anthocyanin occurring in some red wines. Phytochemistry, 1997; 44: 1375-1382.

[70] Mateus N, Pascual-Teresa S, Rivas-Gonzalo JC, Santos-Buelga C, Freitas, V, Structural diversity of anthocyanin-derived pigments in port wines. Food Chemistry. 2002; 76: 335-342. DOI: 10.1016/S0308-8146(01)00281-3

[71] Schwarz M, Jerz G, Winterhalter P, Isolation and structure of pinotin A, a new anthocyanin derivative from pinotage wine. Vitis. 2003; 42(2): 105-106.

[72] Hakånsson AE, Pardon K, Hayasaka Y, De Sa M, Herderich M, Structures and colour properties of new red wine pigments. Tetrahedron Letters, 2003; 44: 4887-4891.

[73] Oliveira J, Azevedo J, Silva MAS, Teixeira N, Cruz L, Mateus N, de Freitas VJ Pyranoanthocyanin dimers: a new family of turquoise blue anthocyanin-derived pigments found in port wine. Journal of Agricultural and Food Chemistry. 2010; 58: 5154-5159.

[74] Mateus N, Silva AMS, Rivas-Gonzalo JC, Santos-Buelga C, Freitas V, A new class of blue anthocyanin-derived pigments isolated from red wines. Journal of Agricultural and Food Chemistry. 2003; 51(7): 1919-1923. DOI: doi.org/10.1021/jf020943a

[75] Mateus N, Oliveira J, Pissarra J, González-Paramás AM, Rivas-Gonzalo JC, Santos-Buelga C, Silva AMS, Freitas V, A new vinylpyranoanthocyanin pigment occurring in aged red wine. Food Chemistry. 2006; 97(4); 689-695. DOI: 10.1016/j.foodchem.2005.05.051

[76] Oliveira J, Santos-Buelga C, Silva AMS, Freitas V, Mateus N, Chromatic and structural features of blue anthocyanin-derived pigments present in Port wine. Analytica Chimica Acta. 2006; 563: 2-9. DOI: 10.1016/j.aca.2005.11.027

[77] Chassaing S, Isorez G, Kueny-Stotz M, Brouillard R, En route to color-stable pyranoflavylium pigments-a systematic study of the reaction between 5-hydroxy-4-methylflavylium

salts and aldehydes. Tetrahedron Letters, 2008; 49: 6999-7004.

[78] Mateus N, Silva A M S, Vercauteren J, de Freitas V, Occurrence of anthocyanin-derived pigments in red wines. Journal of Agricultural and Food Chemistry, 2001; 49: 4836-4840.

[79] Whiting GC, Coggins PA, Organic acid metabolism in cider and perry fermentations. III. Keto-acids in cider-apple juices and ciders. Journal of the Science of Food and Agriculture, 1960; 11:705-709.

[80] Mateus N, Freitas V, Evolution and Stability of Anthocyanin-Derived Pigments during Port Wine Aging. Journal of Agricultural and Food Chemistry. 2001; 49: 5217-5222. DOI: 10.1021/jf0106547.

[81] Romero C, Bakker J, Anthocyanin and colour evolution during maturation of four port wines: effect of pyruvic acid addition. Journal of the Science of Food and Agriculture. 2000; 81: 252-260. DOI: 10.1002/1097-0010(20010115)81:2<252:AID-JSFA810>3.0.CO;2-5

[82] Romero C, Bakker J, Effect of storage temperature and pyruvate on kinetics of anthocyanin degradation, Vitisin A derivative formation, and colour characteristics of model solutions. Journal of Agricultural and Food Chemistry, 2000; 48: 2135-2141. DOI: 10.1021/jf990998l

[83] Monagas M, Núñez V, Bartolomé B, Gómez-Cordovés C, Anthocyanin-derived pigments in Graciano, Tempranillo, and Cabernet Sauvignon wines produced in Spain. American Journal of Enology and Viticulture, 2003; 54: 163-169.

[84] Morata A, Gómez-Cordovés MC, Colomo B, Súarez JA, Pyruvic acid and acetaldehyde production by different strains of *Saccharomyces cerevisiae* with vitisin A and B formation in red wines. Journal of Agricultural and Food Chemistry, 2003; 51:7402-7409.

[85] Asenstorfer RE, Markides AJ, Iland PG, Jones GP, Formation of vitisin A during red wine vinification and maturation. Australian Journal of Grape and Wine Research, 2003; 9:40-46

[86] Mateus N, Silva MAS, Santos-Buelga C, Rivas-Gonzalo JC, De Freitas V, Identification of anthocyanin-flavanol pigments in red wines by NMR and mass spectrometry. Journal of Agricultural and Food Chemistry, 2002; 50: 2110-2116.

[87] Cruz L, Teixeira N, Silva MAS, Mateus N, Borges J, De Freitas V, Role of vinylcatechin in the formation of pyranomalvidin-3-glucoside-(+)-catechin. Journal of Agricultural and Food Chemistry, 2008; 56: 10980-10987.

[88] De Freitas V A P, Glories Y, Monique A, Developmental Changes of Procyanidins in Grapes of Red *Vitis vinifera* Varieties and Their Composition in Respective Wines. American Journal of Enology and Viticulture, 2000; 51: 397-403.

[89] Mateus N, Marques S, Gonçalves AC, Machado J M, De Freitas V, Proanthocyanidin composition of red *Vitis vinifera* varieties from the Douro Valley during ripening: influence of cultivation altitude. American Journal of Enology and Viticulture, 2001; 52: 115-121.

[90] He J, Oliveira J, Silva AM, Mateus N, Freitas V, Oxovitisins: a new class of neutral pyranone-anthocyanin derivatives in red wines. Journal of Agricultural and Food chemistry. 2010; 58(15): 8814-8819. DOI: 10.1021/jf101408q

[91] Oliveira J, de Freitas V, Silva MAS, Mateus N, Reaction between hydroxycinnamic acids and

anthocyanin-pyruvic acid adducts yielding new portisins. Journal of Agricultural and Food Chemistry, 2007; 55: 6349. DOI: 10.1021/jf070968f

[92] Mateus N, Proença S, Ribeiro P, Machado J M, de Freitas V, Grape and wine polyphenolic composition of red *Vitis vinifera* varieties concerning vineyard altitude. Ciencia y Tecnología Alimentaria, 2001; 3: 102-110.

[93] Mateus N, Oliveira J, Haettich-Motta M, de Freitas V, New family of bluish pyranoanthocyanins. Journal of Biomedicine and Biotechnology, 2004; 5: 299.

[94] Rogerson FSS, Freitas VAP, Fortification spirit, a contributor to the aroma complexity of port. Journal of Food Science. 2002; 67: 1564-1569. DOI: 10.1111/j.1365-2621.2002.tb10323.x

[95] Williams AA, Lewis MJ, May HV, The Volatile Flavour Components of Commercial Port Wines. Journal of the Science of Food and Agriculture. 1983; 34: 311-319. DOI: 10.1002/jsfa.2740340316

[96] Vilela A, Monteiro B, Correia E, Sensory profile of Port wines: Categorical Principal Component Analysis, an approach for sensory data treatment. Ciência e Técnica Vitivinícola. 2015; 30:1-8.

[97] Vilela A, Ferreira R, Nunes F, Correia E, Creation and Acceptability of a Fragrance with a Characteristic Tawny Port Wine-Like Aroma. Foods, 2020; 9, 1244; doi:10.3390/foods9091244.

[98] Monteiro B, Vilela A, Correia E, Sensory profile of pink port wines: Development of a flavour lexicon. Flavour and Fragrance Journal. 2014; 29: 50-58.

[99] Ferreira ACS, Pinho PG, Nor-isoprenoids profile during port wine ageing—influence of some technological parameters. Analytica Chimica Acta. 2004; 513: 169-176. DOI: 10.1016/j.aca.2003.12.027

[100] Ferreira ACS, Monteiro J, Oliveira C, Pinho PG, Study of major aromatic compounds in port wines from carotenoid degradation. Food Chemistry. 2008; 110: 83-87. DOI: 10.1016/j.foodchem.2008.01.069

[101] Mendes-Pinto MM, Carotenoid breakdown products the—norisoprenoids—in wine aroma. Archives of Biochemistry and Biophysics. 2009, 483(2): 236-245. DOI: 10.1016/j.abb.2009.01.008

[102] Rogerson FSS, Castro H, Fortunato N, Azevedo Z, Macedo A, Freitas VAP, Chemicals with sweet aroma descriptors found in Portuguese wines from the Douro region: 2,6,6-trimethylcyclohex-2-2,6,6-trimethylcyclohex-2-ene-1,4-dione and diacetyl. Journal of Agricultural and Food Chemistry. 2001; 49: 263-269. DOI: 10.1021/jf000948c

[103] Rogerson FSS, Azevedo Z, Fortunato N, Freitas VAP, 1,3-Dimethoxybenzene, a newly identified component of port wine. Journal of the Science of Food and Agriculture. 2002; 82: 1287-1292. DOI: 10.1002/jsfa.1182

[104] Falqué E, Ferreira AC, Hogg T, Guedes-Pinho P, Determination of aromatic descriptors of Touriga Nacional wines by sensory descriptive analysis. Flavour and Fragrance Journal. 2004, 19(4): 298-302. DOI: 10.1002/ffj.1355

[105] Guedes de Pinho P, Falqué E, Castro M, Silva HO, Machado B, Ferreira AS, Further insights into the floral character of Touriga Nacional wines. Journal of Food Science. 2007, 72(6): S396-S401. DOI: 10.1111/j.1750-3841.2007.00405.x

[106] Freitas VAP, Ramalho PS, Azevedo Z, Macedo A, Identification of some volatile descriptors of the rock-rose like aroma of fortified red wines from Douro Demarcated Region. Journal of Agricultural and Food Chemistry. 1999, 47: 4327-4331. DOI: 10.1021/jf9901035

[107] Guedes de Pinho P, Ferreira ACS, Mendes-Pinto M, Benitez JG, Hogg TA, Determination of carotenoid profiles in grapes, musts, and fortified wines from Douro varieties of *Vitis vinifera*. Journal of Agricultural and Food Chemistry. 2001, 49(11): 5484-5488. DOI: 10.1021/jf010515p

[108] Mendes-Pinto MM, Ferreira ACS, Caris-Veyrot C, Pinho PG, Carotenoid, chrorophyll and chlorophyll-derived compounds in grapes and port wines. Journal of Agricultural and Food Chemistry. 2005; 53: 10034-10041. DOI: 10.1021/jf0503513

[109] Oliveira C, Ferreira AC, Costa P, Guerra J, Pinho PG, Effect of some viticultural parameters on the grape carotenoid profile. Journal of Agricultural and Food Chemistry. 2004; 52: 4178-4184. DOI: 10.1021/jf0498766

[110] Oliveira C, Ferreira ACS, Pinto MM, Hogg T, Alves F, Pinho PG, Carotenoid compounds found in grapes and their relationship to plant water status. Journal of Agricultural and Food Chemistry. 2003; 51: 5967-5971. DOI: 10.1021/jf034275k

[111] Ferreira ACS, Barbe J-C, Bertrand A, Heterocyclic acetals from glycerol and acetaldehyde in port wines: Evolution with aging. Journal of Agricultural and Food Chemistry. 2002, 50: 2560-2564. DOI: 10.1021/jf011391j

[112] Ferreira ACS, Hogg T, Guedes de Pinho P, Identification of key odorants related to the typical aroma of oxidation-spoiled white wines. Journal of Agricultural and Food Chemistry. 2003; 51: 1377-1381. DOI: 10.1021/jf025847o

[113] Ferreira ACS, Barbe JC, Bertrand A, 3-Hydroxy-4,5-dimethyl-2(5H)-furanone: A Key Odorant of the Typical Aroma of Oxidative Aged Port Wine. Journal of Agricultural and Food Chemistry. 2003, 51: 4356-4363. DOI: 10.1021/jf0342932

[114] Martins RC, Monforte AR, Ferreira AS, Port wine oxidation management: A multiparametric kinetic approach. Journal of Agricultural and Food Chemistry. 2013; 61(22): 5371-5379. DOI: 10.1021/jf4005109

[115] Guichard E, Pham TT, Etievant P, Quantitative-determination of sotolon in wines by high-performance liquid-chromatography. Chromatographia. 1993, 37(9-10): 539-542. DOI: 10.1007/BF02275793

[116] Pham TT, Guichard E, Schlich P, Charpentier C, Optimal Conditions for the Formation of Sotolon from alpha-Ketobutyric Acid in the French "Vin Jaune". Journal of Agricultural and Food Chemistry. 1995, 43(10): 2616-2619. DOI: 10.1021/jf00058a012

[117] Pons A, Lavigne V, Landais Y, Darriet P, Dubourdieu D, Identification of a Sotolon Pathway in Dry White Wines. Journal of Agricultural and Food Chemistry. 2010, 58: 7273-7279. DOI: 10.1021/jf100150q

[118] Konig T, Gutsche B, Hartl M, Hubscher R, Schreier P, Schwab W. (3- Hydroxy-4,5-dimethyl-2(5H)-furanone (Sotolon) causing an off-flavor: Elucidation of its formation pathways during storage of citrus soft drinks. Journal of Agricultural and Food Chemistry. 1999, 47(8): 3288-3291. DOI: 10.1021/jf981244u

[119] Ferreira ACS, Rodrigues P, Hogg T, Pinho PG, Influence of Some Technological Parameters on the

Formation of Dimethyl Sulfide, 2-Mercaptoethanol, Methionol, and Dimethyl Sulfone in Port Wines. Journal of Agricultural and Food Chemistry. 2003, 51: 727-732. DOI: 10.1021/jf025934g

[120] Ferreira ACS, Falqué E, Castro M, Silva HO, Machado B, Pinho PG, Identification of key odorants related with high quality Touriga Nacional wine. Developments in Food Science. 2006, 43: 217-220. DOI: 10.1016/S0167-4501(06)80052-X

[121] Ferreira ACS, Avila I, Pinho PG. Sensorial impact of sotolon as the "perceived age" of tawny port wines. In: Frey C, Rouseff R, editors. Natural Flavors and Fragrances. Vol. 908. ACS Symposium Series; 2005. p. 141-159. DOI: 10.1021/bk-2005-0908.ch010

Chapter 7

Salivary Protein-Tannin Interaction: The Binding behind Astringency

Alessandra Rinaldi and Luigi Moio

Abstract

Interactions between salivary proteins and tannins are at the basis of one of the main mechanisms involved in the perception of astringency. Astringency is a tactile sensation evoked in the mouth by plant polyphenol-derived products, such as red wine. It is generally recognised that tannins can provoke negative sensations such as shrinking, drawing, or puckering of the epithelium. On the other hand, the astringency of some red wines can be felt as pleasant mouth feelings of richness, fullness, mouth-coating, and velvet in the mouth. In this chapter, an overview of the research concerned with molecular and sensory mechanisms of astringency was updated. Because of many variables influence the perception of astringency, several methods have been developed to measure the intensity of the sensation. In this context, different indirect assessments were critically evaluated considering the pros and contras and correlated with sensory analysis. We focused the attention on the saliva precipitation index (SPI), based on the binding and precipitation of human saliva with grape and wine tannins, because it has been widely used for many applications in winemaking. A current great challenge is to have an *in vitro* measurement of astringency able to provide information on the fate of wine, from grape to bottle.

Keywords: astringency, salivary proteins, polyphenols, precipitation, methods

1. Introduction

The interaction between plant tannins and macromolecules such as proteins is at the basis of many processes involved in the industry, ecological and agricultural systems [1–3], and food and beverage sensory characteristics. The common factor is the binding between macromolecules and tannins that lead to: (i) the conversion of an animal hide into the leather (tanning or *tannage*); (ii) the plant defence strategies against pathogens [1, 4]; (iii) reduced palatability of high tannin feedings to both terrestrial and marine herbivores and then a reduced interference in the process of digestion [2, 5]; (iv) the perception of astringency in tannin-rich food and beverage [6].

In the tanning process, the tannins bind to the hide's matrix, which is composed primarily of the protein collagen ordered in microcrystalline helical units. The purpose of *tannage* is primarily to increase the hydrothermal stability of the structure of collagen, secondarily to increase biological inertness, and finally, to improve the utility of the hide's physical properties [7].

In higher plants, tannins are primarily reserved as a chemical defence against pathogens. The complex with macromolecules such as cellulose and pectin, send out the exo-enzymes capable of utilising cellulose or pectin, either as a carbon source or for branching cell wall barriers to more nutrient-cytoplasm, depriving of the substrate or binding sites to these substrates. Another important function of tannin complexes is to impede the decomposition of plant litter, also when the leaf is fallen. This provides the delay in decomposition, which allows a constant input or seasonally demanding input of nutrients to the soil [1].

In the other processes, proteins of animal or human saliva interact with tannins of the unripe fruit, forages, or vegetable-derivates such as red wine, tea, and chocolate. Tannin molecules can bind proteins or enzymes at the level of specific amino acids, and modify the folding, the molecular weight, and the core binding site, to form soluble complexes or precipitates, which can alter protein function or inhibit enzyme activity [8]. This binding is at the basis of the astringent sensation experienced when tannins precipitate salivary proteins, and as a result, they lose their ability to lubricate the epithelial membranes of the mouth [6]. This sensation in mouth discourages the animal from feeding the unripe fruit or high-tannin forages and determines the unpleasantness of consumers for some tannin-rich products. These are the reason why, in the last decades, the interest in astringency has been constantly increased in different research areas.

2. Perception of astringency

The term astringency derives from the Latin verb, *ad-stringere* that means tightly bind, strongly join. It refers to the propensity of vegetable tannins to complex with macromolecules, such as proteins and polysaccharides, and alkaloids. Bate-Smith [9] first speculated that astringent sensations were caused by the increase in friction between the mucosal surfaces, which resulted from a reduction in lubrication in the oral cavity as astringent compounds bound salivary proteins. The binding between polyphenols/salivary proteins forms soluble complexes and/or precipitates that can cause the rupture of the salivary pellicle [10], interact with oral cells [11], and stimulate and activate mechanoreceptors (MRs) hold in the mouth [12]. MRs are nerve endings that function like those of the skin, except that they have smaller receptive fields and lower activation thresholds [13]. They are selectively sensitive to different stimulus properties, such as particle size and/or mouth movements, and project such information to the central nervous system [14]. Besides, the activation of G-coupled proteins also seems to be involved in the perception of astringency, activating signal transduction pattern as that of taste recognition [15]. Some brain regions (hippocampus and anterior cingulate cortex) that have been shown to respond to basic tastes were activated by the intensity and pleasantness of astringency [16]. In particular, the right ventral anterior insula that responded to astringent stimuli contributed to the ability to recognise the qualitative features of astringency. The activation of the trigeminal nerve, chorda tympani, and brain regions involved in memory and emotions could explain astringency as a multi perceptual phenomenon.

Whilst the chemical definition of astringency is related to the ability of tannins to bind proteins, in sensory terms, it is described as different and concomitant feelings of drying, puckering, and roughing [17, 18]. Astringency can be defined as a tactile sensation, because: (i) it is perceived on non-gustatory surfaces such as on the soft palate, gingiva, lips [12], (ii) does not show adaptation but also (iii) increases upon repeated ingestion [19], leading to carry-over effects during the tasting. However, side tastes as bitterness, sourness, and sweetness can highly modulate the overall astringency [20]. The sensitivity of MRs to astringents as well as basic tastes may elucidate the complexity of red wine astringency, which has been described by 33 different subqualities [21]. Amongst these "hard," "green," and "rich" have been associated with bitterness, acidity, and high flavour concentration, respectively [22], "harsh," "abrasive," and "drying" have been found to define astringency as a negative sensation, whilst the "complex" and "mouth-coat" subqualities have been associated to a positive impact during tasting [21]. These subqualities were also associated with touch standards when utilised to describe the tactile astringent sensations in the mouth elicited by red wines [23, 24]. The qualitative traits of astringency as "soft", "mouth-coat", and "rich" represented the drivers of liking for Sangiovese wine [25]. Similarly, for Tannat [26], and Côtes du Rhône and Rioja appellations wines [27], the attribute "mouth-coat" contributed to the quality of the wine.

It is also true that the perception of astringency is mediated by psychological factors [28], but salivary protein composition [29] and tannin's structure and composition [30, 31] represent the principal factors. In this regard, numerous reviews have been produced during the past years [32–38].

3. Salivary proteins

Saliva is a biological fluid primarily produced by the three pairs of "major" salivary glands (parotid, submandibular, and sublingual glands) in mouth and by the minor ones by 10% [39]. In the whole, saliva are presently more than 2000 different proteins and peptides [40, 41], which are the result of protein post-translational modifications before being secreted in the mouth [42]. Although saliva is predominantly a watery fluid (99.5%) with a complex mixture of proteins (0.3%; 1–2 mg/mL), ions and other organic compounds (0.2%) are also present. The whole saliva continuously baths the oral cavity and having a pH ranging from 6.2 to 7.4 acts as a buffering system. The saliva is continuously secreted (0.3–7 mL/min) and plays a role in protecting the tooth and mucosal integrity, in antibacterial and antiviral activity, digestion of food, speech, lubrication, taste, and represents a biomarker tool for some diseases [41, 43]. The main families of proteins include enzymes (amylase, carbohydrase, lipase), lactoferrin, high (M1), and low (M2) molecular-weight glycoproteins (mucins), peptides as agglutinins, immunoglobulins, proline-rich proteins (PRPs), cystatins, histatins and statherins [44].

There is evidence that saliva may affect the way we perceive the taste and mouthfeel of foods in various ways [45–47]. During the wine tasting, saliva transports and dissolves the stimuli substances [48]. Saliva constituents are of great importance for establishing protein-tannin interactions. In particular, the PRPs, histatins, mucin, amylase are the main salivary proteins involved in the binding with polyphenols eliciting astringency [49]. The differences between the binding of the same polyphenol to different proteins result from differences in the amino acid sequences [50].

The PRPs account for approximately 70% of the total secretory protein and are subdivided into acidic, basic, and glycosylated PRPs. They are characterised by an abundance of proline, glutamic acid/glutamine, and glycine [51]. The presence of these four amino acids, especially proline, which are the so-called alpha-helix breakers, enables the protein to form secondary structures, which assumes a random coils conformation in solution [10, 52]. This feature may allow PRPs to universally bind various types of polyphenols, mainly tannins with different sizes and structures. Some species, such as humans, rats, and mice, produce PRPs containing about 40% proline [53, 54]. However, some species produce salivary proteins, which are rich in proline but do not show a high affinity to tannins due to extensive glycosylation [54].

Parotid and submandibular secretions also contain several low molecular-weight histidine-rich peptides [55, 56]. Amongst 12 forms, the histatin 1, 3, and 5 are predominant and vary in size from 7 to 38 residues. These peptides show a high content of basic residues, such as lysine, arginine, and histidine [57]. They tightly bind tannins, even if some peptides are devoid of proline [58]. Conversely, others observed high tannin precipitation by histatins thanks to the interactions formed by basic residues and proline [59].

Amongst the low molecular weight salivary proteins, there is a selectivity in binding polyphenols (as PGG, procyanidin trimer, epicatechin, malvidin-3-glucoside): the acidic PRPs considerably form soluble and insoluble complexes with PGG and trimer but not with epicatechin; basic PRPs and glycosylated PRPs seem to not interact with trimer, whilst basic PRPs show a high affinity for epicatechin, malvidin-3-glucoside, and a mixture of both; the statherin shows no selectivity [60, 61].

Mucins are the major constituents of the viscous layer coating hard and soft tissues in the oral cavity. Mucins are generally composed of a peptide core (apomucin) enriched in serine, threonine, and proline residues and carbohydrate side chains (oligosaccharides) that are linked O-glycosidically to threonine or serine. M1 is a polymeric mucin due to the formation of disulfide linkages between cysteine residues in non-glycosylated domains, whilst M2 is a monomer [62]. Average proline content of 10% seems to be responsible for protein-phenol interactions [63].

Amylase is secreted mainly by the parotid gland in both glycosylated and non-glycosylated isoforms [64]. It is an enzyme capable of hydrolysing bonds within amylose and amylopectin and is composed mainly of amino acids like aspartic acid > glutamic acid > arginine [65]. However, amino acids as tyrosine and tryptophan seem to be crucial for interaction with polyphenols [66]. The non-glycosylated form of amylase contains 22 proline and 16 tryptophan amino acid residues in its sequence that enable the binding with polyphenols [50].

4. Tannins

Astringent wines are commonly defined as "tannic" because tannins are the main polyphenolic compounds involved in the sensation of astringency. Swain and Bate-Smith [67] provided the first useful phytochemical definition of tannin, being "water-soluble phenolic compounds, having molecular weights lying between 500 and 3000, which have the ability to precipitate alkaloids, gelatin, and other proteins". Tannins can be classified in condensed tannins, phlorotannins, and hydrolysable tannins. Condensed tannins are large macromolecules that consist of two or more monomeric (+)-catechin or (−)-epicatechin units called procyanidins, whilst prodelphinidins consist of (+)-gallocatechin or (−)-epigallocatechin units. In plants, condensed tannins are found as oligomers (2–10 monomer units) or polymers (>10 monomer units). The number of monomer units in a polymer may be as high as 83 units [68]. The subunit composition varies amongst tannins from grape skins, seeds, and stems [69–71]. The phlorotannins are present in marine brown algae as polymers of phloroglucinol (1,3,5 trihydroxy-benzene) in different ranges of molecular sizes (126 Da–650 kDa). They are analogous to the terrestrial condensed tannins since they do not contain a carbohydrate core [72]. Hydrolysable tannins, structurally perhaps the most complex tannins, comprise three subclasses such as simple gallic acid, poly-galloyl esters of glucose (gallotannins), and esters of ellagic acid (ellagitannins). Derivatives of gallic acid contain one to five galloyl groups that can be esterified to either glucose (e.g., pentagalloyl glucose) or quinic acid (e.g., monogalloyl quinic acid). Gallotannis can contain six or more galloyl groups and can be characterised by having one or more digalloyl groups (e.g., hetpagalloyl glucose). Complex gallotannins have a higher capacity for precipitating proteins than simple galloyl glucoses [73].

Ellagitannins may be divided into six subgroups: hexahydroxydiphenoyl esters, dehydro-hexahydroxydiphenoyl esters and their modifications, nonahydroxytriphenoyl esters (e.g., vescalagin), flavonoellagitannins (e.g., acutissimin A), and oligomers with different degrees of oligomerisation and types of linkages [74].

Tannins are the main responsible for the qualitative aspects of astringency as well for the intensity of the sensation. Grape seed and skin tannins are felt astringent as the mean degree of polymerisation (mDP), and galloylation increased [75]. Their ability to precipitate proteins also increases with mDP up to a given degree of polymerisation [34, 76]. However, monomeric and dimeric flavan-3-ols can induce astringent and bitter sensations [77]. Galloylation of monomers/oligomers and polymers enhances protein precipitation, and its extent depends on the grape variety [78]. The presence of high galloylation seems to be responsible for the coarse perception [75], which in turn can be decreased by a high content of epigallocatechin units on the tannin molecule. On the contrary, it seems that the hydroxylation of B-ring seems to decrease velvety astringency and increase the perception of puckering and drying astringency of wine fractions [79]. Salivary proteins seem to have a higher affinity for condensed tannins than for hydrolysable tannins because of different structural flexibility, size, polarity, affinity constants, and presence of free galloyl groups [80–84]. Oakwood tannins were mainly associated with smooth and mouth-drying sensations at low concentrations [85]. Astringency subqualities such as mouth-coat, full-body, persistent were mainly associated with oak-derived tannin, whilst the velvet, soft, and satin terms were associated with the exotic wood-derived tannin [25].

4.1 Other stimuli

Compounds able to elicit sensations as tastes and mouth feelings are called *stimuli*. Chemically diverse astringents such as complex salts such as aluminium sulfate (alum), acids, and other phenolics, have also been shown to evoke astringency [17, 86]. Five organic acids and one inorganic elicited astringency and astringent subqualities [87], and dryness has also been reported [86, 88]. The addition of malic and lactic acid in red wine at the same pH did not differ significantly in astringency despite the difference in titratable acidity [89]. However, these acids were defined astringent in addition to their sour taste [90]. Wines more abundant in malic acid showed higher reactivity towards saliva proteins and then higher potential astringency than tartaric acid-rich wines at the same pH, probably due to different buffer capacities [91]. The astringency of acids is attributed either to the direct contribution of H^+ ions or to the hydrogen bonding capabilities of the hydroxyl groups on the anion or un-dissociated acid [17]. Denaturation of proteins in the saliva could also affect the binding and dissociation of phenolic compounds and their precipitation. The intensity of astringency linearly increases as a function of pH reduction [19], implying significant precipitation of salivary proteins [92].

Anthocyanins, composed of a sugar bound to the anthocyanidin moiety (cyanidin, peonidin, delphinidin, petunidin, and malvidin), impart colour to the grapes and red wine and can be modified by different enological practices [93]. Controversial is the studies of the influence of anthocyanins on astringency. An anthocyanin fraction added in model wine solution was felt as "rough and chalk," and slightly contributed to the overall astringency probably for contamination of the fraction with unknown phenolic compounds [94]. Successively, the isolated fractions of anthocyanidin–glucosides and anthocyanin coumarates did not influence astringency of wine solutions either the "coarse," "chalk," or "dry" astringent subqualities [95]. However, anthocyanins were able to interact with human salivary proteins forming soluble aggregates [96], and even precipitates, being the cinnamoylated the most reactive fraction (precipitation between 6.5 and 17.5%), also influencing the astringency perception [97]. Pyranoanthocyanins, anthocyanin-derived pigments that can form during red wine ageing, seems to be involved in astringency, since they are able to interact with salivary proteins by phenol, catechol or even flavanols structures, similarly to procyanidins [98].

Flavonols (kaempferol, quercetin, and myricetin) are present in grapes and wine as glycosides (sugar attached). In the plant, they act as a natural sunscreen in the skin of grape berries. In wine, they can be hydrolysed and act as cofactors for colour enhancement. Flavonol glycosides, such as 3-O-glucosides and 3-O-galactosides of quercetin, syringetin, and isorhamnetin, have been reported to be astringent at low detection threshold levels and characterised by a velvety astringency [99]. The addition of quercetin 3-O-glucoside (2 g/L) to wine increased astringency, leading to the formation of complexes with saliva at 200 μM [100]. However, such concentrations are not naturally present in red wine, in which quercetin 3-O-glucoside can range from 2 up to 34 mg/L, depending on the cultivar [101].

Many sensory active non-volatile compounds comprising hydroxybenzoic acids, hydroxycinnamic acids, flavon-3-ol glycosides, and dihydroflavon-3-ol rhamnosides were identified as the key inducers of the astringent mouthfeel of red wines using a molecular sensory approach [99]. The phenolic acids in wines, especially hydroxycinnamic and benzoic acid derivatives, have been reported to be more puckering astringent. These compounds have also been correlated with astringency in free-run and pressed wine [102]. The trans-*p*-coumaric, cis-aconitic, and trans-caftaric acids seem to participate in the astringency of Spanish wines [103].

5. Polyphenol-protein interactions

Given that the carbonyl function of salivary proteins is a very effective hydrogen bond acceptor [104], it would appear that it would play a significant role in bonding to polyphenols hydroxyls [10, 105]. Nowadays, the interaction between proteins and proanthocyanidins is widely recognised to be a combination of hydrogen bonding and hydrophobic effects in the acidic wine matrix. However, covalent bonding is also possible between proteins and polyphenols during oxidation [106] and nucleophilic addition processes [107]. In this chapter, we focused on the non-covalent binding involved in the astringent sensation.

Physico-chemical quantities (binding constants, stoichiometry, and atomic structure of complexes, driving forces for the association) have been utilised to understand the multifaceted sensation of astringency. Many techniques including circular dichroism (CD) [108], isothermal titration microcalorimetry (ITC) [109], fluorescence spectroscopy [50], dynamic light scattering (DLS) [110], and nuclear magnetic resonance (NMR) [111] have been employed to understand the formation mechanism of protein/polyphenol aggregates in solution. Generally, these studies focused on interactions between protein segment from human saliva PRPs proteins family and selected procyanidins, because it represents the easiest way to simulate such a complex phenomenon. They can reveal the hydrophobic interactions formed between the phenolic rings of the procyanidins and proline residues, and the hydrogen bonding between the hydroxyl groups on the phenolic B-ring and hydrogen acceptor sites of the peptide bond [52, 112]. The aggregation of procyanidin with peptide seems to be firstly mediated by hydrophobic forces, and then hydrogen bonding has been postulated to provide directional and robust bonding that stabilises the complex. The peptide is coated by polyphenols, which provides a crosslink between two or more peptides up to a critical point, after which precipitation begins.

The stability of these complexes depends on the tannin dimension and number of free phenolic groups, as well as the nature of the protein involved [81, 109].

The driving factors that determine the binding between tannins and salivary proteins were identified to be the critical micelle concentration value (CMC), tannin structure preferences, and tannin colloidal state [113]. Below the values observed in wine (from 1.5 to 2.9 mM), procyanidins specifically interacted with peptide through hydrophilic recognition. A network of interactions can be formed depending on tannin conformation, and precipitation of the complex can occur, or if an intramolecular staking Π-Π of phenolic groups is preferred, the precipitation is not observed. Above these values, tannins spontaneously tend to form aggregates that, at first through specific interactions bind proteins, and then surrounded by the hydrophobic residues, stabilise the complex by hydrophobic bonding. To summarise, both hydrophilic and hydrophobic interactions contribute to form a complex network, which determines the precipitation of salivary proteins with tannins.

6. Assessments of astringency

A method for measuring astringency remains one of the great analytical challenges in wine chemistry and oenology. The interest in investigating the mechanisms and interactions between polyphenols and proteins can allow us to find the optimal way to simulate and evaluate what happens during the red wine tasting. Quite often, sophisticated techniques rely on the purification of both tannin and protein fractions, the extrusion from the wine content, and the omission of matrix components during reactions, and all contribute to send away astringency from the reality that is: wine polyphenols interacting with salivary proteins in mouth, causing drying sensations.

Several procedures have been carried out during the last decades for measuring tannins. Additionally, analyses of soluble (turbidimetric analysis) and insoluble (precipitation protein assays) protein-polyphenols complex have been developed for assessing astringency. The sensory analysis represents the human response as an analytical tool to evaluate wine perception. Many training and tasting sections are necessary over a long period involving a high number of tasters to form a reliable panel. In the case of astringency, it is complicated to discern amongst tastes and brings on fatigue. A method capable of estimating tannin palatability has to be the most objective as possible and must correlate with sensory data in order to reflect the real phenomenon of wine tasting.

6.1 *Stimuli* analysis: pros and contras

Amongst *stimuli* able to elicit astringency, tannins are the main compounds responsible for this sensation. Tannins are intrinsically amphiphilic molecules with high reactivity, have a diverse range of structures, and are often found in matrices with other phenolic molecules containing similar functional groups. Besides using sophisticated equipment and analytical techniques, there is also a great interest in a relatively simple method.

In the past, many colourimetric techniques were developed to analyse phenolics compounds spectrophotometrically. The first one used the Folin-Denis reagent [114], which was successively modified [115, 116], and lastly into the Folin-Ciocalteau assay [117]. However, they were not specific for tannins but detected any phenolic compound. More specific colour reactions were used to measure condensed tannins and their precursors. Depolymerisation in HCl and n-butanol of proanthocyanidins yield anthocyanidins that can be quantified spectrophotometrically [118, 119]. Others used vanillin reagent for flavanols [6, 120], or *p*-dimethylaminocinnamaldehyde for a more specificity and colour stability [121, 122]. Only the flavonoid-based condensed tannins can be detected with these reagents. As tannins can inhibit the catalytic activity of enzymes [6], many methods used the interaction with proteins in solution to measure the inhibition of different enzymes spectrophotometrically [123].

Other methods, based on the acid-catalysed condensation reactions with benzyl mercaptan (thiolysis) and phloroglucinol (phloroglucinolysis), can determine both the chain length (mDP) and composition by HPLC [124, 125]. Most of our current knowledge about the general composition and structure of grape and wine tannins have been obtained by depolymerisation [126]. Poor yields due to reaction product instability, reactions with non-proanthocyanidin compounds, and side reactions also contribute negatively to the utility of thiolytic methods [124]. The problem with phloroglucinolysis, on the other hand, is that it produces low yields, and only a fraction of the tannin is converted to known flavan-3-ol products [127]. Normal-phase HPLC (NP-HPLC) method has also been developed to quantify the proanthocyanidins into low and high molecular-weight polymers [128]. A simple method based on Fourier transform mid-infrared (FT-MIR) spectroscopy combined with multivariate data analysis, was successfully used to measure the tannin concentration of 86 red wines, previously purified by solid-phase extraction (SPE) [129].

6.2 Precipitation assays: pros and contras

Protein precipitation assays are of particular interest because the interaction of proteins with tannins can be used to model astringency perception [130]. The ability of gelatin to precipitate phenols, including tannins, has been observed since 1934 [131]. The same phenomenon was observed when hide powder or polyvinylpyrrolidone were used in high concentrations [132]. Bate-Smith [130] noted that protein of skin differed from proteins of saliva, which caused the "puckery" sensation induced by tannin. For measuring the relative astringency of tannins, a spectrophotometric technique based on the precipitation of the haemoglobin with tannin was then introduced [130]. Similarly, another spectrophotometric technique measured the inhibition of β-glucosidase after the precipitation with tannic acid and condensed tannins [133]. Alternatively, Hagerman and Butler [134] used bovine serum albumin (BSA) as a precipitant agent, which was successively taken by Harbertson et al. [135] for wine analysis. Glories [136] proposed the gelatin index, in which tannins were precipitated by gelatin protein. This procedure required the measure of proanthocyanidin concentration before and after precipitation with an excess of gelatin. Besides, gelatin is a heterogeneous mixture of proteins, and its composition may change amongst the different commercial products, leading to a source of variability and imprecision of data. For this, some researchers replaced gelatin with ovalbumin [137]. Another tannin assay used the methylcellulose to precipitate tannin (MCP) [138, 139]. The MCP tannin assay is based on the formation of an insoluble polymer-tannin complex, which can be separated by centrifugation. The total phenolic content (absorbance at 280 nm) is measured in control and treated samples. However, if the assays utilise synthetic agent or protein different from saliva, the binding reaction seems not to reproduce the physiological conditions during the wine tasting, because the binding affinity of the protein is not comparable to that of salivary protein. In the case of bovine serum albumin, it has been shown that the salivary protein has a higher affinity for tannin than BSA. In fact, in the presence of an excess of BSA, the tannin preferentially bound the salivary protein. Other proteins, including dietary proteins, may not complex any tannin in the presence of the salivary tannin-binding protein [8]. The use of salivary proteins has been proposed to represent the model system for astringency better. In precipitation assays, fractionated [8, 140] or whole [141, 142] human saliva has been used. Mixing whole saliva and grape polyphenols give rise to a "soft cloudy" precipitate, which gathered after centrifugation on the bottom of the tube so that the supernatant was easily recovered without disturbing this pellet. The binding reaction was performed at 25°C, and the complex formed was successively precipitated by centrifugation at 4°C in order to stop further reactions. The induced precipitation allowed to separate the proteins bound to polyphenols from whose remained in the solution that not reacted with them. Both the nature of condensed tannin [141] and salivary proteins [142] involved in the precipitation were analysed. In both works, the sodium dodecyl sulphate-polyacrylamide gel electrophoresis (SDS-PAGE) of human saliva was carried out, Sarni-Manchado et al. [141], analysed the tannins in the supernatant and pellet. In contrast, Gambuti et al. [142], analysing the supernatant, revealed the proteins mainly reactive with polyphenols by comparison with the control saliva. Evidence of the qualitative and quantitative changes in salivary protein profile after tasting tannin solutions and wines was also made by HPLC [143]. Interactions and precipitation of low molecular weight salivary proteins with procyanidins confirmed the involvement of different families of salivary proteins in the development of astringency [144]. The use of salivary proteins involves the collection of human saliva from different healthy volunteers according to a specific protocol, and it must take into account the salivary flow to limit the effect of individual differences in astringency perception due to subjects' saliva characteristics [145].

6.3 Nephelometry: pros and contras

Nephelometry is a method that allows a direct estimation of the amount of protein/tannin complexes by measuring the scattered light in the solution that results from the gradual formation of a cloudy precipitate corresponding to the soluble aggregate. Chapon [146] proposed this technique by studying the interactions between beer polyphenols and proteins involved in the colloidal instability of beer. Similarly, the haze formed between salivary proteins and polyphenols represents the first step in the development of astringency and can be measured with a turbidimeter [147, 148]. A continuous flow method was also used to study the interactions between grape extracts and wine with BSA at different concentrations [149]. Globular proteins and PRPs were used to measure a relative tannin specific activity of procyanidin oligomers from grape seeds [30], and PRPs showed the strongest affinity. Human salivary proteins have been considered as the most suitable model proteins. For this reason, in turbidity measurement, whole human saliva [148] and mucin, a high molecular weight salivary protein [150], were used as model proteins for astringency assessment. Based on polyphenol/mucin reactivity, a micro-plate assay was also developed [151]. Tannic acid [150], grape seed extracts [151], wine extracts [63], tannin fractions added to model solutions [152] were analysed by nephelometry. The turbidity of the solution, formed by the tannin-protein aggregates, linearly correlated with astringency. However, no direct analysis of wines was carried out. Lastly, instead, wine samples were analysed trough nanotechnology such as localised surface plasmon resonance (LSPR) combined with surface imprinted polymers, as a measure of the interactions of polyphenol with salivary protein and then astringency [153].

6.4 Sensory analysis: pros and contras

The sensory analysis represents the human response to wine tasting. A sensory panel can provide information about the sensory properties of a product, but significant training is required before the panel becomes a reliable sensory instrument. Astringency is a difficult sensory attribute to evaluate, owing to particular characteristics of the sensation. Generally, it is evaluated by tasting but can suffer from individual subjectivity. The feeling can take over 15 seconds to develop fully and is known to build in intensity and become increasingly difficult to clear from the mouth over repeated exposures [19, 154]. Carry-over effects can occur. When wines or tannic solutions are evaluated by a well-trained panel using established sensory methodologies, the panel leader can expect to obtain reliable information about the intensity in the perceived astringency of the samples. Screening, selection, training, and panel maintenance are exercises that help the panel attain proficiency before sample evaluation. Classical methodologies widely applied are descriptive and rating sensory analyses. The first helps to distinguish between samples by a qualitative description of their sensory properties [75] and the second permits to scale samples according to the intensity of the perception. However, time-intensity (TI) is a temporal methodology widely used. This method consists of recording one by one the intensity evolution of given attributes [155]. However, TI showed some limitations because it is time-consuming due to the evaluation of only a few attributes at the same time [156]. Furthermore, carry-over effects can overcome when assessing the temporal perception of an attribute [157]. To overcome these drawbacks, Pineau et al. [156] developed a new method called temporal dominance of sensations (TDS), which consists of identifying and rating sensations perceived as dominant until the perception ends. Before the development of this method, a similar experimental approach was successfully used to describe the temporality of sensations in wines [158]. Astringency, a dynamic sensation, takes many seconds to develop after the basic tastes, and the duration depends on the wine. Notwithstanding, TDS can be difficult when panellists had select the dominant attribute and score its intensity, but proper training can overcome this problem [159].

It is also essential to discuss and familiarise with the terms associated with astringency. A vocabulary of 33 terms has been proposed by a combined panel of experienced tasters and winemakers to describe the mouthfeel characteristics of red wines [160]. The check-all-that-apply (CATA) question that consists of a list of subqualities from which the panellists have to select all the options they consider appropriate to that wine has been utilised for the characterisation of the astringency subqualities of Tannat wine [161]. Recently, a sensory method that combines CATA approach and training in astringency subqualities with touch-standards resulted very useful for investigating the astringency characteristics of red wines [24, 25, 162]. In any case, intense training is necessary to distinguish astringency from other tastes, especially bitterness, and to reveal the different qualitative attributes. Fatigue and loss of *stimuli* memory may occur, particularly with panellists who are unfamiliar with astringency, and when too many samples are presented. Training is also expensive and time-consuming. However, it is necessary to investigate the astringency subqualities of red wines. Sensory analysis is of fundamental importance, but in some cases, it is not possible to perform, so the replacement with an analytical instrument able to measure astringency could help in research as well as in the winery.

6.5 Correlation between sensory and analytical analysis

Because astringency is one of the main attributes for wine quality, winemakers are interested in an analytical and objective method to evaluate it. No method can substitute entirely sensory analysis, but a method that results in a reproducible index has to correlate quite well with it. A statistically significant correlation between the sensorial and analytical methods is necessary.

The gelatin index has represented the almost widely analytical method for estimating astringency in red wine [136]. Besides, it furnished only approximate results [137]. Successively, a positive correlation (R2 = 0.56) between the gelatin index and time-intensity data was obtained only at a low concentration of polyphenols utilising 29 wines judged by 10 panellists [163]. A method that used the ovalbumin in alternatively to gelatin as a precipitation agent was proposed to determine astringency [137]. Ten wines were tested by 10 expert enologists evaluating the astringency on a scale from 1 to 100. The method resulted in more reproducible than the gelatin index and was positively correlated ($R2 = 0.77$) with sensory analysis. This method was also used to assess the astringency of Greek wines, and a good correlation was found ($R2 = 0.93$) [164]. Another predictive model for astringency estimation was based on phenolic compounds and colour analysis of 34 wines by 12 judges on a 9-point intensity scale [165]. Multiple regression generated three possible models to predict astringency from analytical data, the most simple depended on total phenolics and co-pigmented anthocyanins, besides the predicted astringency plotted versus observed astringency resulted in low but acceptable correlation from a sensory perspective.

Monteleone et al. [150] proposed a predictive model by measuring the polyphenol-mucin reactivity in which the capability of polyphenolic extracts to induce astringency was estimated on their ability to develop turbidity in the *in vitro* assay. They found a linear relation between astringency perceived by 30 trained judges and the mucin index for tannic acid model solutions ($R2 = 0.993$) grape seed extracts ($R2 = 0.996$), and phenolic extracts ($R2 = 0.95$) [63].

In a study by Kennedy et al. [166], 40 red wines were evaluated by a panel consisting of three winemakers and two enologists for the astringency intensity scored from zero to 10. The aim was to correlate astringency and tannin concentration measured by different analytical methods: absorption at 280 nm, phloroglucinolysis, gel chromatography, and BSA protein precipitation. The analytical method having the strongest correlations with perceived astringency was the protein precipitation one ($R2 = 0.82$). Protein precipitation represents the method the most similar to the physiological response to astringent *stimuli* and can be used as an *in vitro* tool for understanding how tannin can modulate astringency perception. Generally, it was assumed that the most suitable proteins for evaluating astringency are the salivary PRPs. However, other proteins in whole human saliva were preferentially precipitated by increasing tannin solutions [142]. Successively, the percentage decrease of two salivary proteins after the precipitation with tannins, measured by electrophoresis, represented an indicator of the reactivity of tannin. The saliva precipitation index (SPI) was well correlated with the sensory evaluation of the astringency of 57 red wines ($R2 = 0.97$) made by 18 trained assessors [167].

7. The saliva precipitation index (SPI)

The SPI represents a useful tool to assess the physiological response to astringents, measuring the astringency of red wine indirectly. This index evaluated the precipitation of salivary proteins occurring during the tasting of an astringent stimulus. The SPI, analysing the salivary protein pattern by SDS-PAGE electrophoresis, has been improved considering the in-mouth temperature (37°C) for the binding reaction, the choice of resting saliva, and the ratio saliva:wine. The excess of saliva with respect to wine (2:1) in a static environment permits to measure the binding capacity of tannins better [167]. Successively, to reduce the time and solvents, the chip electrophoresis replaced the SDS-PAGE, providing similar results [168]. In the last years, the SPI has been used for different technological practices proving useful information for winemakers and enologists to manage the style and quality of red wines.

7.1 Applications of SPI in winemaking

7.1.1 Enological practices

In winemaking, the clarification process is fundamental to stabilise and clarify the wine by adding exogenous proteins into wine [169]. Proteins used for fining interact with wine tannins by a mechanism similar to that occurring during the tasting. The interaction protein-tannin, binding, and precipitation determine a decrease in polyphenolic compounds responsible for the sensation of astringency [170]. The SPI was used to evaluate the efficacy of the fining of different proteins at different concentrations in Aglianico [171], and Sangiovese wines [172]. In Aglianico, the gelatin (animal protein) and patatin (plant protein) showed similar efficacy in diminishing wine polyphenols reactive towards salivary proteins, and then astringency, whilst in Sangiovese it depended on the polyphenolic content of the wine. The information provided by SPI was useful to understand that each wine, with peculiar polyphenolic composition, should be treated maintaining the ratio anthocyanins and tannins such as to assure a modulation of astringency and at the same time a correct evolution of the colour during ageing.

A common practice is the utilisation of enological tannins as a substitute for oak barrels to improve colour stability and taste and is authorised by the International Organisation of the Vine and Wine (OIV) for musts and wines clarification [173]. Commercial preparations of tannins of different origins showed different abilities in precipitating salivary proteins: condensed tannins resulted in higher SPI and astringency than hydrolysable tannins. The addition of tannins in wines modify the astringency or not depending on the wine phenolic content. The SPI was useful to understand the effect of tannins addition on wine astringency in order not to compromise overall wine quality [83]. Similarly, after a moderate oxidation (21 mg/L of oxygen equivalent), the addition of 2 g/L of enological tannins did not result in an increase in the reactivity of wine tannins towards salivary proteins after 30 days of treatment. This effect was also shown in the oxidation process in the presence of acetaldehyde [174]. The SPI seems to be sensitive to reaction-products such as polymers of flavanols and anthocyanins formed directly or via a molecular bridge (e.g., acetaldehyde) [31, 175], and new-formed proanthocyanidins [93, 176]. This may explain why during the oxidation of red wines, the SPI followed a different trend from BSA reactive tannins [174, 177, 178].

7.1.2 Ageing

The decrease of astringency with time has been shown to depend on the reduced concentration of tannins due to precipitation [31, 68], but the trend is not strictly related to the age of wine [179]. The astringency of red wine decreases during ageing because of the changes in the structure of tannins due to cleavage reactions generating low molecular weight species [31], polymerisation without the participation of anthocyanins and subsequent precipitation [95], direct or indirect condensation with anthocyanins [180], and the formation of flavan-3-ol sulfonates by SO2 [181]. Wine becomes soft and mellow for the decline of tannin mean degree of polymerisation [182], velvet and mouth-coating for the formation of the polymeric pigments [24], or satin for lower content of flavans and astringent tannins (measured by SPI), and higher formation of polymers [183] after ageing. Studies on Sangiovese wine revealed that the astringency profile changed from an unripe, dry astringency towards rich, full-body, and mouth-coating sensations after about 2 years of ageing [184]. However, pucker sensations can appear if the oxidation is excessive [24, 25]. Astringency subqualities have been able to discriminate wines of different denominations with a chemical age of 3–5 years, more than other wine parameters [25]. Red wine benefits of a moderate oxygenation during ageing favouring changes in tannin structures that, affecting their reactivity towards proteins, can modulate wine astringency. The SPI was utilised to objectively evaluate changes in astringency as a function of oxygen uptake before and after bottling [185]. Although conflicting results were reported for astringency after micro-oxygenation of wines, a significant variation of wine reactivity towards salivary proteins and, then, in wine astringency was observed after 42 months of ageing in bottle only in low pH wines. Moreover, oxygen permeating towards closures determined changes in wine phenolics detectable only using SPI. It was significantly lower when the bottles were sealed with closures at high oxygen transfer rate (OTR). Such differences were not perceived by sensory analysis, demonstrating that SPI can be more sensitive in revealing slight differences in the reactivity of tannins. Lastly, the effect of ageing on the precipitation of salivary proteins is a function of ageing time, wine pH and phenolic composition, and oxygen level in red wine. The decisive role of pH on wine astringency has been confirmed in a recent work of Forino et al. [92], in which the SPI was used to measure wines with different pH levels (3.7–3.2) obtained by adding strong acids or bases, which made the wine unsafe to taste. The binding and precipitation of wine tannins with saliva proteins was favoured at low pH values, and this effect was dominant with respect to the tannins content. Previously, the tartaric acid addition in wine, modifying the pH, resulted in high SPI [186], due to the increase of tannins in the phenolate form, and therefore to an increase of hydrogen bonding with salivary proteins. It is also likely that at low pH increases the accessibility of the binding sites leading to enhanced Van der Waal interactions and hydrogen bonding between proteins and polyphenols [187]. However, other parameters, such as ethanol, fructose, and mannoproteins have been shown to influence astringency and SPI [186]. The effect of mannoproteins on the inhibition of salivary protein precipitation was also showed in Aglianico and Sangiovese wines after 12 months of ageing. The sensory analysis confirmed a reduction in wine astringency. Some mannoproteins interact with tannins forming higher molecular weight structures that prevent the binding with salivary proteins, and thus are not able to elicit astringency [94]. Mannoproteins can also act as steric stabilisers limiting the binding with tannins [112]. Wine polysaccharides inhibit tannin-salivary proteins interaction by a mechanism that involves the formation of protein-tannin complex firstly, probably ruled by hydrophobic interactions and stabilised by hydrogen bonds, and then the polysaccharides can act by a ternary mechanism through the encapsulation of this complex, increasing its solubility. However, the efficiency depends on the polarity of both salivary proteins and tannins [188]. Beyond the molecular mechanism, mannoproteins can highly influence the qualitative sensory perception of astringency, conferring positive subqualities of astringency to red wines [162].

8. Conclusions

Astringency is still a complex phenomenon, and despite the many efforts from researchers, it is not fully understood. However, the different *in vitro* assessments have been shown to be useful in evaluating the wine astringency. They could replace the sensory evaluation when there is no possibility of tasting wines: for low sample availability, when tasting is not permitted (as in the pandemic period due to Covid-19) or unsafe, or when too many samples must be tasted. An analytical method for astringency may be potentially useful not only in research purposes but also in the optimisation of the winemaking process and may help wine producers to improve wine quality.

Conflict of interest

The authors declare no conflict of interest.

Nomenclature

BSA	bovine serum albumin
CATA	check-all-that-apply
CD	circular dichroism
CMC	critical micelle concentration
DLS	dynamic light scattering
FT-MIR	Fourier transform mid-infrared
OIV	International Organisation of the Vine and Wine
ITC	isothermal titration microcalorimetry
LSPR	localised surface plasmon resonance
M1	high molecular-weight mucin
M2	low molecular-weight mucin
mDP	mean degree of polymerisation
MRs	mechanoreceptors
MCP	methylcellulose precipitable-tannin
NP-HPLC	normal-phase HPLC
NMR	nuclear magnetic resonance
OTR	oxygen transfer rate
PRPs	proline-rich proteins
SPI	saliva precipitation index
SDS-PAGE	sodium dodecyl sulphate-polyacrylamide gel electrophoresis
SPE	solid-phase extraction
TDS	temporal dominance of sensations
TI	time-intensity

Author details

Alessandra Rinaldi[1*] and Luigi Moio[2]

1 Biolaffort, Bordeaux, France

2 Department of Agricultural Sciences, University of Napoli "Federico II", Enology Sciences Section, Avellino, Italy

*Address all correspondence to: alessandra.rinaldi@unina.it

References

[1] Zucker WV. Tannins: Does structure determine function? An ecological perspective. The American Naturalist. 1983;**121**(3):335-365. DOI: 10.1086/284065

[2] Stern JL, Hagerman AE, Steinberg PD, Mason PK. Phlorotannin-protein interactions. Journal of Chemical Ecology. 1996;**22**(10):1877-1899. DOI: 10.1007/BF02028510

[3] Aerts RJ, Barry TN, McNabb WC. Polyphenols and agriculture: Beneficial effects of proanthocyanidins in forages. Agriculture, Ecosystems and Environment. 1999;**75**(1-2):1-12. DOI: 10.1016/S0167-8809(99)00062-6

[4] Bernays EA, Driver GC, Bilgener M. Herbivores and plant tannins. In: Advances in Ecological Research. Vol. 19. London, United Kingdom: Academic Press; 1989. pp. 263-302. DOI: 10.1016/S0065-2504(08)60160-9

[5] Mehansho H, Butler LG, Carlson DM. Dietary tannins and salivary proline-rich proteins: Interactions, induction and defense mechanisms. Annual Review of Nutrition. 1987;**7**:423-440

[6] Goldstein JL, Swain T. Changes in tannins in ripening fruits. Phytochemistry. 1963;**2**(4):371-383. DOI: 10.1016/S0031-9422(00)84860-8

[7] Harlan JW, Feairheller SH. In: Friedman M, editor. Protein Crosslinking. Advances in Experimental Medicine and Biology. Chemistry of the Crosslinking of Collagen during Tanning. Boston, MA: Springer; 1977. DOI: 10.1007/978-1-4684-3282-4_27

[8] Austin PJ, Suchar LA, Robbins CT, Hagerman AE. Tannin-binding proteins in saliva of deer and their absence in saliva of sheep and cattle. Journal of Chemical Ecology. 1989;**15**:1335-1347. DOI: 10.1007/BF01014834

[9] Bate-Smith EC. Leuco-anthocyanins 1. Detection and identification of anthocyanidins formed from leuco-anthocyanins in plant tissues. The Biochemical Journal. 1954;**58**(1):122

[10] Hagerman AE, Butler LG. The specificity of proanthocyanidin-protein interactions. The Journal of Biological Chemistry. 1981;**256**:4494-4497

[11] Soares S, Brandão E, Guerreiro C, Mateus N, de Freitas V, Soares S. Development of a new cell-based oral model to study the interaction of oral constituents with food polyphenols. Journal of Agricultural and Food Chemistry. 2019;**67**(46):12833-12843. DOI: 10.1021/acs.jafc.9b05575

[12] Breslin P, Gilmore M, Beauchamp G, Green B. Psychophysical evidence that oral astringency is a tactile sensation. Chemical Senses. 1993;**18**(4):405-417. DOI: 10.1093/chemse/18.4.405

[13] Trulsson M, Essick GK. Low-threshold mechanoreceptive afferents in the human lingual nerve. Journal of Neurophysiology. 1997;**77**(2):737-748. DOI: 10.1152/jn.1997.77.2.737

[14] Chen J, Engelen L. Food Oral Processing: Fundamentals of Eating and Sensory Perception. Oxford, United Kingdom: Wiley-Blackwell; 2012

[15] Schöbel N, Radtke D, Kyereme J, Wollmann N, Cichy A, Obst K, et al. Astringency is a trigeminal sensation that involves the activation of G protein-coupled signaling by phenolic compounds. Chemical Senses. 2014;**39**(6):471-487. DOI: 10.1093/chemse/bju014

[16] Kishi M, Sadachi H, Nakamura J, Tonoike M. Functional magnetic resonance imaging investigation of brain regions associated with astringency. Neuroscience Research.

2017;**122**:9-16. DOI: 10.1016/j.neures.2017.03.009

[17] Lawless HT, Corrigan CJ, Lee CB. Interactions of astringent substances. Chemical Senses. 1994;**19**:141-154. DOI: 10.1093/chemse/19.2.141

[18] ASTM. Standard definitions of terms relating to sensory evaluation of materials and products. In: Kuznicki JT, Rutkiewic AF, Johnson RA, editors. Annual Book of ASTM Standards. Philadelphia, PA: American Society for Testing and Materials; 2004. pp. 5-15

[19] Guinard JX, Pangborn RM, Lewis MJ. The time-course of astringency in wine upon repeated ingestion. American Journal of Enology and Viticulture. 1986;**37**(3):184-189

[20] Fleming EE, Ziegler GR, Hayes JE. Salivary protein levels as a predictor of perceived astringency in model systems and solid foods. Physiology & Behavior. 2016;**163**:56-63. DOI: 10.1016/j.physbeh.2016.04.043

[21] Gawel R, Iland P, Francis I. Characterising the astringency of red wine: A case study. Food Quality and Preference. 2001;**12**(1):83-94. DOI: 10.1016/S0950-3293(00)00033-1

[22] King MC, Cliff MA, Hall J. Effectiveness of the 'Mouth-feel Wheel' for the evaluation of astringent subqualities in British Columbia red wines. Journal of Wine Research. 2003;**14**(2-3):67-78. DOI: 10.1080/09571260410001677932

[23] De Miglio P, Pickering GJ. The influence of ethanol and pH on the taste and mouthfeel sensations elicited by red wine. Journal of Food, Agriculture and Environment. 2008;**6**:143-150

[24] Rinaldi A, Moio L. Effect of enological tannin addition on astringency subqualities and phenolic content of red wines. Journal of Sensory Studies. 2018;**33**(3):e12325. DOI: 10.1111/joss.12325

[25] Rinaldi A, Moine V, Moio L. Astringency subqualities and sensory perception of Tuscan Sangiovese wines. OENO One. 2020;**54**(1):75-85. DOI: 10.20870/oeno-one.2020.54.1.2523

[26] Vidal L, Antúnez L, Rodríguez-Haralambides A, Giménez A, Medina K, Boido E, et al. Relationship between astringency and phenolic composition of commercial Uruguayan Tannat wines: Application of boosted regression trees. Foodservice Research International. 2018;**112**:25-37. DOI: 10.1016/j.foodres.2018.06.024

[27] Sáenz-Navajas M, Ballester J, Pêcher C, Peyron D, Valentin D. Sensory drivers of intrinsic quality of red wines. Foodservice Research International. 2013;**54**(2):1506-1518. DOI: 10.1016/j.foodres.2013.09.048

[28] Martens M. A philosophy for sensory science. Food Quality and Preference. 1999;**10**(4-5):233-244. DOI: 10.1016/S0950-3293(99)00024-5

[29] Lu Y, Bennick A. Interaction of tannin with human salivary proline-rich proteins. Archives of Oral Biology. 1998;**43**(9):717-728. DOI: 10.1016/S0003-9969(98)00040-5

[30] De Freitas V, Mateus N. Structural features of procyanidin interactions with salivary proteins. Journal of Agricultural and Food Chemistry. 2001;**49**(2):940-945. DOI: 10.1021/jf000981z

[31] Cheynier V, Dueñas-Paton M, Salas E, Maury C, Souquet JM, Sarni-Manchado P, et al. Structure and properties of wine pigments and tannins. American Journal of Enology and Viticulture. 2006;**57**(3):298-305

[32] Gawel R. Red wine astringency: A review. Australian Journal of Grape and

Wine Research. 1998;**4**(2):74-95. DOI: 10.1111/j.1755-0238.1998.tb00137.x

[33] de Freitas V, Mateus N. Protein/polyphenol interactions: Past and present contributions. Mechanisms of astringency perception. Current Organic Chemistry. 2012;**16**(6):724-746. DOI: 10.2174/138527212799958002

[34] Bajec MR, Pickering GJ. Astringency: Mechanisms and perception. Critical Reviews in Food Science and Nutrition. 2008;**48**:858-875. DOI: 10.1080/10408390701724223

[35] McRae JM, Kennedy JA. Wine and grape tannin interactions with salivary proteins and their impact on astringency: A review of current research. Molecules. 2011;**16**(3):2348-2364. DOI: 10.3390/molecules16032348

[36] Ma W, Guo A, Zhang Y, Wang H, Liu Y, Li H. A review on astringency and bitterness perception of tannins in wine. Trends in Food Science and Technology. 2014;**40**(1):6-19. DOI: 10.1016/j.tifs.2014.08.001

[37] Soares S, Brandão E, Mateus N, de Freitas V. Sensorial properties of red wine polyphenols: Astringency and bitterness. Critical Reviews in Food Science and Nutrition. 2017;**57**(5):937-948. DOI: 10.1080/10408398.2014.946468

[38] Li SY, Duan CQ. Astringency, bitterness and colour changes in dry red wines before and during oak barrel aging: An updated phenolic perspective review. Critical Reviews in Food Science and Nutrition. 2019;**59**(12):1840-1867. DOI: 10.1080/10408398.2018.1431762

[39] Humphrey SP, Williamson RT. A review of saliva: Normal composition, flow, and function. Journal of Prosthetic Dentistry. 2001;**85**(2):162-169. DOI: 10.1067/mpr.2001.113778

[40] Scarano E, Fiorita A, Picciotti PM, Passali GC, Calò L, Cabras T, et al. Proteomics of saliva: Personal experience. ACTA Otorhinolaryngologica Italica. 2010;**30**(3):125-130

[41] Bandhakavi S, Stone MD, Onsongo G, Van Riper SK, Griffin TJ. A dynamic range compression and three-dimensional peptide fractionation analysis platform expands proteome coverage and the diagnostic potential of whole saliva. Journal of Proteome Research. 2009;**8**(12):5590-5600. DOI: 10.1021/pr900675w

[42] Dawes C. Salivary flow patterns and the health of hard and soft oral tissues. Journal of the American Dental Association. 2008;**139**:18S-24S. DOI: 10.14219/jada.archive.2008.0351

[43] Schipper RG, Silletti E, Vingerhoeds MH. Saliva as research material: Biochemical, physicochemical and practical aspects. Archives of Oral Biology. 2007;**52**(12):1114-1135. DOI: 10.1016/j.archoralbio.2007.06.009

[44] Helmerhorst EJ, Oppenheim FG. Saliva: A dynamic proteome. Journal of Dental Research. 2007;**86**(8):680-693. DOI: 10.1177/154405910708600802

[45] Christensen CM. Role of saliva in human taste perception. In: Meiselman HL, Rivlin RS, editors. Clinical Measurement of Taste and Smell. New York: Macmillan Publishing Co; 1985. pp. 414-428

[46] Fischer U, Boulton RB, Noble AC. Physiological factors contributing to the variability of sensory assessments: Relationship between salivary flow rate and temporal perception of gustatory stimuli. Food Quality and Preference. 1994;**5**:55-64. DOI: 10.1016/0950-3293(94)90008-6

[47] Spielman AI. Interaction of saliva and taste. Journal of Dental Research. 1990;**69**:838-848. DOI: 10.1177/00220345900690030101

[48] Matsuo R. Role of saliva in the maintenance of taste sensitivity. Critical Reviews in Oral Biology and Medicine. 2000;**11**(2):216-229. DOI: 10.1177/10454411000110020501

[49] Lamy E, Torregrossa AM, Castelo PM, Silva FC. Saliva in ingestive behavior research: Association with oral sensory perception and food intake. In: Tvarijonaviciute A, Martínez-Subiela S, López-Jornet P, Lamy E, editors. Saliva in Health and Disease. Cham: Springer; 2020. pp. 23-48. DOI: 10.1007/978-3-030-37681-9_2

[50] Soares S, Mateus N, De Freitas V. Interaction of different polyphenols with bovine serum albumin (BSA) and human salivary α-amylase (HSA) by fluorescence quenching. Journal of Agricultural and Food Chemistry. 2007;**55**:6727-6735. DOI: 10.1021/jf070905x

[51] Hay DI, Bennick A, Schlesinger DH, Minaguchi K, Madapallimattam G, Schluckebier SK. The primary structures of six human salivary acidic proline-rich proteins (PRP-1, PRP-2, PRP-3, PRP-4, PIF-s and PIF-f). The Biochemical Journal. 1988;**255**(1):15-21. DOI: 10.1042/bj2550015

[52] Murray NJ, Williamson MP. Conformational study of a salivary proline-rich protein repeat sequence. European Journal of Biochemistry. 1994;**219**(3):915-921. DOI: 10.1111/j.1432-1033.1994.tb18573.x

[53] Mehansho H, Hagerman A, Clements S, Butler L, Rogler J, Carlson DM. Modulation of proline-rich protein biosynthesis in rat parotid glands by sorghums with high tannin levels. Proceedings of the National Academy of Sciences. 1983;**80**(13):3948-3952. DOI: 10.1073/pnas.80.13.3948

[54] Mole S, Butler LG, Iason G. Defense against dietary tannin in herbivores: A survey for proline rich salivary proteins in mammals. Biochemical Systematics and Ecology. 1990;**18**(4):287-293. DOI: 10.1016/0305-1978(90)90073-O

[55] Oppenheim FG, Yang YC, Diamond RD, Hyslop D, Offner GD, Troxler RF. The primary structure and functional characterisation of the neutral histidine-rich polypeptide from human parotid secretion. The Journal of Biological Chemistry. 1986;**261**(3):1177-1182

[56] Troxler RF, Offner GD, Xu T, Vanderspek JC, Oppenheim FG. Structural relationship between human salivary histatins. Journal of Dental Research. 1990;**69**(1):2-6. DOI: 10.1177/00220345900690010101

[57] Inzitari R, Cabras T, Rossetti DV, Fanali C, Vitali A, Pellegrini M, et al. Detection in human saliva of different statherin and P-B fragments and derivatives. Proteomics. 2006;**6**:6370-6379. DOI: 10.1002/pmic.200600395

[58] Wróblewski K, Muhandiram R, Chakrabartty A, Bennick A. The molecular interaction of human salivary histatins with polyphenolic compounds. European Journal of Biochemistry. 2001;**268**(16):4384-4397. DOI: 10.1046/j.1432-1327.2001.02350.x

[59] Baxter NJ, Lilley TH, Haslam E, Williamson MP. Multiple interactions between polyphenols and a salivary proline-rich protein repeat result in complexation and precipitation. Biochemistry. 1997;**36**(18):5566-5577. DOI: 10.1021/bi9700328

[60] Soares S, Mateus N, De Freitas V. Interaction of different classes of salivary proteins with food tannins. Foodservice Research International. 2012;**49**(2):807-813. DOI: 10.1016/j.foodres.2012.09.008

[61] Soares S, Santos Silva M, García-Estévez I, Brandão E, Fonseca F, Ferreira-da-Silva F, et al. Effect of

malvidin-3-glucoside and epicatechin interaction on their ability to interact with salivary proline-rich proteins. Food Chemistry. 2019;**276**:33-42. DOI: 10.1016/j.foodchem.2018.09.167

[62] Tabak LA. Structure and function of human salivary mucins. Critical Reviews in Oral Biology and Medicine. 1990;**1**(4):229-234. DOI: 10.1177/10454411900010040201

[63] Condelli N, Dinnella C, Cerone A, Monteleone E, Bertuccioli M. Prediction of perceived astringency induced by phenolic compounds II: Criteria for panel selection and preliminary application on wine samples. Food Quality and Preference. 2006;**17**:96-107. DOI: 10.1016/j.foodqual.2005.04.009

[64] Kauffman DL, Watanabe S, Evans JR, Keller PJ. The existence of glycosylated and non-glycosylated forms of human submandibular amylase. Archives of Oral Biology. 1973;**18**(9):1105-1113. DOI: 10.1016/0003-9969(73)90084-8

[65] Muus J. The amino acid composition of human salivary Amylase[1]. Journal of the American Chemical Society. 1954;**76**(20):5163-5165

[66] Gyémánt G, Zajácz Á, Bécsi B, Ragunath C, Ramasubbu N, Erdődi F, et al. Evidence for pentagalloyl glucose binding to human salivary α-amylase through aromatic amino acid residues. Biochimica et Biophysica Acta (BBA) - Proteins and Proteomics. 2009;**1794**:291-296. DOI: 10.1016/j.bbapap.2008.10.012

[67] Swain T, Bate-Smith EC. Flavonoid compounds. In: Florkin M, Mason HS, editors. Comparative Biochemist, a Comprehensive Treaty, Vol. III: Constituents of Life-Part A. New York: Academic Press; 1962. pp. 755-809

[68] Es-Safi N, Fulcrand H, Cheynier V, Moutounet M. Studies on the acetaldehyde-induced condensation of (−)-epicatechin and malvidin 3-o-glucoside in a model solution system. Journal of Agricultural and Food Chemistry. 1999;**47**(5):2096-2102. DOI: 10.1021/jf9806309

[69] Prieur C, Rigaud J, Cheynier V, Moutounet M. Oligomeric and polymeric procyanidins from grape seeds. Phytochemistry. 1994;**36**:781-784. DOI: 10.1016/S0031-9422(00)89817-9

[70] Souquet JM, Cheynier V, Brossaud F, Moutounet M. Polymeric proanthocyanidins from grape skins. Phytochemistry. 1996;**43**:509-512. DOI: 10.1016/0031-9422(96)00301-9

[71] Souquet JM, Labarbe B, Le Guernevé C, Cheynier V, Moutounet M. Phenolic composition of grape stems. Journal of Agricultural and Food Chemistry. 2000;**48**:1076-1080. DOI: 10.1021/jf991171u

[72] Targett NM, Arnold TM. Minireview—Predicting the effects of brown algal phlorotannins on marine herbivores in tropical and temperate oceans. Journal of Phycology. 1998;**34**:195-205. DOI: 10.1046/j.1529-8817.1998.340195.x

[73] McManus JP, Davis KG, Beart JE, Gaffney SH, Lilley TH, Haslam E. Polyphenol interactions. Part 1. Introduction; some observations on the reversible complexation of polyphenols with proteins and polysaccharides. Journal of the Chemical Society, Perkin Transactions. 1985;**2**:1429-1438. DOI: 10.1039/P29850001429

[74] Salminen JP, Karonen M, Sinkkonen J. Chemical ecology of tannins: Recent developments in tannin chemistry reveal new structures and structure–activity patterns. Chemistry--A European Journal. 2011;**17**(10):2806-2816. DOI: 10.1002/chem.201002662

[75] Vidal S, Francis IL, Guyot S, Marnet N, Kwiatkowski M, Gawel R, et al. The mouthfeel properties of grape and apple proanthocyanidins in a wine-like medium. Journal of the Science of Food and Agriculture. 2003;**83**:564-573. DOI: 10.1002/jsfa.1394

[76] Sun B, Md S, Leandro C, Caldeira I, Duarte FL, Spranger I. Reactivity of polymeric proanthocyanidins toward salivary proteins and their contribution to young red wine astringency. Journal of Agricultural and Food Chemistry. 2013;**61**(4):939-946. DOI: 10.1021/jf303704u

[77] Hufnagel JC, Hofmann T. Orosensory-directed identification of astringent mouthfeel and bitter-tasting compounds in red wine. Journal of Agricultural and Food Chemistry. 2008;**56**(4):1376-1386. DOI: 10.1021/jf073031n

[78] Rinaldi A, Jourdes M, Teissedre P, Moio L. A preliminary characterisation of Aglianico (*Vitis vinifera* L. cv.) grape proanthocyanidins and evaluation of their reactivity towards salivary proteins. Food Chemistry. 2014;**164**:142-149. DOI: 10.1016/j.foodchem.2014.05.050

[79] Gonzalo-Diago A, Dizy M, Fernández-Zurbano P. Taste and mouthfeel properties of red wines proanthocyanidins and their relation to the chemical composition. Journal of Agricultural and Food Chemistry. 2013;**61**(37):8861-8870. DOI: 10.1021/jf401041q

[80] Hagerman AE, Rice ME, Ritchard NT. Mechanisms of protein precipitation for two tannins, pentagalloyl glucose and epicatechin$_{16}$ (4→8) catechin (procyanidin). Journal of Agricultural and Food Chemistry. 1998;**46**(7):2590-2595. DOI: 10.1021/jf971097k

[81] Charlton AJ, Baxter NJ, Khan ML, Moir AJG, Haslam E, Davies AP, et al. Polyphenol/peptide binding and precipitation. Journal of Agricultural and Food Chemistry. 2002;**50**(6):1593-1601. DOI: 10.1021/jf010897z

[82] Hofmann T, Glabasnia A, Schwarz B, Wisman KN, Gangwer KA, Hagerman AE. Protein binding and astringent taste of a polymeric procyanidin, 1,2,3,4,6-penta-O-galloyl-β-D-glucopyranose, castalagin, and grandinin. Journal of Agricultural and Food Chemistry. 2006;**54**(25):9503-9509. DOI: 10.1021/jf062272c

[83] Rinaldi A, Gambuti A, Moine-Ledoux V, Moio L. Evaluation of the astringency of commercial tannins by means of the SDS–PAGE-based method. Food Chemistry. 2010;**122**(4):951-956. DOI: 10.1016/j.foodchem.2010.03.105

[84] Dobreva MA, Green RJ, Mueller-Harvey I, Salminen J, Howlin BJ, Frazier RA. Size and molecular flexibility affect the binding of ellagitannins to bovine serum albumin. Journal of Agricultural and Food Chemistry. 2014;**62**(37):9186-9194. DOI: 10.1021/jf502174r

[85] Stark T, Wollmann N, Wenker K, Lösch S, Glabasnia A, Hofmann T. Matrix-calibrated LC-MS/MS quantitation and sensory evaluation of oak ellagitannins and their transformation products in red wines. Journal of Agricultural and Food Chemistry. 2010;**58**(10):6360-6369. DOI: 10.1021/jf100884y

[86] Lee CB, Lawless HT. Time-course of astringent sensations. Chemical Senses. 1991;**16**(3):225-238. DOI: 10.1093/chemse/16.3.225

[87] Corrigan Thomas CJ, Lawless HT. Astringent subqualities in acids. Chemical Senses. 1995;**20**(6):593-600. DOI: 10.1093/chemse/20.6.593

[88] Rubico SM, McDaniel MR. Sensory evaluation of acids by free-choice

profiling. Chemical Senses. 1992;**17**(3):273-289. DOI: 10.1093/chemse/17.3.273

[89] Kallithraka S, Bakker J, Clifford M. Red wine and model wine astringency as affected by malic and lactic acid. Journal of Food Science. 1997;**62**(2):416-420. DOI: 10.1111/j.1365-2621.1997.tb04016.x

[90] Corrigan Thomas CJ, Lawless HT. Astringent subqualities in acids. Chemical Senses. 1995;**20**(6):593-600. DOI: 10.1093/chemse/20.6.593

[91] Picariello L, Rinaldi A, Martino F, Petracca F, Moio L, Gambuti A. Modification of the organic acid profile of grapes due to climate changes alters the stability of red wine phenolics during controlled oxidation. Vitis. 2019;**58**(5):127-133. DOI: 10.5073/vitis.2019.58.special-issue.127-133

[92] Forino M, Picariello L, Rinaldi A, Moio L, Gambuti A. How must pH affects the level of red wine phenols. Lebensmittel-Wissenschaft & Technologie. 2020;**129**:109546. DOI: 10.1016/j.lwt.2020.109546

[93] He F, Liang N, Mu L, Pan Q, Wang J, Reeves MJ, et al. Anthocyanins and their variation in red wines I. Monomeric anthocyanins and their colour expression. Molecules. 2012;**17**(2):1571-1601. DOI: 10.3390/molecules17021571

[94] Vidal S, Francis L, Williams P, Kwiatkowski M, Gawel R, Cheynier V, et al. The mouthfeel properties of polysaccharides and anthocyanins in a wine like medium. Food Chemistry. 2004;**85**(4):519-525. DOI: 10.1016/S0308-8146(03)00084-0

[95] Vidal S, Francis L, Noble A, Kwiatkowski M, Cheynier V, Waters E. Taste and mouthfeel properties of different types of tannin-like polyphenolic compounds and anthocyanins in wine. Analytica Chimica Acta. 2004;**513**(1):57-65. DOI: 10.1016/j.aca.2003.10.017

[96] Ferrer-Gallego R, Soares S, Mateus N, Rivas-Gonzalo J, Escribano-Bailón MT, de Freitas V. New anthocyanin–human salivary protein complexes. Langmuir. 2015;**31**(30):8392-8401. DOI: 10.1021/acs.langmuir.5b01122

[97] Paissoni MA, Waffo-Teguo P, Ma W, Jourdes M, Rolle L, Teissedre P. Chemical and sensorial investigation of in-mouth sensory properties of grape anthocyanins. Scientific Reports. 2018;**8**:17098. DOI: 10.1038/s41598-018-35355-x

[98] García-Estévez I, Cruz L, Oliveira J, Mateus N, de Freitas V, Soares S. First evidences of interaction between pyranoanthocyanins and salivary proline-rich proteins. Food Chemistry. 2017;**228**:574-581. DOI: 10.1016/j.foodchem.2017.02.030

[99] Hufnagel JC, Hofmann T. Orosensory-directed identification of astringent mouthfeel and bitter-tasting compounds in red wine. Journal of Agricultural and Food Chemistry. 2008;**56**(4):1376-1386. DOI: 10.1021/jf073031n

[100] Ferrer-Gallego R, Brás NF, García-Estévez I, Mateus N, Rivas-Gonzalo JC, de Freitas V, et al. Effect of flavonols on wine astringency and their interaction with human saliva. Food Chemistry. 2016;**209**:358-364. DOI: 10.1016/j.foodchem.2016.04.091

[101] Gambuti A, Picariello L, Rinaldi A, Forino M, Blaiotta G, Moine V, et al. New insights into the formation of precipitates of quercetin in Sangiovese wines. Journal of Food Science and Technology. 2020;**57**(7):2602-2611. DOI: 10.1007/s13197-020-04296-7

[102] Rinaldi A, Louazil P, Iturmendi N, Moine V, Moio L. Effect of marc

pressing and geographical area on Sangiovese wine quality. Lebensmittel-Wissenschaft & Technologie. 2020;**118**:108728. DOI: 10.1016/j.lwt.2019.108728

[103] Sáenz-Navajas M, Avizcuri J, Ferreira V, Fernández-Zurbano P. Insights on the chemical basis of the astringency of Spanish red wines. Food Chemistry. 2012;**134**(3):1484-1493. DOI: 10.1016/j.foodchem.2012.03.060

[104] Luck G, Liao H, Murray NJ, Grimmer HR, Warminski EE, Williamson MP, et al. Polyphenol, astringency and proline rich proteins. Phytochemistry. 1994;**37**:357-371

[105] Haslam E. Polyphenol-protein interactions. The Biochemical Journal. 1974;**139**:285-288

[106] Haslam E. Natural polyphenols (vegetable tannins) as drugs: Possible modes of action. Journal of Natural Products. 1996;**59**(2):205-215. DOI: 10.1021/np960040+

[107] Beart JE, Lilley TH, Haslam E. Plant polyphenols—Secondary metabolism and chemical defence: Some observations. Phytochemistry. 1985;**24**(1):33-38. DOI: 10.1016/S0031-9422(00)80802-X

[108] Jöbstl E, O'Connell J, Fairclough JPA, Williamson MP. Molecular model for astringency produced by polyphenol/protein interactions. Biomacromolecules. 2004;**5**(3):942-949. DOI: 10.1021/bm0345110

[109] Frazier RA, Papadopoulou A, Mueller-Harvey I, Kissoon D, Green RJ. Probing protein-tannin interactions by isothermal titration microcalorimetry. Journal of Agricultural and Food Chemistry. 2003;**51**(18):5189-5195. DOI: 10.1021/jf021179v

[110] Poncet-Legrand C, Edelmann A, Putaux J, Cartalade D, Sarni-Manchado P, Vernhet A. Poly(L-proline) interactions with flavan-3-ols units: Influence of the molecular structure and the polyphenol/protein ratio. Food Hydrocolloids. 2006;**20**(5):687-697. DOI: 10.1016/j.foodhyd.2005.06.009

[111] Gonçalves R, Mateus N, Pianet I, Laguerre M, de Freitas V. Mechanisms of tannin-induced trypsin inhibition: A molecular approach. Langmuir. 2011;**27**(21):13122-13129. DOI: 10.1021/la202280c

[112] Poncet-Legrand C, Gautier C, Cheynier V, Imberty A. Interactions between flavan-3-ols and poly(L-proline) studied by isothermal titration calorimetry: Effect of the tannin structure. Journal of Agricultural and Food Chemistry. 2007;**55**(22):9235-9240. DOI: 10.1021/jf071297o

[113] Cala O, Pinaud N, Simon C, Fouquet E, Laguerre M, Dufourc EJ, et al. NMR and molecular modeling of wine tannins binding to saliva proteins: Revisiting astringency from molecular and colloidal prospects. The FASEB Journal. 2010;**24**(11):4281-4290. DOI: 10.1096/fj.10-158741

[114] Folin O, Denis W. On phospho-tungstic-phosphomolybdic compounds as colour reagents. The Journal of Biological Chemistry. 1912;**12**:239-243

[115] Swain T, Hillis WE. The phenolic constituents of *Prunus domestica*. I.—The quantitative analysis of phenolic constituents. Journal of the Science of Food and Agriculture. 1959;**10**(1):63-68. DOI: 10.1002/jsfa.2740100110

[116] Laurent S. Étude comparative de différentes méthodes d'extraction et de dosage des tannins chez quelques ptéridophytes. Archives Internationales de Physiologie et de Biochimie. 1975;**83**(4):735-752. DOI: 10.3109/13813457509081892

[117] Singleton VL, Rossi JA. Colorimetry of total phenolics with

phosphomolybdic-phosphotungstic acid reagents. American Journal of Enology and Viticulture. 1965;**16**(3):144-158

[118] Bate-Smith E. Phytochemistry of proanthocyanidins. Phytochemistry. 1975;**14**(4):1107-1113. DOI: 10.1016/0031-9422(75)85197-1

[119] Porter LJ, Hrstich LN, Chan BG. The conversion of procyanidins and prodelphinidins to cyanidin and delphinidin. Phytochemistry. 1985;**25**(1):223-230. DOI: 10.1016/S0031-9422(00)94533-3

[120] Broadhurst RB, Jones WT. Analysis of condensed tannins using acidified vanillin. Journal of the Science of Food and Agriculture. 1978;**29**(9):788-794. DOI: 10.1002/jsfa.2740290908

[121] McMurrough I, McDowell J. Chromatographic separation and automated analysis of flavanols. Analytical Biochemistry. 1978;**91**(1):92-100. DOI: 10.1016/0003-2697(78)90819-9

[122] Vivas N, Glories Y, Lagune L, Cédric S, Augustin M. Estimation du degrédepolymérisationdesprocyanidines du raisin et du vin par la méthode au ρ-dimethylaminocinnamaldéhyde. OENO One. 1994;**28**(4):319. DOI: 10.20870/oeno-one.1994.28.4.1138

[123] Schofield P, Mbugua D, Pell A. Analysis of condensed tannins: A review. Animal Feed Science and Technology. 2001;**91**(1-2):21-40. DOI: 10.1016/S0377-8401(01)00228-0

[124] Matthews S, Mila I, Scalbert A, Donnelly DM. Extractable and non-extractable proanthocyanidins in barks. Phytochemistry. 1997;**45**(2):405-410. DOI: 10.1016/S0031-9422(96)00873-4

[125] Kennedy JA, Jones GP. Analysis of proanthocyanidin cleavage products following acid-catalysis in the presence of excess phloroglucinol. Journal of Agricultural and Food Chemistry. 2001;**49**(4):1740-1746. DOI: 10.1021/jf001030o

[126] Lorrain B, Ky I, Pechamat L, Teissedre P. Evolution of analysis of polyhenols from grapes, wines, and extracts. Molecules. 2013;**18**(1):1076-1100. DOI: 10.3390/molecules18011076

[127] McRae JM, Falconer RJ, Kennedy JA. Thermodynamics of grape and wine tannin interaction with polyproline: Implications for red wine astringency. Journal of Agricultural and Food Chemistry. 2010;**58**(23):12510-12518. DOI: 10.1021/jf1030967

[128] Kennedy JA, Waterhouse AL. Analysis of pigmented high-molecular-mass grape phenolics using ion-pair, normal-phase high-performance liquid chromatography. Journal of Chromatography. A. 2000;**866**(1):25-34. DOI: 10.1016/S0021-9673(99)01038-9

[129] Fernández K, Agosin E. Quantitative analysis of red wine tannins using fourier-transform mid-infrared spectrometry. Journal of Agricultural and Food Chemistry. 2007;**55**(18):7294-7300. DOI: 10.1021/jf071193d

[130] Bate-Smith E. Haemanalysis of tannins: The concept of relative astringency. Phytochemistry. 1973;**12**(4):907-912. DOI: 10.1016/0031-9422(73)80701-0

[131] Nierenstein M. The Natural Organic Tannins. London: J&A Churchill Ltd; 1934

[132] Swain T. Tannins and lignins. In: Rosenthal GA, Janzen DH, editors. Herbivores: Their Interactions with Secondary Plant Metabolites. New York: Academic Press; 1979. pp. 657-682

[133] Goldstein JL, Swain T. The inhibition of enzymes by tannins. Phytochemistry. 1965;**4**:185-192

[134] Hagerman AE, Butler LG. Protein precipitation method for the quantitative determination of tannins. Journal of Agricultural and Food Chemistry. 1978;**26**(4):809-812. DOI: 10.1021/jf60218a027

[135] Harbertson JF, Picciotto EA, Adams DO. Measurement of polymeric pigments in grape berry extract sand wines using a protein precipitation assay combined with bisulfite bleaching. American Journal of Enology and Viticulture. 2003;**54**(4):301-306

[136] Glories Y. La couleur des vins rouges. 1°2° partie. Connaissance Vigne Vin. 1984;**18**:253-271

[137] Llaudy MC, Canals R, Canals J, Rozés N, Arola L, Zamora F. New method for evaluating astringency in red wine. Journal of Agricultural and Food Chemistry. 2004;**52**(4):742-746. DOI: 10.1021/jf034795f

[138] Sarneckis CJ, Dambergs RG, Jones P, Mercurio M, Herderich MJ, Smith PA. Quantification of condensed tannins by precipitation with methyl cellulose: Development and validation of an optimised tool for grape and wine analysis. Australian Journal of Grape and Wine Research. 2008;**12**(1): 39-49. DOI: 10.1111/j.1755-0238.2006.tb00042.x

[139] Mercurio MD, Dambergs RG, Herderich MJ, Smith PA. High throughput analysis of red wine and grape phenolics adaptation and validation of methyl cellulose precipitable tannin assay and modified Somers color assay to a rapid 96 well plate format. Journal of Agricultural and Food Chemistry. 2007;**55**(12): 4651-4657. DOI: 10.1021/jf063674n

[140] Yan Q, Bennick A. Identification of histatins as tannin-binding proteins in human saliva. The Biochemical Journal. 1995;**311**(1):341-347. DOI: 10.1042/bj3110341

[141] Sarni-Manchado P, Cheynier V, Moutounet M. Interactions of grape seed tannins with salivary proteins. Journal of Agricultural and Food Chemistry. 1999;**47**(1):42-47. DOI: 10.1021/jf9805146

[142] Gambuti A, Rinaldi A, Pessina R, Moio L. Evaluation of Aglianico grape skin and seed polyphenol astringency by SDS–PAGE electrophoresis of salivary proteins after the binding reaction. Food Chemistry. 2006;**97**(4):614-620. DOI: 10.1016/j.foodchem.2005.05.038

[143] Kallithraka S, Bakker J, Clifford MN. Evidence that salivary proteins are involved in astringency. Journal of Sensory Studies. 1998;**13**(1):29-43. DOI: 10.1111/j.1745-459X.1998.tb00073.x

[144] Soares S, Vitorino R, Osório H, Fernandes A, Venâncio A, Mateus N, et al. Reactivity of human salivary proteins families toward food polyphenols. Journal of Agricultural and Food Chemistry. 2011;**59**(10):5535-5547. DOI: 10.1021/jf104975d

[145] Dinnella C, Recchia A, Vincenzi S, Tuorila H, Monteleone E. Temporary modification of salivary protein profile and individual responses to repeated phenolic astringent stimuli. Chemical Senses. 2010;**35**(1):75-85. DOI: 10.1093/chemse/bjp084

[146] Chapon L. Nephelometry as a method for studying the relations between polyphenols and proteins. Journal of the Institute of Brewing. 1993;**99**(1):49-56. DOI: 10.1002/j.2050-0416.1993.tb01146.x

[147] de Freitas V, Mateus N. Nephelometric study of salivary protein-tannin aggregates. Journal of the Science of Food and Agriculture. 2002;**82**(1):113-119. DOI: 10.1002/jsfa.1016

[148] Horne J, Hayes J, Lawless HT. Turbidity as a measure of salivary

protein reactions with astringent substances. Chemical Senses. 2002;**27**(7):653-659. DOI: 10.1093/chemse/27.7.653

[149] Carvalho E, Mateus N, De Freitas V. Flow nephelometric analysis of protein–tannin interactions. Analytica Chimica Acta. 2004;**513**(1):97-101. DOI: 10.1016/j.aca.2003.10.010

[150] Monteleone E, Condelli N, Dinnella C, Bertuccioli M. Prediction of perceived astringency induced by phenolic compounds. Food Quality and Preference. 2004;**15**(7-8):761-769. DOI: 10.1016/j.foodqual.2004.06.002

[151] Fia G, Dinnella C, Bertuccioli M, Monteleone E. Prediction of grape polyphenol astringency by means of a fluorimetric micro-plate assay. Food Chemistry. 2009;**113**(1):325-330. DOI: 10.1016/j.foodchem.2008.07.058

[152] Rébénaque P, Rawyler A, Boldi M, Deneulin P. Comparison between sensory and nephelometric evaluations of tannin fractions obtained by ultrafiltration of red wines. Chemosensory Perception. 2015;**8**(1):33-43. DOI: 10.1007/s12078-015-9175-x

[153] Guerreiro JRL, Teixeira N, de Freitas V, Sales MGF, Sutherland DS. A saliva molecular imprinted localised surface plasmon resonance biosensor for wine astringency estimation. Food Chemistry. 2017;**233**:457-466. DOI: 10.1016/j.foodchem.2017.04.051

[154] Lyman BJ, Green BG. Oral astringency: Effects of repeated exposure and interactions with sweeteners. Chemical Senses. 1990;**15**(2):151-164. DOI: 10.1093/chemse/15.2.151

[155] Lee WE, Pangborn RM. Time-intensity: The temporal aspects of sensory perception. Food Technology. 1986;**40**(11):71-82

[156] Pineau N, Schlich P, Cordelle S, Mathonnière C, Issanchou S, Imbert A, et al. Temporal dominance of sensations: Construction of the TDS curves and comparison with time–intensity. Food Quality and Preference. 2009;**20**(6):450-455. DOI: 10.1016/j.foodqual.2009.04.005

[157] Clark CC, Lawless HT. Limiting response alternatives in time-intensity scaling: An examination of the halo-dumping effect. Chemical Senses. 1994;**19**(6):583-594. DOI: 10.1093/chemse/19.6.583

[158] Pessina R, Patron C, Pineau N, Piombino P, Moio L, Schlich P. Measuring temporality of sensations in wine. In: European Conference on Sensory Science of Food and Beverages "A Sense of Identity". 2004. pp. 26-29

[159] Schlich P. Temporal dominance of sensations (TDS): A new deal for temporal sensory analysis. Current Opinion in Food Science. 2017;**15**:38-42. DOI: 10.1016/j.cofs.2017.05.003

[160] Gawel R, Oberholster A, Il F. A 'Mouth-feel Wheel': Terminology for communicating the mouthfeel characteristics of red wine. Australian Journal of Grape and Wine Research. 2000;**6**(3):203-207. DOI: 10.1111/j.1755-0238.2000.tb00180.x

[161] Vidal L, Antúnez L, Giménez A, Medina K, Boido E, Ares G. Sensory characterisation of the astringency of commercial Uruguayan Tannat wines. Foodservice Research International. 2017;**102**:425-434. DOI: 10.1016/j.foodres.2017.09.022

[162] Rinaldi A, Coppola M, Moio L. Aging of Aglianico and Sangiovese wine on mannoproteins: Effect on astringency and colour. Lebensmittel-Wissenschaft & Technologie. 2019;**105**:233-241. DOI: 10.1016/j.lwt.2019.02.034

[163] Goldner M, Zamora M. Effect of polyphenol concentrations on astringency perception and its correlation with gelatin index of red wine. Journal of Sensory Studies. 2010;**25**(5):761-777. DOI: 10.1111/j.1745-459X.2010.00304.x

[164] Kallithraka S, Kim D, Tsakiris A, Paraskevopoulos I, Soleas G. Sensory assessment and chemical measurement of astringency of Greek wines: Correlations with analytical polyphenolic composition. Food Chemistry. 2011;**126**(4):1953-1958. DOI: 10.1016/j.foodchem.2010.12.045

[165] Cliff M, Brau N, King MC, Mazza G. Development of predictive models for astringency from anthocyanin, phenolic and color analyses of British Columbia red wines. Journal International des Sciences de la Vigne et du Vin. 2002;**36**:21-30

[166] Kennedy JA, Ferrier J, Harbertson JF, des Gachons CP. Analysis of tannins in red wine using multiple methods: Correlation with perceived astringency. American Journal of Enology and Viticulture. 2006;**57**(4):481-485

[167] Rinaldi A, Gambuti A, Moio L. Application of the SPI (saliva precipitation index) to the evaluation of red wine astringency. Food Chemistry. 2012;**135**(4):2498-2504. DOI: 10.1016/j.foodchem.2012.07.031

[168] Rinaldi A, Iturmendi N, Gambuti A, Jourdes M, Teissedre P, Moio L. Chip electrophoresis as a novel approach to measure the polyphenols reactivity toward human saliva. Electrophoresis. 2014;**35**(11):1735-1741. DOI: 10.1002/elps.201300622

[169] Ribéreau-Gayon P, Glories Y, Maujean A, Dubourdieu D. The chemistry of wine stabilisation and treatments. In: Handbook of Enology. Chichester, United Kingdom: John Wiley and Sons; 2006

[170] Maury C, Sarni-Manchado P, Lefebvre S, Cheynier V, Moutounet M. Influence of fining with different molecular weight gelatins on proanthocyanidin composition and perception of wines. American Journal of Enology and Viticulture. 2001;**52**(2):140-145

[171] Gambuti A, Rinaldi A, Moio L. Use of patatin, a protein extracted from potato as alternative to animal proteins in fining of red wine. European Food Research and Technology. 2012;**235**(4):753-765. DOI: 10.1007/s00217-012-1791-y

[172] Rinaldi A, Errichiello F, Moio L. Alternative fining of Sangiovese: Effect on polyphenolic and sensory characteristics. Australian Journal of Grape and Wine Research. 2020

[173] OIV. International Code of Oenological Practices. Paris, France: Organisation Internationale de la Vigne et du Vin; 2012. p. 1

[174] Sheridan MK, Elias RJ. Exogenous acetaldehyde as a tool for modulating wine color and astringency during fermentation. Food Chemistry. 2015;**177**:17-22. DOI: 10.1016/j.foodchem.2014.12.077

[175] Saucier C, Little D, Glories Y. First evidence of acetaldehyde-flavanol condensation products in red wine. American Journal of Enology and Viticulture. 1997;**48**(3):370-373

[176] Fulcrand H, Atanasova V, Salas E, Cheynier V. The fate of anthocyanins in wine: Are there determining factors? In: Red Wine Color. ACS Symposium Series. Washington, USA: American Chemical Society; Vol. 886. 2004. pp. 68-88. DOI: 10.1021/bk-2004-0886.ch006

[177] Picariello L, Gambuti A, Petracca F, Rinaldi A, Moio L. Enological tannins affect acetaldehyde evolution,

colour stability and tannin reactivity during forced oxidation of red wine. International Journal of Food Science and Technology. 2018;**53**(1):228-236. DOI: 10.1111/ijfs.13577

[178] Gambuti A, Picariello L, Rinaldi A, Moio L. Evolution of Sangiovese wines with varied tannin and anthocyanin ratios during oxidative aging. Frontiers in Chemistry. 2018;**6**(63):1-11. DOI: 10.3389/fchem.2018.00063

[179] McRae JM, Dambergs RG, Kassara S, Parker M, Jeffery DW, Herderich MJ, et al. Phenolic compositions of 50 and 30 year sequences of Australian red wines: The impact of wine age. Journal of Agricultural and Food Chemistry. 2012;**60**(40):10093-10102. DOI: 10.1021/jf301571q

[180] Weber F, Greve K, Durner D, Fischer U, Winterhalter P. Sensory and chemical characterisation of phenolic polymers from red wine obtained by gel permeation chromatography. American Journal of Enology and Viticulture. 2013;**64**(1):15-25. DOI: 10.5344/ajev.2012.12074

[181] Ma L, Watrelot AA, Addison B, Waterhouse AL. Condensed tannin reacts with SO_2 during wine aging, yielding flavan-3-ol sulfonates. Journal of Agricultural and Food Chemistry. 2018;**66**(35):9259-9268. DOI: 10.1021/acs.jafc.8b01996

[182] Chira K, Pacella N, Jourdes M, Teissedre P. Chemical and sensory evaluation of Bordeaux wines (cabernet-sauvignon and merlot) and correlation with wine age. Food Chemistry. 2011;**126**(4):1971-1977. DOI: 10.1016/j.foodchem.2010.12.056

[183] Gambuti A, Picariello L, Rinaldi A, Ugliano M, Moio L. Impact of 5 years bottle aging under controlled oxygen exposure on sulfur dioxide and phenolic composition of tannin-rich red wines. OENO One. 2020

[184] Rinaldi A, Gonzalez A, Moio L. The sensory profile of astringency: Application on Sangiovese wines. In: Proceedings of 11th International Symposium of Oenology of Bordeaux and 11th Edition of the Symposium In Vino Analytica Scientia (ŒnoIVAS 2019). France: Bordeaux; 25-28 June 2019

[185] Gambuti A, Siani T, Picariello L, Rinaldi A, Lisanti MT, Ugliano M, et al. Oxygen exposure of tannins-rich red wines during bottle aging. Influence on phenolics and color, astringency markers and sensory attributes. European Food Research and Technology. 2017;**243**(4):669-680. DOI: 10.1007/s00217-016-2780-3

[186] Rinaldi A, Gambuti A, Moio L. Precipitation of salivary proteins after the interaction with wine: The effect of ethanol, pH, fructose, and mannoproteins. Journal of Food Science. 2012;**77**(4):C485-C490. DOI: 10.1111/j.1750-3841.2012.02639.x

[187] Poncet-Legrand C, Cartalade D, Putaux J, Cheynier V, Vernhet A. Flavan-3-ol aggregation in model ethanolic solutions: Incidence of polyphenol structure, concentration, ethanol content, and ionic strength. Langmuir. 2003;**19**(25):10563-10572. DOI: 10.1021/la034927z

[188] Brandão E, Silva MS, García-Estévez I, Williams P, Mateus N, Doco T, et al. The role of wine polysaccharides on salivary protein-tannin interaction: A molecular approach. Carbohydrate Polymers. 2017;**177**:77-85. DOI: 10.1016/j.carbpol.2017.08.075

Chapter 8

Wine Stabilisation: An Overview of Defects and Treatments

Fernanda Cosme, Luís Filipe-Ribeiro and Fernando M. Nunes

Abstract

Wine is widely consumed due to its distinctive sensory characteristics. However, during wine production and storage, several defects can appear. These can be the result of unwanted microbiological activity or due to the unbalanced levels of some compounds resulting from an unbalanced grape chemical composition or inadequate winemaking practices and storage conditions. The main purpose of wine stabilisation is the removal of wine defects, either visual, olfactive, gustative, or tactile, the increase in wine safety and stability by fining and filtration operations, avoiding the occurrence of some usual wine precipitations after bottling. Although the best strategy is to prevent the appearance of wine defects, when present, several fining agents or additives, and technologies are available today with different performances and impact on wine quality. By physicochemical and sensory analysis, the defect is detected, and if the objective is removing them, some laboratory trials are performed to achieve a better treatment approach. This review overviews the principal wine defects and treatments available today and in the near future. Generally, the future trend is the use of more sustainable and environmentally friendly fining agents and technologies, looking for treatments with better performance and specificity.

Keywords: wine defects, stabilisation, wine treatments, fining agents

1. Introduction

All around the world in wine production many sensory defects, physicochemical instabilities, and a few toxic compounds can appear, which usually results in a decrease of the wine quality and/or safety, being responsible for economic losses for the wine industry. Nowadays, the origin of most wine defects and processing conditions that favour their formation are well identified. At the same time, many viticultural and technological solutions are available to completely or partially avoid their formation, and various additives and fining agents are allowed to be used to avoid their formation or remove the compounds, or their precursors, responsible for the instability [1]. Frequently, the question is not how to remove the defect or their precursors, but how to remove it without changing the wine sensory profile.

The most troublesome defects for wine producers are those occurring after wine bottling, as the intervention at this stage is rather limited and therefore preventive actions are the most efficient and sometimes the only strategy. After wine bottling the main external factor that can influence wine stability and the kinetics of the chemical reactions and interactions of the wine constituents is the temperature that can significantly affect for example the solubility of tartrate salts, induce colour changes in red and white wines, and the formation of protein haze. Also, the redox potential, mainly determined by the levels of dissolved oxygen in the wine can significantly influence the wine shelf life. Before bottling, depending on the wine defect to be treated two main stabilisation strategies are currently used: subtractive and non-subtractive. For the first approach fining agents are used that can interact with the compound(s) responsible for the wine defect or their precursors removing them from the wine. In the second type of strategy, an additive is added to the wine that interacts with the compound(s) responsible for the wine defect affecting their ability to form crystals as in the case of carboxymethylcellulose in the tartaric instability [2] or decrease the vapour pressure of the compound(s) responsible for an undesirable odour for example like chitosan in the case of the ethylphenols responsible for the '*Brett character*' [3].

In the next sections (2, 3, 4, and 5) the main wine defects and their stabilisation treatments currently allowed and used will be overviewed as well as the stabilisation treatments currently under research. For a matter of simplicity and systematisation, the wines defect that can occur in wine production are divided into four main groups according to their impact on the wine sensory quality and safety: (1) visual sensory defects; (2) off-odours and taints; (3) taste and tactile defects (4) safety-related defects. The wine visual defects are concerned with changes in wine limpidity and undesirable colour changes that can occur during wine production and especially after wine bottling. These defects can be due to the formation of precipitates related to solubility issues, the formation of precipitates related to the colloidal instability of some wines components, the reaction between wine components, or due to the oxidation of wine phenolic compounds.

Wine aroma significantly determines consumer acceptability [4], being extremely complex as it is the result of the cumulative effect of a diverse group of volatile compounds present at levels ranging from fractions of ng/L up to mg/L. These compounds can interact with the olfactory epithelium to generate a sensory perception [5, 6]. The levels of these volatiles are dependent on both viticultural [7] and oenological practices [8, 9]. Although several hundreds of these volatile compounds can be present in wines [10], only a few are present at levels above the perception threshold and thus being responsible for characteristic odours [11]. During wine production, several off-odours and taints can cause severe quality problems. Off-odours are considered to be the occurrence of any atypical odours resulting from compounds formed by the deterioration of the wine, including chemical reactions and microbial spoilage; whereas, taints result from external contamination of the wine as a result of exposure of grapes to contaminating environments or migration of compounds form packaging materials contaminated by either synthetic chemicals or chemicals produced or transformed by microbial action [12]. Sometimes, compounds that impart a positive aroma character in one instance may cause an off-odour when present in the wrong context or at high levels. Besides, faults in one wine may not be undesirable in another: for example, the complex oxidised bouquets of Sherries, the fusel odours of Port wine, and the baked character of Madeira wine. Some faults, such as a barnyard odour, generated by ethylphenols, may be considered pleasingly 'rustic,' or part of the terroir character of certain wines. The evident presence of ethyl acetate in the aroma of wine is also usually considered a fault. However, in expensive Sauternes, it appears to be acceptable (or ignored). Nevertheless, there is general agreement among most wine professionals as to what constitutes an aromatic fault in table wines. These wine defects, their origin, and their sensory impact are discussed in Section 2. Balancing the oral sensations of wine is one of the most demanding tasks for a winemaker since a distinguishing feature of superior wines is the harmony achieved among these seemingly simple sensations. Indeed, imbalances created by excessive acidity, astringency, or bitterness, among others, are often the first deficiencies noted by a panellist [13]. Since it is very important to understand the factors that affect the sensory properties responsible for wine taste and tactile imbalance, to allow the wine industry to be able to control these sensory characteristics of the final product from the first production stages. The main wine taste and tactile defects are astringency, bitterness, and acidity imbalance. Wine astringency is caused by excess proanthocyanidins in young red wine, it is considered a tactile sensation [14], responsible for the 'drying', 'roughing' and 'puckering'. The ageing process reduces astringency due to oxidation and will be less evident in aged wines. Bitterness is a taste perception related to phenolic compounds with low molecular weights such as flavonol aglycones, especially myricetin or quercetin [15] as well as to monomeric or small phenolic flavanols [16]. Bitterness, astringency, and acidity could change depending on the oenological practice. In Section 4 the sensory impact of excessive astringency, bitterness and acidity will be highlighted, as well as the stabilisation/correction strategies.

In Section 5, the formation of the main potential toxic compounds that affect wine safety is overviewed. These compounds are present in wines due to the activity of bacteria and moulds in the wine production and due to the bad hygienic sanity of the grapes used for winemaking, respectively.

2. Origin of visual sensory defects and strategies for wine stabilisation

The two most important visual sensory defects that affect wine quality occur during the storage of bottled wine. These are the tartrate instabilities and protein instability. Other instabilities that can occur include red wine colour colloidal instability, pinking of certain white wines produced from white grape varieties, browning of white and rose wines, and oxidation of all wines as well as metallic instabilities such as iron and copper instability.

2.1 Tartrate instabilities

The crystallisation of potassium hydrogen tartrate or calcium tartrate salts may occur in wines where their concentration product exceeds their solubility product [17, 18]. The formation of these crystals results in the formation of deposits at the bottom vat and sometimes in the bottled wine. Prevention of tartaric precipitation in bottled wine is essential as consumers associate its occurrence with poor production conditions or an exogen wine material. Methods to prevent tartaric crystals precipitation include metatartaric acid, cold stabilisation, and electrodialysis [19]. The addition of mannoproteins obtained from the hydrolysis of the yeast cell wall was authorised by the European Community since 2005 [19]. Mannoproteins inhibit the crystallisation of tartrate salts by lowering the crystallisation temperature [20] preventing the occurrence of precipitates in wine [21]. Arabic gum can also have some effectiveness to prevent tartaric instability, as they are protective colloids [22], and more recently sodium carboxymethylcellulose for white wine (since 2009) and potassium polyaspartate (since 2015) were authorised [19].

2.2 Protein instability

Precipitation of soluble proteins in bottled wines can also occur and this results in the formation of an amorphous haze or deposit. This defect occurs frequently in white wines or wines with low polyphenol content. It is rare in wines with relatively high levels of flavonoid phenols, particularly tannins, which complex and precipitates proteins during wine production. The most important proteins that have been related to wine protein instability are pathogenesis-related proteins of *Vitis vinifera* that include the chitinases and thaumatin-like proteins as described by Tian et al. [23]. The formation of wine protein haze is a multifactorial process with several factors known to influence the process, such as storage or wine ageing temperature, pH, ionic strength, wine protein composition, organic acids, ethanol, phenolic compounds, metals, and sulphate content; however, other important factors remain unidentified, such as the non-proteinaceous component(s) usually named X factor [24]. These proteins can be slowly denatured and aggregate throughout wine storage, forming a light-dispersing haze; therefore, this phenomenon needs to be prevented by removing them from the wine, usually by fining, before wine bottling [25]. Bentonite fining is the most used process to avoid protein instability in white wine, with the dose used being preferentially determined previously by stability tests [25]. However, bentonite fining can have a detrimental effect on wine quality, for example, by the removal of colour and aroma compounds [26]. Therefore, alternative techniques to bentonite fining have been studied, such as ultrafiltration, the addition of proteolytic enzymes, flash pasteurisation, other adsorbents, zirconium oxide, natural zeolites, chitin and chitosan, carrageenan and the use of some mannoproteins [25].

2.3 Colour colloidal instability

Colour instability of some red wines can cause product depreciation. In the last years, excessive precipitation of colouring matter has been observed at the bottom of the storage tanks and even in bottled wines [27]. This precipitation occurs along with the natural evolution of red wine during the storage period and results in a considerable modification of its colour and limpidity that is undesirable in terms of visual perception and loss in taste and flavour [22]. This precipitate may happen after a few months of ageing. It may be sometimes gelatinous and strongly red coloured. If the red wine is aged and bottled, the precipitate may occur later in the form of a thin leaf, lining the inner side of the bottles [28]. A method commonly used to reduce wine turbidity and stabilise the colouring matter is fining by the addition of proteins (albumin, casein, or gelatine) that promotes flocculation or precipitation before bottling [29]. However, in some cases, the precipitate formation can still be observed in the later stages of the ageing process. The addition of arabic gum could prevents wine colour instability [22].

2.4 Metallic instabilities

Two metal-dependent instabilities can also result in the formation of wine turbidity or deposits. Excess levels of iron (5–20 mg/L) in wine (white or red) can lead to the formation of precipitates with phosphates and tannins resulting in instabilities (iron (III) phosphate [white casse] or tannate [blue casse]). Application of arabic gum or casein/potassium caseinate or citric acid (< 1 g/L) is used to prevent this instability [30]. Copper instability occurs only in white wines, initially as a white haze, and later as a reddish-brown amorphous precipitate, develop upon storage of bottled wine with excess copper (> 0.5 mg/L), under strong reducing conditions and in the presence of SO_2. This metal instability is caused by a reaction of metal traces, mainly copper from machinery, pesticide residues, or treatment with copper sulphate for the treatment of reductive off-flavours [31]. The OIV recommends a maximum copper content of 1.0 mg/L in wines [32]. The application of arabic gum is used to prevent this instability [30]. The protective colloids prevent metal precipitation but do not eliminate the copper. Exchange resins such as polyvinylimidazole-polyvinylpyrrolidone copolymers with selective binding of metals such as copper or iron have been developed [33]. The occurrence of iron and copper-related precipitations are much less usual due to the reduction of the level of these metals in the wines all-around the world by the use of stainless steel vats and wood barrels in the wine production, vat taps, and plastic polymers in the winery hoses, press machines and filling machines.

2.5 Enzymatic and non-enzymatic oxidation

One of the most frequent oenological problems in winemaking is premature wine oxidation, especially the oxidative spoilage of young white wines causing wine browning [34]. During winemaking and bottle-ageing wine, components react with oxygen [35]. Moderate oxidation of red wines phenolic compounds can contribute positively to the red wine colour stabilisation and decrease wine astringency, nevertheless, excessive oxidation can have negative effects on wine quality [36]. Wine oxidation generally results in wine colour changes, an important sensory attribute that is the first to be appreciated by consumers. Today the market wants white wines with a citrine colour, almost colourless, except for those white wines fermented in oak barrels or wines with some ageing time. In rosé wines, many colours can be found on the market, since the 'Provence style', with a slight salmon colour, until rosé wines with the colour of open coloured red wines like 'Palhete', wines produced with white and red grapes and with some maceration. In the red wines, many styles and colours can be found, from the faint colour of Pinot Noir wines to the wines produced with Alicante Bouschet or Vinhão grape varieties that yield wines with intense red colours. The fast colour change in a white or rosé bottled wine is normally the result of an oxidative problem.

The deleterious browning reaction in must and wine occurs due to the oxidation of phenolic compounds and can start as soon the grapes are crushed due to the polyphenol oxidase activity. Polyphenol oxidase with tyrosinase and catecholase activity are natural enzymes present in grape berry. They can catalyse the oxidation of monophenols to *o*-diphenols and further oxidation to orthoquinone. In wine hydroxycinnamates and flavanols, such as caffeoyltartaric acids and catechin, respectively [37] are oxidised to the corresponding quinones. Further reaction of the quinones can result in the formation of a brown colour, especially that of catechin than can yield by dimerisation the yellow dehydrodicatechin B [38] Another problem can arise when grapes are affected with *Botrytis cinerea* [39] and the resulting must become contaminated with laccase enzyme. Laccase catalyses the one-electron oxidation of a broad range of compounds including substituted phenolics to the corresponding radicals [40]. Wine phenolic acids, catechins, anthocyanins, tannins, and stilbenes are converted into the corresponding quinones, which often react further to dark coloured polymers [41]. The latter are generally insoluble in water and precipitate out from must and wine. Grape polyphenol oxidase is sensitive to low concentrations of SO_2 being inactivated, but laccase is more resistant to SO_2, and it may be present in the final wine [22], while polyphenol oxidase rarely survives the fermentation process [42]. After fermentation, with the enzyme removed or inactivated, oxidation reactions in white wine are based on non-enzymatic pathways, where Fe (II) is oxidised to Fe (III), producing hydrogen peroxide, and the following reaction where Fe (III) coordinates with catechols and oxidises them to semiquinones [43]. Then the semiquinones disproportionate to form reactive electrophilic quinones and these reactive compounds have a key impact on wine chemistry, by degrading several colour and flavour substances [35]. Reactions of oxidation products with flavonoids are well known, and some of the products are pigmented. When tartaric acid is oxidised to glyoxal, the resulting bridged product continues to react, creating a xanthylium product that absorbs in the visible region, and may contribute to the yellow hue of oxidised wines [44].

To avoid the fast colour evolution, the winemakers use SO_2 that due to their antioxidant and antioxidasic properties protect wine colour [45]. Unlike grape oxidases, which are inhibited by sulphites even at low levels, fungal laccases tend to be more resistant. The most effective treatment to eliminate the laccase activity in the must is heat treatment (2 min, 75°C). Ascorbic acid reduces and recycles quinones back to their original catechol forms, being generally used in pre-bottling. The presence of other nucleophiles, such as glutathione, 3-sulfanylhexanol and H_2S, leads to the formation of additional products on different positions of the benzene ring [45], and such reactions should also prevent browning since the quinone is being quenched. There are also many technological solutions that when used can results in a more stable wine colour as, in white wines, a fast liquid/solid separation in the press machines, reducing phenolic acids by wine fining with PVPP, potassium caseinate/casein, isinglass, gelatine, patatin and pea protein. Winemakers need to be especially cautious when handling a cold wine, such as during cold stabilisation. Oxygen is more soluble at lower wine temperatures. However, the oxidation reaction speeds up when the temperature rises. As the cold wine warms up the greater amount of dissolved oxygen will contribute to serious wine oxidation. To minimise the adverse effects of oxidation during wine racking the winemakers employ several techniques such as, using SO_2, using gentle pumps that minimise aeration, and checking hoses and fittings for leaks, and flushing hoses and containers with inert gas before wine racking. In modern winemaking, the inert gases are often used to minimise oxygen pickup in the head space of partially filled containers and during wine racking. The common inert gases used include; nitrogen, CO_2, argon, and a mixture of these gases in various proportions. For economic reasons, the use of nitrogen and CO_2 seems to be more common. To provide an inert gas cover over the wine surface in a partially filled container, CO_2 or argon should be used. These gases are denser than the air and form an inert layer devoid of oxygen. The danger of oxygen exposure is greater during the wine racking. To minimise oxygen/air contact the system is purged with the inert gas. In the process of purging the inert gas is passed through the system such as hoses, transfer lines, equipment, and the receiving tank to displace air. The wine is then racked under an inert atmosphere.

2.6 Pinking

The development of a salmon-red blush colour in white wines produced exclusively from white grape varieties is known as pinking, and the phenomenon is observed occasionally. It is perceived as an undesirable phenomenon by both wine consumers and the industry. Although with seasonal and regional variations, pinking has been observed worldwide, with predominance in white wines produced from *V. vinifera* L. grape varieties such as Chardonnay, Chenin Blanc, Crouchen, Muscat Gordo Blanco, Palomino, Riesling, Sauvignon Blanc, Sémillon, Sultana, and Thompson Seedless [46]. Pinking is mainly observed when white wines are produced under reducing conditions [47]. The pinking phenomenon is frequently observed after the bottling and storage of white wines or after alcoholic fermentation (AF) [48]. In wines made from Síria white grape variety, it was shown that the compounds responsible for the appearance of the salmon colour after bottling were due to the presence of small amounts of anthocyanins in the wine that could also be detected both in the pulp and in the skin of the white grapes [46].

Although it cannot be excluded that other compounds can be responsible for the appearance of a salmon colour in white wines from other grape varieties, the presence of the small number of anthocyanins in Chardonnay, Sauvignon Blanc, and Riesling has been shown [49]. To avoid the pinking problem there are various preventive or curative oenological treatments, including adding PVPP or PVPP associated with bentonite or increasing the redox potential using ascorbic acid in the pre-bottling stage [46].

3. Off-odours and taints and strategies for their mitigation and wine stabilisation

Uncontrolled or undesirable microbiological activity developed in the wine can be responsible for several wine spoilage problems. These defects are diverse in origin and chemical compounds involved impact as well on the wine sensory quality.

3.1 Wine off-odours

One of the main problems that can occur is the development of high levels of volatile acidity, mainly acetic acid (I, **Figure 1**). Acetic acid can be formed at the beginning of wine production (in grapes), during fermentation, and in the bottled wine as a bacterial or yeast metabolite [50]. High volatile acidity is associated with bad SO_2 management or extreme wine exposure to oxygen that stimulate the growth of aerobic acetic acid bacteria (AAB), that increases acetic acid. This results in an olfactory sensory defect known as vinegar off-odour. Vinegary wines are typically sharply acidic with an irritating odour. Ideally, the content of acetic acid should not exceed 0.7 g/L in wine. Several methodologies, aiming to decrease excessive volatile acidity of acidic wines have been proposed [50], such as microbial stabilisation of the acidic wine followed by blending with other wines, reverse osmosis, nanofiltration, and biological removal of acetic acid through refermentation [22].

Acetaldehyde (ethanal) (II, **Figure 1**) in wine can impart some undesirable flavours, when above a certain level. The average values of acetaldehyde in white wine are about 80 mg/L, in red wine 30 mg/L and for Sherries wine 300 mg/L [51]. Acetaldehyde is an intermediate product of yeast fermentation; however, it is more commonly associated with ethanol oxidation, catalysed by the enzyme ethanol

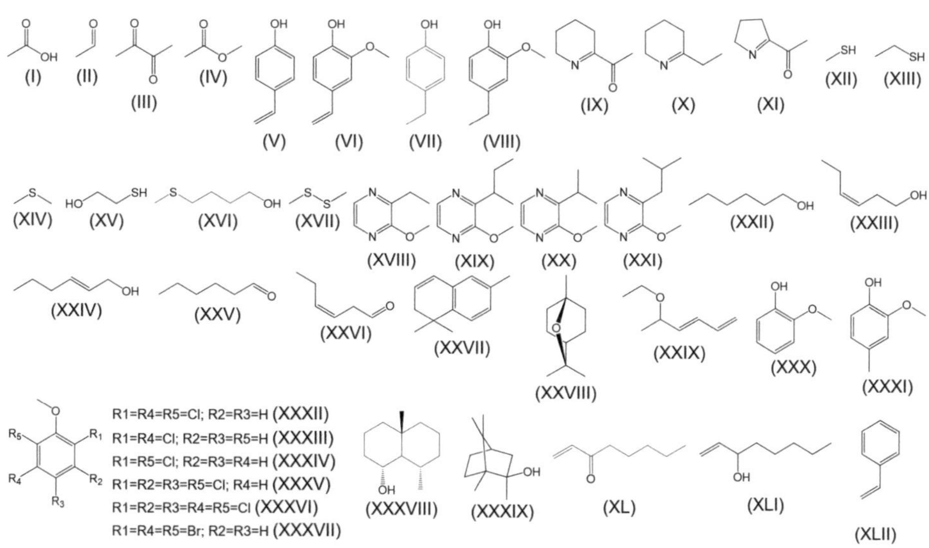

Figure 1.
Structure of the main wine off-odours and taints.

dehydrogenase. Moreover, acetaldehyde can be formed by non-enzymatic oxidation throughout the storage and ageing of wine [52]. During wine oxidation, iron (II) reduces oxygen to the hydroperoxyl radical, which converts wine *ortho*-diphenols phenols into quinones and H_2O_2. Ferrous ion associated with H_2O_2 generates hydroxyl radical that can react with ethanol to yield acetaldehyde [53]. The sensory threshold for acetaldehyde in red wines is typically in the range of 40–100 mg/L [54]. If present at low levels gives a pleasant fruity aroma, but at high levels, it possesses a pungent irritating odour [55]. Indeed, excess acetaldehyde produces a 'green,' 'grassy,' 'nutty,' 'sherry-like,' 'bruised apple,' or even 'vegetative' off-flavour [30, 56]. The level of acetaldehyde in wine can be reduced by appropriate yeast strain selection, as well as the prevention of oxidation during the winemaking process [57]. The reduction of acetaldehyde can also be done by wine lactic acid bacteria (LAB) of the genera *Lactobacillus* and *Oenococcus* which can degrade free and SO_2-bound acetaldehyde [58]. Acetaldehyde also strongly binds to SO_2, reducing the free acetaldehyde content, and thus the perception of its aroma in wines [2].

Diacetyl (2,3-butanedione, III in **Figure 1**), is usually found in low levels, as a result of yeast metabolism (<1 mg/L), but it is principally formed during malolactic fermentation (MLF), by the metabolism of citric acid, which is usually naturally present in wines at levels between 0.1–0.7 g/L [59]. If present in an excessive content sufficient to affect wine's flavour, is usually considered as a fault, generating a buttery, nutty or toasty, lactic off-odour. The detection threshold for diacetyl in a 10% aqueous ethanol is 0.1 mg/L [11]. However, the diacetyl detection threshold is dependent on the wine matrix. It has been reported as 0.2 mg/L in white wine (Chardonnay) and from 0.9 mg/L (Pinot Noir) to 2.8 mg/L (Cabernet Sauvignon) in red wines [60]. Also, diacetyl quickly bounds SO_2, and the free and bound forms of diacetyl are in chemical equilibrium, depending on the pH, the level of SO_2, and the presence of other SO_2 binding components, such as acetaldehyde, α-ketoglutaric acid, and pyruvic acid are important [61]. It is assumed that only the unbound form of diacetyl is sensorially active. According to Nielsen and Richelieu [61] the addition of 80 mg SO_2, which is within the range used in the wine industry, reduced the free diacetyl content (20 mg/L) by 75%.

All wines contain a few tens of mg/L of ethyl acetate (30–60 mg/L, IV in **Figure 1**) produced by yeast, higher levels indicate AAB activity, formed by esterification between acetic acid and ethanol. This compound at low levels in wine (<50 mg/L) may not be unpleasant, contributing to 'fruity' aroma properties and add complexity to the wine, but at levels >150 mg/L ethyl acetate can confer an unpleasant 'fingernail polish' aroma [62]. Ethyl acetate has a perception threshold in the wine of around 160–180 mg/L, which is much lower than that of acetic acid (750 mg/L) [22]. The deleterious effect of ethyl acetate can be in part reduced by ageing [63] but, after 6 months of bottle ageing, the ethyl acetate levels (140–180 mg/L) affect the wine flavour, giving wines a hot flavour which reinforces the impression of bitterness on the aftertaste [22]. It is usually more perceived in white wine than red wines. Factors that can influence ethyl acetate formation include the yeast strain used during the AF as well as the temperature of fermentation and SO_2 levels. Ethyl acetate is also produced by AAB and is related to dissolved oxygen levels in the wine [64].

Vinylphenols and ethylphenols are collectively known as volatile phenols (VPs). Vinylphenols (4-vinylphenol and 4-vinylguaiacol, V, and VI in **Figure 1**, respectively) are produced by the yeast *S. cerevisiae*, LAB such as *L. plantarum*, and *Dekkera/Brettanomyces* yeasts [65]. Their impact on wine quality is almost exclusively observed in white wines, as these wines can contain significant quantities of vinylphenols which, beyond a certain content (limit threshold = 725 µg/L of 4-vinylguaiacol+4-vinylphenol (1:1)), can be responsible for a depreciating 'phenolic' or 'pharmaceutic' characteristic [66]. *S. cerevisiae* possesses a cinnamate carboxylase enzyme which can transform by non-oxidative decarboxylation, the phenolic acids *p*-coumaric and ferulic acids, into corresponding vinylphenols. This activity is only expressed during AF and with a variable intensity depending on the yeast strain. Although *Dekkera/Brettanomyces* yeasts can also produce vinylphenols they are more likely to reduce the available vinylphenols to ethyl derivatives. It has been shown that *Dekkera/Brettanomyces* is the only known microorganism that under winemaking conditions can produce significant amounts of VPs [67]. The ethylphenols are formed by these yeasts through decarboxylation of the corresponding hydroxycinnamic acids to vinylphenols, and subsequent reduction to ethylphenols, yielding 4-ethylphenol (VII in **Figure 1**) from *p*-coumaric acid and 4-ethylguaiacol (VIII in **Figure 1**) from ferulic acid [67, 68]. Some attributes, such as animal, stable, horse sweat was designated by the widespread term '*Brett-character*' in oenology [69]. The perception threshold of EPs (4-ethylphenols, designated as 4-ethylphenol and 4-ethylguaiacol) is influenced by the wine matrix. The values reported by Chatonnet et al. [68] 440 μg/L for 4-EP and 135 μg/L for 4-EG were found in a model solution. In red wines, the 4-ethylphenol presents a detection threshold of 230 μg/L [70] while the combination of 4-ethylphenol with 4-ethylguaiacol shows a threshold of 400 μg/L [68]. Nowadays, perhaps it is the most problematic sensory defect in red wine production around the world, with million litres being be contaminated each year [71]. In the last years, research has been performed to remove these negative VPs from contaminated red wines [67] and efficient treatments include activated carbons (ACs) and fungal chitosan to avoid the growth of contaminated yeast or to reduce the head space volatility of these negative VPs [3, 72]. New materials have been evaluated for their removal aiming to decrease the negative impact of the former treatments on wine quality. Of the new material that includes molecularly imprinted polymers [73], chitosan [3] and degassed and ethanol impregnated cork powder [74], that can remove about 70% of ethylphenols allowing a significant recovery of the wine's fruit and floral character [74]. This material is cheap and easily prepared from cork powder wastes, being natural with good biodegradability, and low environmental impact.

The formation of mousy off-flavours can occur during (MLF) either by the action of LAB (particularly heterofermentative strains) or *Dekkera/Brettanomyces* yeast. This off-flavour can be associate to three compounds, namely the N – heterocyclic volatile bases 2-acetyltetrahydropyridine (sensory threshold in water =1.6 μg/L, IX, **Figure 1**), 2-ethyltetrahydropyridine (odour threshold in wine = 150 μg/L, X, **Figure 1**, [75] and 2-acetylpyrroline (detection threshold in water = 0.1 μg/L, XI, **Figure 1**, [76], being the first one produced at the highest levels. *Dekkera/Brettanomyces* are capable of producing at least two of these compounds, whereas LAB are capable to produce all the three [77]. Although the biosynthetic pathway for the mousy off-flavour compounds formation in wine is unknown, the conditions necessary for its production have been established. L-lysine and L-ornithine are the precursors of the heterocycle ring of the three mousy compounds, and ethanol and acetaldehyde are responsible for the acetyl side chain. The presence or absence of certain metal ions and oxygen has a substantial effect on off-flavour production [77]. However, there is still not know efficient treatment to remove the mousy off-flavour from wines [78]. At present research studies are being performed on the use of molecular imprinting technology for developing materials with the capacity to selectively remove the mousy off-odour [79]. Therefore, at present, it is necessary to prevent the biosynthesis of the mousy off-odour-forming compounds, by the elimination or strict control of the yeast and bacteria responsible for their formation. This can be achieved by implementing effective microbial control strategies in the winery [77].

Aroma properties evocative of rotten eggs, cabbage, garlic, putrefaction are termed 'reduction'. These aroma attributes are generally considered to contribute negatively to overall wine sensory quality and are considered to be related to different low molecular weight volatile sulphur compounds, such as H_2S, (odour threshold in red wine 1.1 μg/L), methyl mercaptan (methanethiol, odour threshold in red wine 1.8 μg/L, XII, **Figure 1**), ethyl mercaptan (ethanethiol, odour threshold in red wine 1.1 μg/L, XIII, **Figure 1**), and dimethyl sulphide (odour threshold in red wine 25 μg/L, XIV, **Figure 1**) [80]. Yeast fermentation is frequently associated with the occurrence of reductive off-odours, mainly linked to the formation of H_2S and mercaptan by the yeast as mentioned by Pereira et al. [81]. As nitrogen availability is considered one of the main factors for H_2S production by yeast, a strategy that could be adopted is the addition of yeast assimilable nitrogen to supplement fermentation [80]. The production of H_2S during the AF is normal and the quantity produced is dependent on multifactorial factors such, yeast DNA, grape juice turbidity, level of assimilable nitrogen in the grape juice, levels of methionine and cysteine, fermentation temperature, high levels of SO_2, and sulphates. This type of aroma sometimes masks completely the positive varietal and fermentative aroma, however, H_2S is very volatile and usually, simple wine aeration is enough to remove them or can be precipitated with copper sulphate or copper citrate. The excessive aeration of the wine in the presence of H_2S could lead, by oxidation, to the production of heavy thiols that could be exceedingly difficult to remove from the wine. On the other hand, mercaptans and the other sulphides, are more intractable. Mercaptans impart off-odours reminiscent of rotten onions and disulphides are formed under similar reductive conditions and generate cooked-cabbage odours. Related compounds, such as 2-mercaptoethanol (XV, **Figure 1**) and 4-(methyl thiol) butanol (XVI, **Figure 1**), produce intense barnyard and chive–garlic odours, respectively.

Light-struck refers to a reduced-sulphur odour that can develop in wine during exposure to light [62]. This defect is associated with the formation of volatile sulphur compounds with unpleasant aroma notes, formed by the methionine degradation catalysed by the photochemically activated riboflavin. Methanethiol (XII, **Figure 1**) and dimethyl disulphide (XVII, **Figure 1**) are the main compounds responsible for the light-struck taste in white wine termed as 'cooked cabbage' [82, 83]. Exposure of wine to light at wavelengths close to 370 or 442 nm is particularly effective in inducing the light-struck taste [84], manly when clear glass bottles are used [85]. The preventive strategies are the most efficient as this defect generally develops after wine bottling, and these are mainly related to the reduction of the riboflavin levels in grape juice and wine. There are classic and authorised fining agents, such as bentonite and AC (activated carbon) that can be used to remove with relative efficiency riboflavin from white wine [86]. After application, if bentonite the average residual riboflavin was 60% [86, 87]. Also during the AF, the selection of low riboflavin-producing yeasts can be used as it was shown that it is yeast strain-dependent [86, 87].

Several herbaceous off-odours may be detected in wines. The presence of excessive sensations of herbaceous off-odour results in a decrease in the fruit notes, normally not appreciated by consumers. The source of this off-odour can generally be due to the presence of alkylmethoxypyrazines or aldehydes and alcohols with C6. The main alkylmethoxypyrazines found in grapes, musts, and wines are 3-ethyl-2-methoxypyrazine (ETMP, XVIII, **Figure 1**); 3- sec-butyl-2-methoxypyrazine (SBMP, XIX, **Figure 1**), 3-isopropyl 2 methoxypyrazine (IPMP, XX, **Figure 1**); and 3- isobutyl-2-methoxypyrazine (IBMP, XXI, **Figure 1**), conferring aromatic notes described as 'green pepper', or 'tomato leaf'.

Alkylmethoxypyrazines represent a narrow, delineated group of extremely powerful odorants characterised by extremely low sensory perception thresholds (1–2 ng/L in distilled water [88]; being present in green plant tissues, including grapes [89]. The content of methoxypyrazine in the wine depends primarily on grape composition [90], being observed a complex relationship between viticultural practices and varietal aroma, being difficult to predict the final wine aroma because of the multiple compounds and pathways involved. This vegetative character is most commonly, although not exclusively, associated with Sauvignon Blanc, Cabernet Sauvignon, and other Bordeaux varietals [91]. IPMP may also be present in certain grapes and thus found in the derived wine as a varietal character. The excessive green bell pepper aroma found in red wines containing IBMP is generally considered unfavourable to wine quality. However, the presence of this compound at low levels is often noted to augment the quality of certain wines obtained from red varieties (Cabernet Franc, Cabernet Sauvignon, Carménère, Merlot) or white varieties (Sauvignon Blanc, Sémillon) by adding to the intrinsic flavour complexity of these varietals [92]. The presence of IBMP can be a positive quality factor when it is not dominant but is in balance and complemented by other herbaceous and fruity aromas [93].

Aldehydes and alcohols with 6 carbon atoms are volatile, odorous molecules that can contribute to the herbaceous aroma in the wine. Their cut-grass-like aroma is the characteristic odour of freshly damaged green leaves; therefore, these compounds are often referred to as green leaf volatiles [94] and may also impart a bitter flavour [95]. The C6 alcohols frequently found in grapes include hexanol (XXII, **Figure 1**), (Z)-3-hexenol (XXIII, **Figure 1**), and (E)-2-hexenol (XXIV, **Figure 1**). (E)-2-hexenol, (E)-3-hexenol may also be found in wine at levels of μg/L [96]. The C6 aldehydes commonly identified in grapes are hexanal (XXV, **Figure 1**) and (E)-2-hexenal (XXVI, **Figure 1**); also C7 aldehydes have been found, but at lower content concerning C6 aldehydes [97]. At low levels (< 0.5 mg/L threshold), these C6 volatiles compounds contribute positively to the overall aroma of the wine. These C6 compounds may be present in a free volatile form or in bound form, as glycosides [98]. They are mainly generated through the enzymatic breakdown of C18 polyunsaturated fatty acids contained in plant membranes. The C6 aldehydes and alcohols derive from the oxidation of grape polyunsaturated fatty acids such as oleic acid, linoleic acid, and linolenic acid initiated by the lipoxygenase pathway when the berries are crushed [99]. Their levels in must can be in the order of several hundreds of μg/L [100] or even more than 13,000 μg/L [101], with very variable odour thresholds (400–8000 μg/L) [11]. Their levels depend on several factors, including the grape variety and ripeness, treatments before fermentation, and temperature/duration of contact with the skins.

1,1,6- trimethyl-1,2-dihydronaphthalene (TDN, XXVII, **Figure 1**) exhibits kerosene- and petrol-like off-flavour when present at high levels. Precursors of TDN are carotenoid derived compounds originating from the grapes [102]. These precursors are slowly converted to TDN in the wine acidic medium. Kerosene/petrol aroma usually becomes perceivable after several years of wine storage. TDN is an ambiguous aroma compound, defining the varietal character of Riesling wine but also constituting a repelling taint [103] Comparing wines made of various grape varieties, a perceivable amount of TDN is found mostly in Riesling wines. The recognition threshold of TDN has been reported by Simpson [104] to be in the range of 20 μg/L, while Sacks et al. [105] determined a detection threshold of 2 μg/L. Exposing the grapes to more sunlight by defoliation increases both TDN levels [106]. Low pH and bottle ageing will increase their content likewise due to hydrolytic cleavage of the TDN precursors [102, 106].

3.2 Wine taints

1,8-Cineole (eucalyptol, XXVIII, **Figure 1**) is known to give the perception of eucalyptus and minty flavour, not negative sensory notes by themselves, the reason why there is a discussion if it should be considered as a positive or as a taint in red wine. The sensory perception threshold is very low, of 1.1 μg/L, and a recognition threshold of 3.2 μg/L in Californian Merlot wine [107]. Farina et al. [108] reported similar threshold values in Uruguayan Tannat wine. According to Saliba et al. [109], the mechanism by which 1,8-cineole occurs in the finished wine is not well understood and three mechanisms have been suggested: the compound develops from chemical precursors during the winemaking and bottle ageing processes, namely by chemical transformation of limonene and α-terpineol [108]; that grapes naturally produce the compound during berry development [110]; 1,8-cineole is introduced via another source for example from trees.

Geranium taint is due to the presence of 2-ethoxy-3,5-hexadiene (XXIX, **Figure 1**) in wine, which has an odour reminiscent of crushed geranium leaves. It is originated from the reduction of sorbic acid carried out by the LAB. The reduction product sorbitol under wine conditions isomerises to 3,5-hexadiene-2-ol that after reaction with ethanol generates the 2-ethoxyhexa-3,5-diene which has a sensory threshold of about 0.1 mg/L [111].

Grapevines and grape exposure to smoke from firers can result in wines with undesirable sensory characters, such as 'smoky', 'burnt', 'ashy' or 'medicinal', usually described as 'smoke taint'. Smoke taint markers in grapes and wine are the VPs, guaiacol (XXX, **Figure 1**), and 4-methylguaiacol (XXXI, **Figure 1**) [112]. Kennison et al. [113], showed that trace levels ($\leq$1 μg/L) of guaiacol, 4-methylguaiacol, 4-ethylguaiacol and 4-ethylphenol were detected in grape juice derived from grapes harvested from grapevines exposed to smoke; but significant quantities of these phenols were released when the grape juice was fermented, or hydrolysed with strong acid or β-glucosidase enzymes. These compounds are known to exhibit 'smoky', 'phenolish', 'sharp', and 'sweet' aromas [113]. Guaiacol causes a phenolic and medicinal taint in a contaminated wine [114], its flavour threshold is 0.030 mg/L in an aqueous solution containing 12% ethanol. An aroma threshold of 0.020 mg/L in a dry white wine was reported by Simpson et al. [115] and a detection threshold in the water of 5.5 μg/L and white wine of 95 μg/L and red wine, 75 μg/L [112].

2,4,6-trichloroanisole (TCA, XXXII, **Figure 1**) is probably the most known compound associated with wine defect, being the key compound responsible for the 'cork taint' in wines [116]. It is very easy to recognise because of its low sensory threshold, which is from 0.03 to 1–2 ng/L in water and 4 ng/L in white wine for trained assessors. However, the threshold values in wine depend strongly on the kind of wine, the wine style, and the experience of the panellist [117]. 'Cork taint' is mostly described as a musty, mouldy, or earthy smell, being sometimes also described as burnt rubber, smoky or even camphor. Other chloroanisoles, such as 2,4-dichloroanisole (2,4-DCA, XXXIII, **Figure 1**), 2,6-dichloroanisole (2,6-DCA, XXXIV, **Figure 1**), 2,3,4,6-tetrachloroanisole (TeCA, XXXV, **Figure 1**) and pentachloroanisole (PCA, XXXVI, **Figure 1**) can also contribute to the 'cork taint' but they do not play a dominant role in this fault. 2,4,6-tribromoanisole (TBA, XVII, **Figure 1**) can also have a significant role in the musty/mouldy of-odour of wines [118]. Moreover, the aroma masking effect of TCA or TBA can be perceived in the wines at levels even lower than perception thresholds (4 ng/L) [118]. As cork is a natural product from the cork oak it is subject to microbial contamination and its quality is dependent on good agricultural practices and quality control during processing, transport, and storage. Chlorophenolic biocides (nowadays forbidden but accumulated in the environment) are the common precursors which can be transformed by certain fungi to TCA and different chloroanisoles. Other pathways of chloroanisoles formation usually include reactions of chlorination and methylation of compounds naturally present in wooden and cork materials [119]. At the same time 2,4,6-tribromophenol (TBP), which application is not restricted present, can play the role of TBA precursor and increases the risk of musty/mouldy taint in wines [118]. Moulds are considered the most significant causative organisms of cork taint, with implicated genera including *Penicillium*, *Aspergillus*, *Cladosporium*, *Monilia*, *Paecilomyces* and *Trichoderma*. Nevertheless, the process of wine contamination by haloanisoles is complex. Since cork stoppers are the most known source of these compounds, the musty/mouldy fault was named 'cork taint'. However, succeeding studies demonstrated that musty/mouldy defects in wine are not originated exclusively from the naturally contaminated cork materials [118, 120]. TCA and other haloanisoles can be formed in different wooden parts inside the cellar (barrels, ceiling constructions, pallets) and subsequently released into the air. Hereafter, the contamination of winery equipment and 'clean' cork stoppers occurs and haloanisoles can be transmitted further to the wine. Besides, cork taint-related flavours were found in wines that were barrel samples or not closed with natural cork stoppers, indicating that natural cork stoppers are not the only source of mouldy off-flavours [118, 120]. Therefore, depending on the compound causing the cork taint, the consumer has a different impression of the problem. Several compounds with similar negative flavour attributes were discovered in mouldy and musty smelling wines that were not affected by TCA (geosmin, 2-methyl-isoborneol, octane-3-one, pyrazines, etc.) [121]. That said, misguided hygiene practices have historically been part of the cork-taint problem. Cleaning using chlorinated bleach was common in wineries until a link to cork taint was found. Contact between barrels and bleach on cellar floors was a particular pathway for TCA to strike. Flame-retardant paints and fungicides were found to taint wine with TBA. Barrelled wines were particularly badly hit, and some facilities had to be rebuilt. Nowadays most wineries know to avoid chemicals containing tribromophenols. Heat-treated wood is more common, and barrels are rarely cleaned with chlorine. Different approaches were made regarding the removal of TCA and TBA from tainted wines; either by fining with ACs and filtered afterwards, or polyethylene was added as an adsorbent to the wine [121].

Along with TCA and TBA, geosmin (XXXVIII, **Figure 1**), 2-methylisoborneol (MIB XXXIX, **Figure 1**), 1-octen-3-one (XL, **Figure 1**) and 1-octen-3-ol (XLI, **Figure 1**) are compounds that are closely linked with the growth of moulds [116]. Their presence in wines can impart typical earth, mushroom, fungal and mouldy flavour [122, 123]. The mushroom aroma is associated mainly with 1-octen-3-one and 1-octen-3-ol, whereas the earthy aroma is attributed to (−)-geosmin and an earthy-camphor aroma to 2-methylisoborneol. Geosmin may result from the development in grapes picked in unfavourable weather conditions by microorganisms. It is a chiral compound and the (−) form is much more odoriferous than the (+) form. (−) Geosmin is also the only enantiomer to have been identified in pure cultures of *Streptomyces* sp. and *Penicillium* sp. strains isolated from rotten grapes [121]. Geosmin olfactory detection threshold depends on the wine matrix: In water: 10 ng/L [122]; In wine: white wine, 60–65 ng/L [122]; red wine, 80–90 ng/L [122]. MIB is a metabolite of *Botrytis cinerea*, some *Penicillium* spp. and some *Streptomyces* spp. MIB and 1-octen-3-one have also been found in musts made from rotten grapes but not in the corresponding wines, indicating that they are not stable during AF [123]. The findings of both compounds in bottled wine can therefore be linked to the cork stopper and the growing of mould on the cork during the manufacturing process. MIB olfactory detection threshold has been determined as: 0.012 μg/L (in water La Guerche et al. [123]; 0.04 μg/L [123]; in red wine). 1-octen-3-ol (olfactory detection threshold in the water of 2 μg/L and red wine, 40 μg/L, La Guerche et al. [123], however, has also been found on rotten grapes and the musts made from them and is stable during AF so that an occurrence of this compound in a wine can be caused by mould growth on the grapes as well as by contaminated cork stoppers [124]. Lisanti et al. [125] showed that in the red wine the potassium caseinate and grape seed oil treatments decreased the level of geosmin by 14% and 83%, respectively, while in the white wine, the AC and the grape seed oil were able to decrease the level of geosmin by 23% and 81%, respectively. However, after estimating the olfactory impact of the volatile compounds by OAVs (concentration/odour perception threshold), only the treatment with grape seed oil was able to decrease the relative contribution of geosmin in the profile of the odour active compounds, in both wines.

Wine can accidentally be contaminated with styrene when trace amounts of the styrene (XLII, **Figure 1**) are released during wine storage in polyester tanks reinforced with fibre glass [126]. Also, occasionally styrene contamination has been detected in wine in contact with synthetic closures [127]. The taste threshold for styrene in water has been reported as 22 μg/L [128] but may be higher in wine. An amount higher than 100 μg/L (the generally accepted threshold of sensory perception), styrene can modify the wine sensory characteristics by imparting a taste of plastic and adhesive. Wagner et al. [129] found in German wines values ranging from 0 to 19 μg/L.

4. Origin of taste and tactile sensory defects and strategies for wine stabilisation

The wine imbalances by acidity, astringency, or bitterness, are often the first defects noted in the sensory perception of wine quality [13].

4.1 Acidity

Organic acids are the main responsible for sourness and able of modifying this sourness sensation in wines producing a pleasant and refreshing sensation [130]. However, when present at high levels they are responsible for an unpleasant acidity. Therefore, it is generally accepted that too much acidity will taste excessively sour and sharp, while wines with too little acidity will taste flabby and flat and present a less defined flavour profile [131]. Organic acids contribute to the tartness and mouth-feel properties of wine. Tartaric acid is the main organic acid in wine, which, at high levels (>5 g/L), is responsible for an unpleasant taste. Other acids include malic, citric, fumaric, succinic, pyruvic, α-ketoglutaric, lactic, and acetic [3]. However, different organic acids have different sensory properties, and the impact of organic acids is therefore not only linked to total acidity and pH, but to the specific levels of each acid in the wine [132]. The perceived sourness was imparted by L-tartaric acid, D-galacturonic acid, acetic acid, succinic acid, L-malic acid, and L-lactic acid and was slightly suppressed by the levels of chlorides of potassium, magnesium, and ammonium [16]. Acidity adjustment is the reduction or increase in titratable acidity so that the resulting wine will be acceptable. Acidity adjustment can be performed by the addition of an approved acid, the chemical deacidification with approved salts, and using ion exchange resins, either cation, anion or both, electromembrane processes and by biological deacidification. Tartaric acid is commonly used to increase the titratable acidity and reduce the pH in the wine industry, because of its stability and the fact that yeast and other microorganisms are unable to metabolise it at wine pH [133].

The reduction of titratable acidity by the addition of carbonate salts such as calcium carbonate, can be done in one of two ways, the first, is a direct addition which is not recommended as it results in wines which are unstable with respect to calcium tartrate, the second is to treat only a wine portion. This process causes the pH to increase up increase to 4 or 4.5 at the end of the addition. The tartaric and malic acids are primarily in the ionised forms. The precipitation of their calcium salts is favoured and this also lowers the calcium levels. This method is referred to as the double-salt and it may be used to reduce the total acidity of high-acid grape musts before fermentation. The precipitation is primarily that of calcium tartrate and under certain circumstances the coprecipitation of calcium malate [2]. Ion exchange resins, either cation exchange alone or as a combination of anion and cation exchange, can also be used to change wine acidity [19]. In red wine for example the acid reduction can be achieved by using LAB strains. MLF refers to the conversion of malic acid to lactic acid and CO_2. This secondary fermentation usually takes place after the AF. The benefits of MLF is the acidity reduction and simultaneously add the complexity of aroma and taste and provides a more microbiological stable wine [2].

4.2 Astringency

One of the most important sensations and a quality attribute is astringency. Gawel et al. [134] presented a structured vocabulary derived by a panel of experienced wine tasters that describe the astringent sub-qualities of red wines, such as velvety, drying, puckering, or roughing. Astringency is mainly a tactile sensation [135] not a taste because it can be perceived in regions of the oral cavity where there is no taste receptor [96, 136, 137]. The major mechanism proposed to astringency perception is the interaction and precipitation of salivary glycoproteins, namely by tannins generating a loss of lubrication [136]. Vidal et al. [138] showed in model solutions that astringency perception of proanthocyanidins increases with their mean degree of polymerisation (mDP) and their percentage of galloylation [139]. Oligomeric proanthocyanidins have been described as inducing lower roughness than the more polymerised molecules, whereas an increase in galloylation has been associated with a higher perceived drying and roughing astringency [139]. However, other wine phenolic compounds, such as flavonols, phenolic acids, or anthocyanins, can also play an important role in astringency development [139].

4.3 Bitterness

Bitter perception in wines is related to phenolic compounds with low molecular weights as well as to monomeric or small phenolic flavanols [16]. Concerning the latter, they have been described for a long time as the main contributors to the bitterness generated by flavonoid phenols [140]. Monomeric flavonoid phenols are primarily bitter but as the molecular weight increases upon polymerisation, astringency increases more rapidly than bitterness. It has also been shown that chiral difference between the two major wine monomeric flavanols produces a significant difference in temporal perception of bitterness: (−)-epicatechin is significantly bitterer and has a significantly longer duration of bitterness in the mouth than (+)-catechin [140].

Protein fining agents could induce some sensory changes. Astringency and bitterness of wine can decline due to its interaction with tannins. The fining process directly occurs from the precipitation of proanthocyanidins by these protein fining agents and it is influenced by the chemical characteristics of the protein used. The interactions between proanthocyanidins and protein fining agents depend on molecular weight, amino acid composition and surface charge density of the proteins used [141–144].

Different proteins are used for wine fining such as gelatine, egg albumin, isinglass, and casein/potassium caseinate. Different types of gelatine remove different amounts of proanthocyanidins (9–16%) depending on the wine phenolic composition and structural characteristics of the proanthocyanidins and on the gelatine composition and characteristics [142, 143]. It has been generally thought that proteins bind primarily high polymerised tannins as well as high galloylated tannins, and therefore are preferentially removed [141], but some recent work showed that each of the different proteins (gelatine, egg albumin, isinglass, casein) and different size fractions of the same protein class interact differentially with different sizes of tannins [142, 143]. Regardless, allergen labelling may make wine fining with any of the animal-derived products impractical although some effort has been made to evaluate plant-derived proteins [144]. Recent studies of wine astringency demonstrated that tannins must be different two-fold for a trained panel to be able to successfully differentiate the wines [145]. Further, since some of the polymeric pigments can precipitate with protein there is the risk of losing stable colour [146]. As mentioned previously a higher astringency intensity is directly associated with a higher concentration of proanthocyanidins with a higher mean degree of polymerisation [147]. During ageing, astringency perception becomes softer, the reasons for the change in wine astringency could involve a decrease in proanthocyanidin concentration accompanied by a decrease in proanthocyanidins structural changes [148].

Therefore, the phenolic composition could be modulated during the winemaking steps (maceration/fermentation, stabilisation (fining) and ageing) and consequently, it allows the modulation of wine astringency and/or astringency sub-qualities as well as the wine bitterness.

5. Origin of potentially toxic compounds and strategies to improve wine safety

In fermented beverages in which a variety of microorganisms exist it may be inevitably the production of toxic products as a result of their metabolism and side reactions, including ethyl carbamate (I, **Figure 2**), biogenic amines (II, **Figure 2**) mycotoxins, namely ochratoxin A (III, **Figure 2**) and aflatoxin B1 (IV, **Figure 2**). They are generally generated due to the incomplete metabolism of nitrogen-containing compounds during the fermentation process [149].

5.1 Ethyl carbamate

Ethyl carbamate (EC), also known as urethane, is an ethyl ester of carbamic acid that can be found in several fermented beverages [150] including wines [151]. EC levels in wines can range from n.d.-19 μg/L in white wine, n.d.-54 μg/L in red wine, 14–50 μg/L in fortified wine, and n.d.-58 μg/L in sherry-type wine [152]. EC is classified as a 'probable human carcinogen' by the IARC since 2007 (group 2A) [153]. Although currently there is no harmonised maximum level for EC, some countries have established their criteria for example in Canada the maximum level is 30 μg/L for wine, the Canadian guidelines were adopted by other countries such as Czech Republic, Brazil, France, Germany, and Switzerland. South Korea also set the maximum limit of 30 μg/L only for table wine. For fortified wine, the maximum level of

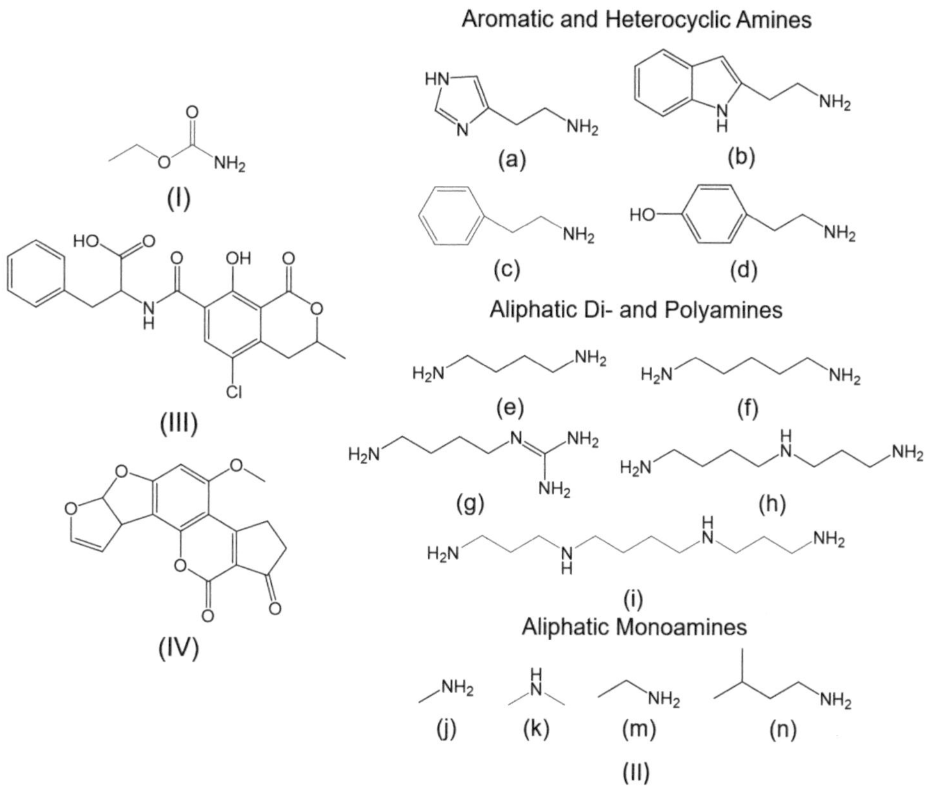

Figure 2.
Main wine potentially toxic compounds. (I) EC; (II) BAs – (a) histamine; (b) tryptamine; (c) phenylethylamine; (d) tyramine; (e) putrescine; (f) cadaverine; (g) agmatine; (h) spermine; (i) spermidine; (j) methylamine; (k) dimethylamine; (m) ethylamine; (i) isopentylamine; (III) OTA; (IV) AB1.

EC is 100 µg/L in Canada and the Czech Republic, and of 60 µg/L in the US [154]. EC can be produced from at least five precursors, namely urea, citrulline, carbamyl phosphate, cyanic acid and diethyl pyrocarbonate. In turn, urea and citrulline can be respectively generated by yeast and LAB by metabolisation of arginine, a major amino acid found in grape juice and wine [155]. The fermentation conditions, such as pH, temperature, ethanol level, light irradiation, oxygen, storage time, yeast or LAB strains can also affect the formation of EC [156]. For example, lowering the temperature during fermentation and storage, lowering the pH, lowering the ethanol content, and addition of diammonium phosphate as a yeast nutritional supplement reduces the EC concentration. The development of techniques for EC elimination from alcoholic beverages [155] has attracted considerable attention, and enzymatic decomposition methods have been widely employed given their safety and environmentally friendly nature. Two enzymes are used, namely, urease, which is commercially available and can degrade urea, the major precursor of EC [157], and urethanase, which can directly catalyse EC degradation [158]. To reduce EC concentration in wine, the use of acid urease seems to be the most appropriate way to suppress EC formation [19, 159]. Moreover, the efficiency of commercial acid urease treatment varies with several factors, including pH, temperature, the presence, and concentration of inhibitors (malate, ethanol, phenolic compounds), and type of wine [157]. Therefore, the immobilisation of acid urease in chitosan beads enhances the protection against inhibitors, increases the stability of the enzyme, and has the advantage of facilitating enzyme recycling and consequently reducing the cost of its use. EC content can also be effectively reduced by decreasing the generation of its precursors. Significant advances have been made via genetic technologies in modifying fermentation strains that produce less EC precursors. Genetic modification approaches have the potential to provide safe, affordable, and effective methods to decrease EC formation. Several studies have shown that the modification of catalytic enzymes, such as urea carboxylase, arginase, and allophanate hydrolase, showed the ability to reduce the concentration of EC [156]. Additionally, the modification of urea permease and amino acid permease, which are regulated by several factors and directly affect the generation of EC precursors have been explored [160]. Since the metabolism pathways related to urea have been fully considered for the high-efficiency minimisation of EC, enhancing the gene expression of $DUR_{1,2}$ and DUR_3, which encode urea degradation enzymes and permease, respectively, is considered to be a viable strategy. In this way, the modification of permease has led to the construction of functionally enhanced urea-importing wine yeast cells, which can continuously express the DUR_3 gene and reduce EC level in Chardonnay wine by 81% [161].

5.2 Biogenic amines

Biogenic amines (BAs) are low molecular weight organic bases that have adverse physiological effects on humans when absorbed at high levels [162]. BAs are formed by decarboxylation of the corresponding amino acids by microorganisms such as LAB [162]. *Pediococcus,* as well as *Lactobacillus,* have been implicated in the production of BAs in wines that have undergone spontaneous MLF [162]. The final BAs levels in wine depend on the availability of the precursor amino acids and the BAs producing bacteria [162]. The only currently available simple and efficient solution to avoid or minimise BAs formation in wine is the use of MLF starter cultures [163]. After inoculation, the selected strain becomes dominant during MLF. Another recommendation is to avoid the practices that increase amino acid and peptide levels in musts. This also implies that LAB have more substrates, not only for producing BAs but also to survive better and longer after MLF [164]. Although there are currently no official values for the maximum limits for histamine and other BAs, the maximum value imposed for the levels of histamine, has been established through wine purchase and sale contracts, with the German companies demanding a maximum level of 2 mg/L of histamine. For these reasons, strategies for the reduction/elimination of BAs in wine are necessary, especially histamine. Until now, there are not useful treatments for reducing BAs levels especially, in red wines. However, it has been shown that uncommon wine LAB strains have amine oxidase activities that degrade histamine, tyrosine, and putrescine [165]. Also, non-*Saccharomyces* yeasts, such as *Schizosaccharomyces*, can decrease malic acid content in wine, they can be an excellent alternative to LAB, avoiding the MLF. Also, the use of *Schizosaccharomyces* reduces the risk of BAs production [166]. According to the OIV Resolution [19], only bentonite is applied in already contaminated wines to reduce the content of the BAs in the final wine [167]. Bentonite has a negative surface charge density, being able to exchange the cations adsorbed on its surface by the wine BAs. However, due to the negative impact on wine aroma combined with the high wine losses due to the high volume of lees, bentonite is presently not an adequate solution and other options need to be studied. Till now, there is no effective way of removing BAs in the finished wine.

5.3 Mycotoxins

Mycotoxins are secondary metabolites produced by several fungi that grow in food, including wine, under particular circumstances, with ochratoxin A (OTA) being one of the most important [168]. In 1993, the International Agency for

DOI: http://dx.doi.org/10.5772/intechopen.95245

Research on Cancer (IARC) classified OTA as possibly carcinogenic for humans (group 2B) [169]. Since 2006, the maximum limit for OTA in wine is 2 μg/kg [170]. OTA has been related to wine contaminations since 1996 [171] and after that, the occurrence of OTA in wine samples has been described in several works. Blesa et al. [172] found an OTA incidence in wines of 53% in 521 red wines, 69% in 98 rosé wines, and 61% in 301 white wines. These data show that it is important to prevent and control the occurrence of this mycotoxin in wines. To eliminate this toxin, several chemicals, microbiological and physical methods have been described [168]. Nevertheless, in the case of wines, effective removal processes are limited, and at present, the use of adsorbents is the most common. OTA content can reduce in the final wine from 70 to 32% by fermenting with selected yeast [173], *Non-Saccharomyces*, such as *Schizosaccharomyces*, are also promising in reducing the OTA content by about 70% during fermentation [173]. Several fining agents have been evaluated concerning their ability to remove OTA from wines, and it was found that AC presented a good adsorption capacity for OTA [174]. Filtration before bottling about 0.45 μm of wine can easily reduce the final content in OTA by about 80% [174], but none of them is 100% efficient in removing OTA from wines.

Aflatoxins are a group of highly toxic secondary metabolites produced by fungi of the genus *Aspergillus* [175]. AFB1 is the most predominant and toxic aflatoxin. It is classified as a Group 1 human carcinogen (IARC) [169]. Aflatoxins and Aflatoxins-producing strains [176] have been detected in grape and musts [177]. The presence of aAFB1 in wines is caused by fungi that grow on grapes in the vineyards. In the literature, there are few studies regarding aflatoxins contamination in wines [177]. One of them is from Di Stefano et al. [178] that studied the occurrence of aflatoxins in 30 sweet wines from five winemaking in the Sicilian regions, Italy. The presence of aflatoxin in wines has been documented in recent years, largely because of the adaptability of aflatoxigenic fungi, such as *A. flavus*. At present, the EU has not set a maximum allowable limit for aflatoxins in wine, but this does not mean that the problem can be ignored. Therefore, it is essential to develop technological solutions to reduce/eliminate the levels of aflatoxins in wines. Recently the work from Cosme et al. [179] shows the high efficiency of bentonite in the removal of aflatoxin B1 and B2.

6. Conclusions

As overviewed in this chapter, there are many physicochemical wine instabilities and defects that can appear during wine production. Some defects can significantly decrease the wine's sensory quality. However, today, there are several treatments and solutions to avoid them or reduce their impact on wine quality. The best strategy is always a preventive approach. Nevertheless, some of the defects are intrinsic to the grape composition and/or wine production process therefore they must be removed or minimised before bottling. The available treatments, either by using fining agents, additives, or other technological solutions, are generally effective, although they are sometimes not the perfect solution as they can also impact on positive wine sensory attributes and/or have a detrimental environmental impact. Current research trend is focused on the development of fining agents, additives, or technological solutions with improved specificity that will allow the removal of the defect without changing the other wine characteristics, and at the same time to explore low cost, natural or renewable materials that will allow a lower environmental impact of the stabilisation process.

Author details

Fernanda Cosme*, Luís Filipe-Ribeiro and Fernando M. Nunes
Chemistry Research Centre—Vila Real, (CQ-VR), Food and Wine Chemistry Laboratory, University of Trás-os-Montes e Alto Douro, Vila Real, Portugal

*Address all correspondence to: fcosme@utad.pt

References

[1] Lefebvre S, Sieczkowski N, Vidal F, La sécurité alimentaire en oenologie: cas des proteins végétales. Revue Française d'Oenologie, 210; 23-30.

[2] Boulton RB, Singleton VL, Bisson LF, Kunkee RE, (Eds.), 1996. Principles and Practices of Winemaking. 1th ed. Chapman & Hall, New York, USA.

[3] Filipe-Ribeiro L, Cosme F, Nunes FM, Reducing the negative sensory impact of volatile phenols in red wine with different chitosans: Effect of structure on efficiency. Food Chemistry, 2018; 242, 591-600.

[4] King ES, Kievit RL, Curtin C, Swiegers JH, Pretourius IS, Bastian SEP, Francis IL, The effect of multiple yeast inoculations on Sauvignon Blanc wine aroma composition, sensory properties and consumer preference. Food Chemistry, 2010; 122 (3), 618-626.

[5] Robinson AL, Boss PK, Solomon PS, Trengove RD, Heymann H, Ebeler SE, Origins of grape and wine aroma. Part 1. Chemical components and viticultural impacts. American Journal of Enology and Viticulture, 2014; 65(1), 1-24.

[6] Robinson AL, Boss PK, Solomon PS, Trengove RD, Heymann H, Ebeler SE, Origins of grape and wine aroma. Part 2. Chemical and Sensory Analysis. American Journal of Enology and Viticulture, 2014; 65(1), 25-42.

[7] Jackson DI, Lombard PB, Environmental and management practices affecting grape composition and wine quality–A review. American Journal of Enology and Viticulture, 1993; 44, 409-430.

[8] de Revel G, Martin N, Pripis-Nicolau L, Lonvaud-Funel A, Bertrand A, Contribution to the knowledge of malolactic fermentation influence on wine aroma. Journal of Agricultural and Food Chemistry, 1999; 47, 4003-4008.

[9] Rapp A, Mandery H, Wine aroma. Experienta, 1986; 42, 873-884.

[10] Maarse H, Vischer CA, Volatile Compounds in Food. Alcoholic Beverages. Qualitative and Quantitative Data. Zeist, Neth.: 1989, TNO-CIVO. 6th ed.

[11] Guth H, Quantitation and sensory studies of character impact odorants of different white wine varieties. Journal of Agricultural and Food Chemistry, 1997; 45, 3027-3032.

[12] Whitfield FB, Food off-flavours: cause and effect. In: Developments in Food Flavours (edited by G. G. Birch & M. G. Lindley). 1986; 249-273. London: Elsevier.

[13] Martin N, Sweet/sour balance in champagne wine and dependence on taste/odour interactions. Food Quality and Preference, 2002; 13, 295-305.

[14] Breslin PAS, Gilmore MM, Beauchamp GK, Green BG, Psychophysical evidence that oral astringency is a tactile sensation. Chemical Senses, 1993; 18, 405-417.

[15] Sáenz-Navajas M-P, Ferreira V, Dizy M, Fernández-Zurbano P, Characterization of taste-active fractions in red wine combining HPLC fractionation, sensory analysis and ultra performance liquid chromatography coupled with mass spectrometry detection. Analytica Chimica Acta, 2010; 673, 151-159.

[16] Hufnagel JC, Hofmann T, Orosensory-directed identification of astringent mouthfeel and bitter-tasting compounds in red wine. Journal of

Agriculture and Food Chemistry, 2008; 56, 1376-1386.

[17] Berg HW, Keefer RM, Analytical determination of tartrate stability in wine. I. Potassium bitartrate. American Journal of Enology and Viticulture, 1958; 9, 180-193.

[18] Berg HW, Keefer RM, Analytical determination of tartrate stability in wine. II. Calcium tartrate. American Journal of Enology and Viticulture, 1959; 10, 105-109.

[19] OIV, Codex Oenologique International; Internationale de la Vigne et du Vin: Paris, France, 2019.

[20] Moine-Ledoux V, Dubourdieu D, Rôle des mannoprotéines de levures vis-à-vis de la stabilisation tartrique des vins. Bulletin de l'OIV, 2002; 75, 471-482.

[21] Waters EJ, Alexander G, Muhlack R, Pocock KF, Colby C, O'Neill BK, Høj PB, Jones P, Preventing protein haze in bottled white wine. Australian Journal of Grape and Wine Research, 2005; 11, 215-225.

[22] Ribéreau-Gayon P, Glories Y, Maujean A, Dubourdieu D, Handbook of Enology. Volume 2: The Chemistry of Wine Stabilization and Treatments; JohnWiley and Sons Inc.: Hoboken, NJ, USA, 2006.

[23] Tian B, Harrison R, Pathogenesis-Related Proteins in Wine and White Wine Protein Stabilization. Chapter, Book In: Winemaking - Stabilization, Aging Chemistry and Biochemistry. Editors Cosme F., Nunes, F.M., Filipe-Ribeiro, L. 2020, IntechOpen, DOI: http://dx.doi.org/10.5772/intechopen.92445.

[24] Chagas R, Lourenço AM, Monteiro S, Ferreira RB, Ferreira LM, Is caffeic acid, as the major metabolite present in Moscatel wine protein haze hydrolysate, involved in protein haze formation? Food Research International, 2017; 98, 103-109.

[25] Cosme F, Fernandes C, Ribeiro T, Filipe-Ribeiro L, Nunes FM, White Wine Protein Instability: Mechanism, Quality Control and Technological Alternatives for Wine Stabilisation—An Overview. Beverages, 2020; 6, 19; doi:10.3390/beverages6010019.

[26] Lambri M, Dordoni R, Silva A, Faveri DM, Effect of bentonite fining on odor-active compounds in two different white wine styles. American Journal of Enology and Viticulture, 2010; 61, 225-233.

[27] Canas S, Leandro MC, Spranger MI, The precipitation of colouring matter in red wines. Contribution to the study of the deposits composition Ciência e Técnica Vitivinícola, 2000; 15 (1) 27-40.

[28] Waters EJ, Peng ZK, Pocock KF, Jones GP, Clarke P, Williams PJ, Solid-state 13C NMR investigation into insoluble deposits adhering to the inner glass-surface of bottled red wine. Journal of Agricultural and Food Chemistry, 1994; 42 (8), 1761-1766.

[29] Glories Y, Le froid et la matière colorante des vins rouges. In Revue Française d'Œnologie, 1979; 73, 37-39.

[30] Zoecklein B, Fugelsang K, Gump B, Nury F, Phenolic compounds and wine color. In Wine Analysis and Production; Chapman and Hall: New York, 1995.

[31] Clark AC, Wilkes EN, Scollary GR, Chemistry of copper in white wine: A review. Australian Journal of Grape and Wine Research, 2015; 21, 339-350.

[32] Volpe MG, la Cara F, Volpe F, de Mattia A, Serino V, Petitto F, Zavalloni C, Limone F, Pellecchia R, de PriscoPP, Di Stasi M, Heavy metal

uptake in the enological food chain. Food Chemistry, 2009; 117, 553-560.

[33] Schubert M, Glomb MA, Analysis and chemistry of migrants from wine fining polymers. Journal of Agricultural and Food Chemistry, 2010; 58, 8300-8304.

[34] Chinnici F, Sonni F, Natali N, Riponi C, Oxidative evolution of (+)-catechin in model white wine solutions containing sulfur dioxide, ascorbic acid or gallotannins, Food Research International, 2013; 51, 59-65.

[35] Danilewicz JC, Interaction of polyphenols, oxygen, and sulfite in wine. American Journal of Enology and Viticulture, 2008; 59, 346A-346A.

[36] Lopes P, Silva MA, Pons A, Tominaga T, Lavigne V, Saucier C, Darriet P, Teissedre P-L, Dubourdieu D, Impact of oxygen dissolved at bottling and transmitted through closures on the composition and sensory properties of a sauvignon blanc wine during bottle storage. Journal of Agricultural and Food Chemistry, 2009; 57, 478, 10261-10270.

[37] Singleton VL, Zaya J, Trousdale E, Salgues M, Caftaric Acid in Grapes and Conversion to a Reaction-Product during Processing. Vitis, 1984; 23, 113-120.

[38] Guyot S, Vercauteren J, Cheynier V, Structural determination of colourless and yellow dimers resulting from (+)-catechin coupling catalysed by grape polyphenoloxidase, Phytochemistry, 1996; 42, 1279-1288.

[39] Mayer AM, Polyphenol oxidases in plants and fungi: going place? A review. Phytochemistry, 2006; 67,2318-2331.

[40] Reiss R, Ihssen J, Richter M, Eichhorn E, Schilling B, Thöny-Meyer L, Laccase *versus* laccase-like multi-copper oxidase: a comparative study of similar enzymes with diverse substrate spectra. PLoS One, 2013; 8(6), e65633.

[41] Fowler ZL, Baron CM, Panepinto JC, Koffas MAG, Melanization of flavonoids by fungal and bacterial laccases. Yeast, 2011; 28, 181-188.

[42] Cheynier V, Basire N, Rigaud J. Mechanism of transcaffeoyltartaric acid and catechin oxidation in model solutions containing grape polyphenoloxidase. Journal of Agricultural and Food Chemistry, 1989; 37(4), 1069-1071.

[43] Danilewicz JC, Fe(II):Fe(III) Ratio and Redox Status of White Wines, American Journal of Enology and Viticulture, 2016; 67, 146-152.

[44] Es-Safi N-E, Cheynier V, Moutounet M, Effect of copper on oxidation of (+)-catechin in a model solution system. International Journal of Food Science and Technology, 2003; 38, 153-163.

[45] Nikolantonaki M, Magiatis P, Waterhouse AL, Measuring protection of aromatic wine thiols from oxidation by competitive reactions vs wine preservatives with *ortho*-quinones. Food Chemistry, 2014; 163, 61-67.

[46] Andrea-Silva J, Cosme F, Filipe-Ribeiro L, Moreira ASP, Malheiro AC, Coimbra MA, Domingues MRM, Nunes FM, Origin of the pinking phenomenon of white wines. Journal of Agricultural and Food Chemistry, 2014; 62, 5651-5659.

[47] Du Toit W, Marais J, Pretorius I, Du Toit M, Oxygen in must and wine: a review. South African Journal for Enology and Viticulture, 2006; 27, 76-94.

[48] Simpson R, Miller G, Orr L, Oxidative pinking of whites wines: recent observations. Australian Food Technology Australian, 1982; 34, 46-47.

[49] Arapitsas P, Oliveira J, Mattivi F, Do white grapes really exist? Food Research International, 2015; 69, s 21-25.

[50] Vilela-Moura A, Schuller D, Mendes-Faia A, Silva RF, Chaves SR, Sousa MJ, Côrte-Real M, The impact of acetate metabolism on yeast fermentative performance and wine quality: reduction of volatile acidity of grape-musts and wines: minireview. Applied Microbiology and Biotechnology, 2011; 89, 271-280

[51] McCloskey LP, Mahaney P, An enzymatic assay for acetaldehyde in grape juice and wine. American Journal of Enology and Viticulture, 1981, 32, 159-162.

[52] Sheridan MK, Elias RJ, Exogenous acetaldehyde as a tool for modulating wine color and astringency during fermentation. Food Chemistry, 2015; 177, 17-22.

[53] Picariello L, Gambuti A, Petracca F, Rinaldi A, Moio L, Enological tannins affect acetaldehyde evolution, colour stability and tannin reactivity during forced oxidation of red wine. International Journal of Food Science and Technology, 2018, 53, 228-236.

[54] Peynaud E, The Taste of Wine, 2nd ed.; Wiley: New York, 1996.

[55] Miyake T, Shibamoto T, Quantitative analysis of acetaldehyde in foods and beverages. Journal of Agricultural and Food Chemistry, 1993, 41, 1968-1970.

[56] Kotseridis Y, Baumes R, Identifi cation of impact odorants in Bordeaux red grape juices, in the commercial yeast used for its fermentation, and in the produced wine. Journal of Agricultural and Food Chemistry, 2000; 48, 400-406

[57] Romano P, Suzzi G, Turbanti L, Polsinelli M, Acetaldehyde production in *Saccharomyces cerevisiae* wine yeasts. FEMS Microbiology Letters, 1994; 118, 213-218.

[58] Osborne JP, Dubé Morneau A, Mira de Orduña R, Degradation of free and sulfur-dioxide-bound acetaldehyde by malolactic lactic bacteria in white wine. Journal of Applied Microbiology, 2006; 101, 474-479.

[59] Amerine MA, Ough CS, 'Methods for analysis of musts and wines' 1980; (John Wiley & Sons: New York).

[60] Martineau B, Acree TE, Henick-Kling T, Effect of wine type on the detection threshold for diacetyl. Food Research International, 1995; 28, 139-143.

[61] Nielsen JC, Richelieu M, Control of Flavor Development in Wine during and after Malolactic Fermentation by Oenococcus oeni. Applied Environmental Microbiology, 1999; 65, 2, 740-745.

[62] Jackson RS, Wine Science. Principles and Applications 2000, (Academic Press: London, UK).

[63] Lilly M, Bauer FF, Lambrechts MG, Swiegers JH, Cozzolino D, Pretorius IS, The effect of increased yeast alcohol acetyl transferase and esterase activity on the flavor profiles of wine and distillates. Yeast, 2006; 23, 641-659.

[64] Nykänen L, Nykänen I, Production of esters by different yeast strains in sugar fermentations. Journal of the Institute of Brewing, 1977, 83, 30-31.

[65] Chatonnet P, Coralie V, Dubourdieu D, Influence of polyphenolic components of red wines on the microbial synthesis of volatile phenols. American Journal of Enology and Viticulture, 1997; 48, 443-448.

[66] Chatonnet P, Dubourdieu D, Boidron JN, Lavigne V, Synthesis of volatile phenols by *Saccharomyces*

cerevisiae in wines. Journal of the Science of Food and Agriculture, 1993; 62(2),191-202.

[67] Milheiro J, Filipe-Ribeiro L, Vilela A, Cosme F, Nunes MF, 4-Ethylphenol, 4-ethylguaiacol and 4-ethylcatechol in red wines: Microbial formation, prevention, remediation and overview of analytical approaches, Critical Reviews in Food Science and Nutrition, 2019; 59, 9, 1367-1391.

[68] Chatonnet P, Dubourdieu D, Boidron JN, Pons M, The origin of ethylphenols in wines. Journal of the Science of Food and Agriculture, 1992; 60, 165-178.

[69] Portugal C, Saenz Y, Rojo-Bezares B, Zaragaza M, Torres C, Cacho J, Ruiz-Larrea F. Brettanomyces susceptibility to antimicrobial agents used in winemaking: In vitro and practical approaches. European Food Research and Technology, 2014; 238, 641-652.

[70] Chatonnet P, Boidron JN, Pons M, Élevage des vins rouges en fûts de chêne: évolution de certains composés volatiles et de leur impact aromatique. Science des Aliments, 1990; 10, 565-587.

[71] Licker JL, Acree TE, Henick-Kling T, What Is "Brett" (*Brettanomyces*) Flavor?: A Preliminary Investigation. ACS Symposium Series, 1998; 714, 96-115.

[72] Filipe-Ribeiro L, Milheiro J, Matos CC, Cosme F, Nunes FM, Reduction of 4-ethylphenol and 4-ethylguaiacol in red wine by activated carbons with different physicochemical characteristics: Impact on wine quality. Food Chemistry, 2017; 229, 242-251.

[73] Filipe-Ribeiro L, Cosme F, Nunes FM, New molecularly imprinted polymers for reducing negative volatile phenols in red wine with low impact on wine colour, Food Research International, 2020.

[74] Filipe-Ribeiro L, Cosme F, Nunes FM. A simple method to improve cork powder waste adsorption properties: Valorisation as a new sustainable wine fining agent. ACS Sustainable Chemistry & Engineering, 2019; 7, 1105-1112.

[75] Craig JT, Heresztyn T, 2-Ethyl-3,4,5,6-tetrahydropyridine An assessment of its possible contribution to the mousy off-flavor of wines. American Journal of Enology and Viticulture, 1984; 35, 46-48.

[76] Buttery RG, Ling LC, Juilano BO, 2-Acetyl-1-pyrroline: an important aroma component of cooked rice. Chemistry & Industry (London) 1982; 958-959.

[77] Snowdon EM, Bowyer MC, Grbin PR, Bowyer PK. Mousy off-flavor: a review. Journal of Agricultural and Food Chemistry, 2006; 54(18), 6465-6474. doi:10.1021/jf0528613

[78] Costello PJ, Henschke PA, Mousy off-flavour of wine: precursors and biosynthesis of the causative Nheterocycles 2-ethyl tetrahydropyridine, 2-acetyltetrahydropyridine, and 2-acetyl-1-pyrroline by Lactobacillus hilgardii DSM 20176. Journal of Agricultural and Food Chemistry, 2002; 50, 7079-7087.

[79] Snowdon EM, Bowyer PK, Grbin PR, Bowyer MC, The removal of mousy off-flavour from wine using molecular imprint technology. In Proceedings of the Twelfth Australian Wine Industry Technical Conference, July 24-29, 2004; Blair, R. J., Williams, P. J., Pretorius, I. S., Eds.; Winetitles: Adelaide, Australia, 2004; pp 301-302

[80] Ugliano M, Siebert T, Mercurio M, Capone D, Henschke PAJ, Effect of nitrogen supplementation and Saccharomyces species on hydrogen sulfide and other volatile sulfur

compounds in shiraz fermentation and wine. Journal of Agricultural and Food Chemistry, 2008; 56, 9175-9182.

[81] Pereira AG, Fraga M, Garcia-Oliveira P, Carpena M, Jimenez-Lopez C, Lourenço-Lopes C, Barros L, Ferreira ICFR, Prieto MA, Simal-Gandara J, Management of Wine Aroma Compounds: Principal Basis and Future Perspectives. Chapter, Book -Winemaking - Stabilization, Aging Chemistry and Biochemistry. Editors Cosme F., Nunes, F.M., Filipe-Ribeiro, L. 2020, IntechOpen, DOI: http://dx.doi.org/10.5772/intechopen.92973

[82] Maujean A, Haye M, Feuillat M, Contribution a l'étude des 'goûts de lumière' dans le vin de Champagne. II. Influence de la lumière sur le potentiel d'oxydoreduction. Correlation avec la teneur en thiols du vin. Connaissance de la Vigne et du Vin 1978; 12, 277-290.

[83] Maujean A, Seguin N, Contribution à l'étude des goûts de lumière dans les vins de Champagne. 4. Approaches a une solution oenologique des moyens de prévention des goûts de lumière. Science des Aliments 1983; 3, 603-613.

[84] Maujean A, Seguin N, Contribution à l'étude des goûts de lumière dans les vins de Champagne. 3. Les réactions photochimiques responsables des goûts de lumière dans le vin de Champagne. Sciences des Aliments, 1983; 3, 589-601 ISSN: 0240-8813

[85] Dias DA, Smith TA, Ghiggino KP, Scollary GR, The role of light, temperature, and wine bottle colour on pigment enhancement in white wine. Food Chemistry, 2012; 135, 2934-2941.

[86] Pichler, U, Analisi della riboflavina nei vini bianchi e influenza della sua concentrazione. L'Enotecnico 1996; 32, 57-62.

[87] Fracassetti D, Limbo S, Pellegrino, Tirelli A, Light-induced reactions of methionine and riboflavin in model wine: Effects of hydrolysable tannins and sulfur dioxide. Food Chemistry, 2019; 298,124952.

[88] Bayonove C, Cordonnier R, Dubois P, Study of an aromatic characteristic fraction of Cabernet Sauvignon grape variety, identification of 2-methoxy-3-isobutyl-pyrazine. Comptes Rendus Hebdomadaires Des Seances De L'Academie Des Sciences Serie D 1975; 281, 75.

[89] Roujou de Boubee D, Van Leeuwen C, Dubourdieu D, Organoleptic impact of 2-methoxy-3-isobutylpyrazine on red Bordeaux and Loire wines effect of environmental conditions on concentrations in grapes during ripening. Journal of Agricultural and Food Chemistry, 2000; 48, 4830.

[90] Allen MS, Lacey MJ, Harris RLN, Brown WV, Contribution of methoxypyrazines to Sauvignon Blanc wine aroma. American Journal of Enology and Viticulture, 1991; 42, 109.

[91] Allen MS, Lacey MJ, Boyd S, Methoxypyrazines in red wines: occurrence of 2-methoxy-3-(1-methylethyl) pyrazine. Journal of Agricultural and Food Chemistry, 1995; 43, 769.

[92] Belancic A, Agosin E, Methoxypyrazines in grapes and wines of *Vitis vinifera* cv. Carmenere. American Journal of Enology and Viticulture, 2007; 58, 462.

[93] Sala C, Busto O, Guasch J, Zamora F, Influence of vine training and sunlight exposure on the 3-alkyl-2- methoxypyrazines content in musts and wines from the *Vitis vinifera* variety Cabernet Sauvignon. Journal of Agricultural and Food Chemistry, 2004; 52(11), 3492

[94] Harsch MJ, Benkwitz F, Frost A, Colonna-Ceccaldi B,

Garner RC, Salmon JM, New precursor of 3-mercaptohexan-1-ol in grape juice: thiol-forming potential and kinetics during early stages of must fermentation. Journal of Agricultural and Food Chemistry, 2013; 61, 3703.

[95] Ferrandino A, Schubert A, Carlomagno A, Varietal and prefermentative volatiles during ripening of *Vitis vinifera* cv Nebbiolo berries from three growing areas. Food Chemistry, 2012; 135 (4), 2340.

[96] Soares S, Brandão E, Mateus N, de Freitas V, Sensorial properties of red wine polyphenols: astringency and bitterness. Critical Reviews in Food Science and Nutrition, 2017; 57, 937-948.

[97] Gigot C, Ongena M, Fauconnier M-L, Wathelet J-P, Du Jardin P, Thonart P, The lipoxygenase metabolic pathway in plants: potential for industrial production of natural green leaf volatiles. Biotechnologie, Agronomie, Société et Environnement, 2010; 14 (3), 451.

[98] Kalua CM, Boss PK, Sample preparation optimization in wine and grapes: dilution and sample/headspace volume equilibrium theory for headspace solid-phase microextraction. Journal of Chromatography A, 2008; 1192 (1), 25.

[99] Drawert F, Heimann W, Emberger R, Tressl R, EnzymatischeBildung von Hexen-(2)-al-(1), Hexanal und deren Vorstufen. Liebigs Annalen der Chemie, 1966; 694, 200.

[100] Moreno J, Peinado R, "Enological Chemistry" 2012; First ed. Academic Press, USA.

[101] Ferreira V, López R, Cacho JF, Quantitative determination of the odorants of young red wines from different grape varieties. Journal of the Science of Food and Agriculture, 2000; 80 (11), 1659.

[102] Simpson RF, Miller GC, Aroma composition of aged Riesling wines. Vitis, 1983; 22, 51-63.

[103] Sponholz WR, Hühn T, Factors influencing the aging of the Riesling wine. Clonal material and used yeast strain, Wein-Wissenschaft, 1997; 52 (2), 103-108

[104] Simpson RF, 1,1,6-Trimethyl-1,2-dihydronaphthalene: an important contributor to the bottle aged bouquet of wine. Chemistry and Industry, 1978; 1, 37.

[105] Sacks GL, Gates MJ, Ferry FX, Lavin EH, Kurtz AJ, Acree T, Sensory threshold of 1,1,6-trimethyl-1,2-dihydronaphthalene (TDN) and concentrations in young Riesling and non-Riesling wines. Journal of Agricultural and Food Chemistry, 2012; 60, 2998-3004.

[106] Gerdes SM, Winterhalter P, Elbeler SE, Effect of sunlight exposure on norisoprenoid formation in white Riesling grapes, Carotenoid Derived-Aroma Compounds, Chapter 19, 2002; 262-272, doi:http://dx.doi.org/10.1021/bk-2002-0802.c.

[107] Herve E, Price S, Burns G, Eucalyptol in wines showing a Eucalyptus" aroma. Poster Paper. Actualities Oenologiques 2003" VIIme Symposium International d'Oenologie, 19-21 June 2003; Bordeaux, France.

[108] Farina L, Boido,E, Carrau F, Versini G, Dellacassa E, Terpene compounds as possible precursors of 1,8-cineole in red grapes and wines. Journal of Agricultural and Food Chemistry, 2005; 53, 1633-1636.

[109] Saliba AJ, Bullock J, Hardie WJ, Consumer rejection threshold for 1,8-cineole (eucalyptol) in Australian red wine. Food Quality and Preference, 2009; 20, 500-504.

[110] Poitou X, Thibon C, Darriet P, 1,8-Cineole in French red wines: evidence for a contribution related to its various origins. Journal of Agricultural and Food Chemistry, 2017; 65 (2), 383-393.

[111] du Toit M, Pretorius IS, Microbial spoilage and preservation of wine: using weapon from nature's own arsenal – a review, South African Journal for Enology and Viticulture, 2000; 21(special issue), 74-96.

[112] Kennison KR, Wilkinson KL, Williams HG, Smith JH, Gibberd MR, Smoke derived taint in wine: Effect of postharvest smoke exposure of grapes on the chemical composition and sensory characteristics of wine. Journal of Agricultural and Food Chemistry, 2007; 55, 10897-10901.

[113] Kennison KR, Gibberd MR, Pollnitz AP, Wilkinson KL, Smoke-derived taint in wine: the release of smoke-derived volatile phenols during fermentation of Merlot juice following grapevine exposure to smoke. Journal of Agricultural and Food Chemistry, 2008; 56, 7379-7383.

[114] Boidron JN, Chatonnet P, Pons M, Influence du bois sur certaines substances odorantes des vins. Connaissance de la Vigne et du Vin, 1988; 22, 275-294.

[115] Simpson RF, Amon JM, Daw AJ, Off-flavour in wine caused by guaiacol. Food Australia Journal. 1986; 38(1), 31-33.

[116] Sefton MA, Simpson RF, Compounds causing cork taint and the factors affecting their transfer from natural cork closures to wine – a review, Australian Journal of Grape and Wine Research, 2005; 11, 226-240.

[117] Mazzoleni V, Maggi L, Effect of wine style on the perception of 2,4,6-trichloroanisole, a compound related to cork taint in wine. Food Research International, 2007; 40(6), 694-699.

[118] Chatonnet P, Bonnet S, Boutou S, Labadie M-D, Identification and responsibility of 2,4,6-tribromoanisole in musty, corked odors in wine. Journal of Agricultural and Food Chemistry, 2004; 52, 1255-1262.

[119] Simpson RF, Sefton MA, Origin and fate of 2,4,6-trichloroanisole in cork bark and wine corks, Australian Journal of Grape and Wine Research, 2007; 13, 106-116.

[120] Chatonnet P, Guimberteau G, Dubourdieu D, Boidron JN, Nature et origin des odeurs de « moisi » dans la caves. Incidences sur la contamination des vin, Journal International des Sciences de la Vigne et du Vin, 1994; 28, 131-151.

[121] Jung R, Schaefer V, Christmann M, Hey M, Fischer S, Rauhut D, Removal of 2,4,6-trichloroanisole (TCA) and 2,4,6 – tribromoanisole (TBA) from wine. Mitteilung Klosterneuburg, 2008; 58, 58-67.

[122] Darriet P, Pons M, Lamy S, Dubourdieu D, Identification and quantification of geosmin, a powerful earthy odorant contaminating wines. Journal of Agricultural and Food Chemistry, 2000; 48, 4835-4838.

[123] La Guerche S, Dauphin B, Pons M, Blancard D and Darriet P, Characterisation of some mushroom and earthy off-odors microbially induced by the development of rot on grapes. Journal of Agricultural and Food Chemistry, 2006; 54, 9193-9200.

[124] Darriet P, Lamy S, La Guerche S, Pons M, Dubourdieu D, Blancard D, Steliopoulos P,·Mosandl A, Stereodifferentiation of geosmin in wine. European Food Research and Technology, 2001; 213,122-125.

[125] Lisanti MT, Gambuti A, Genovese A Piombino P, Moio L, Earthy off-flavour in wine: Evaluation of remedial treatments for geosmin contamination. Food Chemistry, 2014; 154,171-178.

[126] Farines V, Mary T, Kelly A, Determination of styrene in wine by headspace solid phase microextraction in wine ad gas chromatography mass. Conference paper, XXXIème Congrès Mondial de la Vigne et du Vin. Verona, 2008, Italy.

[127] Godden P, Francis L, Field J, Gishen M, Coulter A, Valente P,Høj P, Robinson E, Wine bottle closures: physical characteristics and effect on composition and sensory properties of a Semillon wine 1. Performance up to 20 months post-bottling. Australian Journal of Grape and Wine Research, 2001; 7, 64-105.

[128] Linson JPH, Legger-Huysman A, Roozen JP, Sensory analysis of polystyrene packaging material taint in cocoa powder for drinks and chocolate flakes. Food Additives and Contaminants, 1991; 8, 1-7.

[129] Wagner S, Jakob L, Rapp A, Niebergall H, Investigations about the native content of polyols in wine and the possible influence of the technology of wine-making. Wein-Wissenschaft, 1994; 49, 247-253.

[130] Ebeler SE, Analytical chemistry: Unlocking the secrets of wine flavour. Food Reviews International, 2001; 17, 45-64.

[131] Mato I, Suárez-Luque S, Huidobro JF, A review of the analytical methods to determine organic acids in grape juices and wines. Food Research International, 2005; 38, 1175-1188.

[132] Chidi BS, Bauer FF, Rossouw D, Organic Acid Metabolism and the Impact of Fermentation Practices on Wine Acidity: A Review. South African Journal for Enology and Viticulture, 2018; 39, 2,

[133] Volschenk H, Van Vuuren HJJ, Viljoen-Bloom M, Malic acid in wine: Origin, function and metabolism during vinification. South African Journal for Enology and Viticulture, 2006; 27, 123-136.

[134] Gawel R, Iland PG, Francis IL, Characterizing the astringency of red wine: A case study. Food Quality and Preference, 2001; 12, 83-94.

[135] Vivas N, Glories Y, Lagune L, Saucier C, Augustin MJ, Estimation du degré de polymérisation des procyanidines du raisin et du vin par la méthode au ρ-dimethylamino-cinnamaldéhyde. Journal International des Sciences de la Vigne et du Vin, 1994; 28, 319.

[136] Rinaldi A, Moio L, Salivary protein-tannin interaction: the binding behind astringency. Chapter, Book In: Winemaking - Stabilization, Aging Chemistry and Biochemistry. Editors Cosme F., Nunes, F.M., Filipe-Ribeiro, L. 2020, IntechOpen,

[137] Green, BG, Oral astringency: A tactile component of flavor. Acta Psychologica, 1993; 84, 119.

[138] Vidal S, Francis L, Guyot S, Marnet N, Kwiatkowski M, Gawel R, Cheynier V, Waters EJ, The mouth-feel properties of grape and apple proanthocyanidins in a wine like medium. Journal of the Science of Food and Agriculture, 2003; 83, 564-573.

[139] Hornedo-Ortega, R, González-Centeno MR, Chira K, Jourdes M, Teissedre P-L, Phenolic Compounds of Grapes and Wines: Key Compounds and Implications in Sensory Perception Chapter, Book In: Winemaking - Stabilization, Aging Chemistry and Biochemistry. Editors Cosme F., Nunes, F.M., Filipe-Ribeiro,

2020; L. IntechOpen, DOI: 10.5772/intechopen.93127.

[140] Noble AC, Astringency and bitterness of flavonoid phenols. Chemistry of Taste: Mechanisms, Behaviors and Mimics, 2002; 825, 192-201.

[141] Maury C, Sarni-Manchado P, Lefebvre S, Cheynier V, Moutounet M, Influence of fining with different molecular weight gelatins on proanthocyanidin composition and perception of wines. American Journal of Enology and Viticulture, 2001; 52, 140-145.

[142] Cosme F, Ricardo-da-Silva JM, Laureano O, Behavior of Various Proteins on Wine Fining: Effect on Different Molecular Weight Proanthocyanidin Fractions of Red Wine. American Journal of Enology and Viticulture, 2009; 112, 197-204.

[143] Cosme F, Ricardo-da-Silva JM, Laureano O, Interactions between protein fining agents and proanthocyanidins in white wine. Food Chemistry, 2008; 106,536-544

[144] Cosme F, Ricardo-da-Silva JM, Laureano O, Protein fining agents: characterization and red wine fining assay. Italian Journal of Food Science, 2007; 19, 39-56.

[145] Smith PA, McRae JM, Bindon KA, Impact of winemaking practices on the concentration and composition of tannins in red wine. Australian Journal of Grape and Wine Research, 2015; 21, 601-614.

[146] Landon J, Weller K, Harbertson J, Ross CF, Chemical and Sensory Evaluation of Astringency in Washington State Red Wines American Journal of Enology and Viticulture, 2008; 59(2), 153-158

[147] Habertson JF, Picciotto EA Measurement of Polymeric Pigments in Grape Berry Extracts and Wines Using a Protein Precipitation Assay Combined with Bisulfite Bleaching. American Journal of Enology and Viticulture, 2003; 54,4, 301-306.

[148] Soares S, Mateus N, de Freitas V, Interaction of different polyphenols with Bovine Serum Albumin (BSA) and Human Salivary α-Amylase (HSA) by fluorescence quenching. Journal of Agricultural and Food Chemistry, 2007; 55, 6726-6735.

[149] Chira K, Pacella N, Jourdes M, Teissedre PL, Chemical and sensory evaluation of Bordeaux wines (Cabernet-Sauvignon and Merlot) and correlation with wine age. Food Chemistry, 2011; 126, 1971-1977.

[150] Bindon KA, Smith PA, Comparison of the affinity and selectivity of insoluble fibres and commercial proteins for wine proanthocyanidins. Food Chemistry, 2013; 136, 917-928.

[151] Thibon C, Marullo P, Claisse O, Cullin C, Dubourdieu D, Tominaga T. Nitrogen catabolic repression controls the release of volatile thiols by *Saccharomyces cerevisiae* during wine fermentation. FEMS Yeast Research 2008; 8(7),1076-86.

[152] Dennis MJ, Howarth N, Key PE, Pointer M, Massey RC. Investigation of ethyl carbamate levels in some fermented foods and alcoholic beverages. Food Additives & Contaminants, 1989;6,383-9.

[153] IARC (International Agency for Research on Cancer) International agency for research. 96, alcoholic beverage consumption and ethyl carbamate (urethane) 6-13 February 2007 1-5. World Health Organization; Lyon, France: (2010).

[154] Ryu D, Choi B, Kim E, Park S, Paeng H, Kim C, Lee J, Yoon H, Koh E,

Determination of Ethyl Carbamate in Alcoholic Beverages and Fermented Foods Sold in Korea. Toxicological Research, 2015; 31(3), 289-297.

[155] Jiao ZH, Dong YC, Chen QH, Ethyl carbamate in fermented beverages: Presence, analytical chemistry, formation mechanism, and mitigation proposals. Comprehensive Reviews in Food Science and Food Safety, 2014; 13 (4), 611-626.

[156] Uthurry C, Lepe J, Lombardero J, Garcia Del Hierro J, Ethyl carbamate production by selected yeasts and lactic acid bacteria in red wine. Food Chemistry, 2006; 94(2), 262-70.

[157] Fidaleo M, Esti M, Moresi M, Assessment of urea degradation rate in model wine solutions by acid urease from Lactobacillus fermentum. Journal of Agricultural and Food Chemistry, 2006; 54 (17), 6226-6235.

[158] Mohapatra BR, Bapuji M, Characterization of urethanase from Micrococcus species associated with the marine sponge (Spirasfrella species). Letters in Applied Microbiology, 1997; 25 (6), 393-396.

[159] US Food and Drug Administration. 1997. Ethyl carbamate preventative action manual. Center for food safety and applied nutrition. https://www.fda.gov/food/chemicals/ethyl-carbamate-preventative-action-manual

[160] Cooper TG, Kovari L, Sumrada RA, Park H-D, Luche R, Kovari I, Nitrogen catabolite repression of arginase (CAR1) expression in *Saccharomyces cerevisiae* is derived from regulated inducer exclusion. Journal of Bacteriology, 1992; 174(1),48-55.

[161] An D, Ough C, Urea Excretion And Uptake By Wine Yeasts As Affected By Various Factors. American Journal of Enology and Viticulture, 1993; 44 1:35-40.

[162] Silla Santos M H, 'Biogenic amines: their importance in food', International Journal of Food Microbiology, 1996; 29, 213-231.

[163] Arena M E, Manca de Nadra M C, 'Biogenic amine production by Lactobacillus', Journal of Applied Microbiology, 2001; 90, 158-162.

[164] Ortega-Heras M, Perez-Magarino S, del-Villar-Garrachón V, Gonzalez-Huerta C, Moro González LC, Guadarrama Rodríguez A, Villanueva Sánchez S, Gallo González R, Martín de la Helguera S, Study of the effect of vintage, maturity degree, and irrigation on the amino acid and biogenic amine content of a white wine from the Verdejo variety. Journal of the Science of Food and Agriculture, 2014;94, 2073-82.

[165] González Marco A, Jiménez Moreno N, Ancín Azpilicueta C, Influence of addition of yeast autolysate on the formation of amines in wine. Journal of the Science of Food and Agriculture, 2006; 86, 2221-7.

[166] García-Ruíz A, González-Rompinelli EM, Bartolomé B, Moreno-Arribas MV, Potential of wine associated lactic acid bacteria to degrade biogenic amines. International Journal of Food Microbiology, 2011; 148, 115-20.

[167] Grossmann M, Smit I, Loehnertz O, Ansorge A, Biogenic amines and grapes: Effect of microbes and fining agents. In: Proceeding of international symposium of microbiology and food safety of wine. 2007 Vilafranca, Spain: 20-21, November 2007.

[168] Quintela S, Villarán MC, de Armentia IL, Pisters R, Lane DA, Elejalde E, Ochratoxin A removal from red wine by several oenological fining agents: Bentonite, egg albumin, allergen-free adsorbents, chitin and chitosan. Food Additives

& Contaminants, 2012; 468 29, 1168-1174.

[169] IARC, Monographs on the Evaluation of Carcinogenic Risks to Humans. 1993, 56. IARC. Geneva. Switzerland.

[170] E.C. European Commission. Commission Regulation (EC) No. 1881/2006 of 19 December 2006 setting maximum levels for certain contaminants in foodstuffs (consolidated version from 12/12/2014). Off J Eur Union. p. L364/5-L364/24. (2014).

[171] Zimmerli B, Dick R, Ochratoxin A in table wine and grape-juice: occurrence and risk assessment. Food Additives & Contaminants, 1996; 13, 655-668.

[172] Blesa J, Soriano J M, Molto J C, Manes J, Factors affecting the presence of Ochratoxin A in wines. Critical Reviews in Food Science and Nutrition, 2006; 46, 473-478.

[173] Cecchini F, Morassut M, García-Moruno,E, Di Stefano R, Influence of yeast strain on ochratoxin A content during fermentation of white and red must. Food Microbiolology, 2006; 23, 411-417.

[174] Olivares-Marín M, Del Prete V, Garcia-Moruno E, Fernández-GonzálezC,Macías-GarcíaA, Gómez-Serrano V, The Development of an Activated Carbon from Cherry Stones and its use in the Removal of Ochratoxin A from Red Wine. Food Control, 2009; 20, 298-303

[175] Williams JH, Phillips TD, Jolly PE, Stiles JK, Jolly CM, Aggarwal D, Human aflatoxicosis in developing countries: a review of toxicology, exposure, potential health consequences and interventions. The American Journal of Clinical Nutrition, 2004, 80 (5), 1106-1122.

[176] Fredj SMB, Chebil S, Lebrihi A, Lasram S, Ghorbel A, Mliki A, Occurrence of pathogenic fungal species in Tunisian vineyards. International Journal of Food Microbiology, 2007; 113, 245-250.

[177] Chunmei J, Junling S, Qìan H, Yanlin L, Occurrence of toxin producing fungi in intact and rotten table and wine grapes and related influencing factors. Food Control, 2013; 31,5-13.

[178] Di Stefano V, Pitonzo R, Avellone G, Di Fiore A, Monte L, Ogorka AZT, Determination of aflatoxins and ochratoxins in Sicilian sweet wines by High-Performance Liquid Chromatography with fluorometric detection and immunoaffinity cleanup. Food Analytical Methods, 2015; 8, 569-577.

[179] Cosme F, Inês A, Ferreira B, Silva D, Filipe-Ribeiro L, Abrunhosa L, Nunes FM, Elimination of aflatoxins B1 and B2 in white and red wines by bentonite fining. Efficiency and impact on wine quality. Foods, 2020, 9, 1789.

Chapter 9

Pathogenesis-Related Proteins in Wine and White Wine Protein Stabilization

Bin Tian and Roland Harrison

Abstract

Protein stabilisation in white wine is of great concern to winemakers as denaturation of wine proteins may cause haze formation, which is usually considered a wine fault. Pathogenesis-related (PR) proteins derived from grapes are the major soluble proteins remaining in the finished wine, which are mainly responsible for haze formation. The development of PR proteins in grapes during ripening and the extraction of PR proteins from grapes into juice can largely affect the concentration of PR proteins in the final wine, which consequently influence wine protein stability. Bentonite fining is the most common method to remove proteins in white wine before bottling, but it can cause the loss of wine volume and the removal of beneficial aromas. Thus, a number of alternative methods have been proposed for their potential to replace bentonite fining.

Keywords: bentonite fining, haze formation, pathogenesis-related proteins, phenolics, stabilisation, wine

1. Introduction

Wines, like many other natural food products, contain varying amounts of different nitrogenous substances, the most important of which are proteins [1]. Proteins found in wine are mostly derived from grape berries. Using immunological methods, the specific polyclonal antibodies raised against the total proteins of a Portuguese Malvasia Fina monovarietal wine have been applied to analyse the origin of the wine proteins [2] which are entirely from the berry pulp. Three years later, a similar immune detection study was conducted [3] and three various polyclonal antibodies raised against must, yeast, and bacteria proteins were applied to analyse the origin of proteins in Chardonnay wine. The results indicated that most of the wine proteins came from grapes and many of them were glycoproteins, but there were also some proteins from the yeast. Yeast may affect the wine protein composition in two ways: by transferring proteins to the wine during the process of yeast autolysis and/or hydrolysing the must proteins via the exocellular protease present in the yeasts [1]. Furthermore, the analysis of a Sauvignon Blanc wine using nano-high performance liquid chromatography (HPLC)/tandem mass spectrometry showed that within the 20 identified proteins there were two proteins from bacteria and one from fungi, which could be attributed to sources in the vineyard including natural infections and improper handling during harvest [4].

IntechOpen

Grape variety	Sample type	MW	pI	Reference
Sauvignon Blanc	Wine	14.6–77.1 kDa		[4]
Riesling and Gewürztraminer	Grape	11.2–190 kDa		[5]
	Wine	11.2–65 kDa	4.1–8.0	[5]
Chardonnay, Verdeca and Pinot Noir	Wine	6–200 kDa	3.6–9.0	[6]
Macabeo, Xarel-lo, Parellada and Malvar	Wine	14–94 kDa	3.0–5.6	[7]
Muscadine	Grape	19–100 kDa	5.6–7.6	[8]
	Wine	12–50 kDa	4.6–8.8	[9]

Table 1.
Proteins reported in grapes and wines from different grape varieties: molecular weight (MW) and isoelectric point (pI).

In previous studies, proteins from grapes and wines have been reported with molecular weight (MW) in the range 6–200 kDa and isoelectric points (pI) in a range 3–9 kDa, as shown in **Table 1** [4–9]. However, the majority of wine proteins have MW and pI in a low range (20–30 and 4.1–5.8 kDa, respectively) and possess a net positive charge at the pH of the wine [2, 5, 10]. Studies on fractions of wine proteins using denaturing polyacrylamide gel electrophoresis have shown that the wine protein fraction is mainly composed of only a few polypeptides with MW ranging from 15 to 30 kDa, but a more detailed examination of whole protein fraction indicates a very large number of distinct polypeptides, exhibiting similar MW but subtle differences in electrical charges [11]. In that study, the authors also revealed via highly specific antibodies and N-terminal sequencing analysis that most wine polypeptides were structurally similar, suggesting the existence of a common precursor to most or all of the wine proteins which could generate all of the detected polypeptides by limited proteolysis.

In a study on Muscat of Alexandria wine [12], two major proteins with MW of 24 and 32 kDa, respectively, by SDS-PAGE are found with significant contribution to protein haze, and the 24 kDa protein produced about 50% more haze than the 32 kDa protein. The N-terminal sequence of the protein with MW of 24 kDa showed homology to thaumatin and to a number of plant thaumatin-like proteins, and the N-terminal sequence of enzyme digested peptides of the protein with MW of 32 kDa showed homology to plant chitinases [13]. Another study analysed the two main wine proteins present in sodium dodecyl sulphate capillary gel electrophoresis (SDS-CGE), which were determined with MW at 22 and 26 kDa, respectively, being concluded as corresponding to thaumatin-like proteins (TLPs) and chitinases [14]. Both thaumatin-like proteins and chitinases in wine are pathogenesis-related (PR) proteins derived from grapes.

2. Pathogenesis-related proteins in grapes and juice

Pathogenesis-related proteins are a group of plant proteins induced in pathological or related situations [15]. They were first discovered in tobacco as a result of a hypersensitive reaction to tobacco mosaic virus (TMV) [16]. PR proteins are typically acidic, of low molecular weight and highly resistant to proteolytic degradation and to low pH values. On the basis of similarities in amino acid sequences, serological relationship and/or enzymatic or biological activity, 11 families have been recognised and classified for tobacco and tomato [17]. Some of these PR protein family members have also been found in grapevine. The two prominent soluble proteins accumulated in grapes during ripening have been identified as chitinases (PR-3 family) and thaumatin-like proteins (PR-5 family) [18, 19]. However, in early studies, the β-1, 3-glucanases (PR-2 family), a potential indicator of pathogen attack, were not found in grape juice and/or berry extracts [19–22]. With the accomplishment of grapevine genome sequencing programs in 2007 [23, 24] and the development of technology in protein analysis, proteomic analysis of grapevine has significantly improved knowledge of grape proteins and produced a better understanding of their characteristics [25]. These have consequently shown that there are more PR protein family members found in grapevines, such as osmotins (PR-5 family), β-1, 3-glucanases (PR-2 family) and the PR-10 proteins [26–28]. The two major PR proteins in wine, thaumatin-like proteins and chitinases, have been found present in both grape skin and pulp but not in grape seed [29].

Protein content in grape berries generally increases during ripening [30–33]. The accumulation of PR proteins in grape berries during ripening has been observed [33, 34], with *véraison* being the trigger for gene expression. The expression of PR genes in grapes can also be modulated by the classical PR protein inducers such as wounding, chemical elicitors, pathogen attack and abiotic stress [18, 35]. Although the level of PR proteins in grape berries increases, the diversity of PR proteins decreases during grape ripening [36]. In addition, the level and proportion of PR proteins in grapes are dependent on the cultivar, region, climate and viticultural practices [36–41]. Therefore, the actual protein composition in ripe grape berries is a result of the interactions between environmental conditions and intrinsic factors.

Sunlight-exposed fruits presented generally higher total soluble solids, anthocyanins and phenolic compounds and lower titratable acidity, malate and berry weight than non-exposed or canopy-shaded fruits [42–46]. One study on Riesling must show that the total amino acid concentration was significantly lower for fruits exposed to ambient UV-B levels than the low UV-B treatment and reduced UV-B affected amino acid composition, causing higher levels of arginine and glutamine, the main sources of amino acid for yeast metabolism [47]. In a later study [48], UV exclusion resulted in a lower concentration of not only phenolic compounds such as tannins but also PR proteins in grape skin. Interestingly, UV exclusion showed no effect on the PR proteins in the grape pulp.

Fungal infection can significantly influence the concentration of PR proteins in grapes. Grey mould caused by *Botrytis cinerea* is one of the main fungal diseases found in grapevines. A study that compared the juice from healthy grapes against *Botrytis* [49] showed that most proteins normally present in the healthy juice, namely, those between 20 and 30 kDa and a major glycoprotein at 62/64 kDa disappeared in the *Botrytis* infected juice. These results suggested that some proteinases secreted by *Botrytis cinerea* could degrade grape proteins. Another study on *Botrytis cinerea* infection on Chardonnay and Semillon grapes has also revealed that the concentrations of both PR proteins and total proteins in botrytised grape juice decreased compared to the juice from healthy grapes [50]. Conversely, powdery mildew infection on grape berries has been documented as increasing levels of PR proteins [48, 50, 51]. The strongly induced expression of some PR genes such as Vv*Chi*3 (coding for an acidic class III chitinase), Vv*Glu*b (coding for a basic class I glucanase) and Vv*Tl*2 (coding for a thaumatin-like protein) has been reported in powdery mildew infected grape berries [35]. A recent study [38] also showed that a number of proteins were induced in leaf tissues of Cabernet Sauvignon in response to powdery mildew infection, suggesting that Cabernet Sauvignon is able to initiate a basal defence but is unable to restrict fungal growth or slow down disease progression.

Extraction of PR proteins from grapes into juice can be greatly influenced by harvesting and grape processing conditions. Studies carried out in Australia [52, 53] showed that the juice obtained from mechanical harvesting coupled with long-distance transport had a higher concentration of PR proteins than juice obtained from hand harvesting fruit, which is likely due to the long skin contact during transport. A more recent study [54] conducted in New Zealand showed that Sauvignon Blanc juices from machine harvesting followed by 3 h skin contact had a significantly lower concentration of proteins, including PR proteins, than those from hand harvesting followed by 3 h skin contact. It was likely due to the greater juice yield in machine harvesting treatment and the interactions between proteins and phenolic compounds. In the following study [55], the authors confirmed that longer skin contact can increase the extraction of PR protein but the final concentration of PR proteins in juice can be modulated by the co-extracted phenolic compounds.

3. Protein haze formation in white wine

Protein haze can appear in bottled white wine as shown in **Figure 1** if unstable proteins are not removed before the wine bottling. Although research studies have investigated the protein stabilisation in white wines [12, 13, 56–59], the precise mechanism of protein haze formation still remains incompletely understood. One hypothesis [60] is that the first step in protein haze formation in wines is protein denaturation, a process accelerated by heating, after which the denatured proteins aggregate into large enough particles to be visually detected as haze, a process that may be affected either positively or negatively by non-protein wine components. In a recent study [61], two different mechanisms were proposed responsible for the heat-induced precipitation of the Arinto wine proteins: (1) at the higher pH values, it appears to result from isoelectric precipitation of proteins; (2) at the lower pH values, it seems to be associated with the presence of the non-protein wine components.

As the main soluble proteins remain in the finished wine, the slow denaturation of PR proteins is thought to lead to protein aggregation, flocculation into a hazy suspension and formation of precipitates. A study using purified thaumatin-like proteins and chitinases from grape juice [62] suggested that chitinases are the primary cause of heat-induced haze formation and their concentration was directly correlated to the turbidity of heat-induced haze formation, but conversely, thaumatin-like proteins seemed to have no measurable impact on turbidity. This result was confirmed by a latter study [54] in which chitinases were found to have a good linear correlation with protein stabilisation in Sauvignon Blanc wine. Different protein haze formation behaviour between thaumatin-like proteins and chitinases could be due to the difference in the protein structure of these two types of proteins. The thaumatin-like proteins start unfolding (or denaturing) at 62°C, but most of the proteins will refold again when the temperature drops down [63]. In contrast, chitinases have a lower unfolding temperature and this denaturation is irreversible. Thus, once chitinases are unfolded, they may aggregate and precipitate out of the solution.

A study used a reconstitution method [64] to investigate the heat-induced aggregation behaviour of purified wine proteins and showed that the chitinases were the protein most prone to aggregate and the one that formed the largest particles. It is important to note that in the reconstitution experiment, four thaumatin-like protein isoforms, chitinases, phenolics and polysaccharides in a Chardonnay wine were isolated individually, and the wine stripped of these compounds was used as a base to reconstitute each of the proteins alone or in combination with the isolated phenolics and/or polysaccharides. Although phenolics and polysaccharides did not show a significant impact on aggregation behaviour of chitinases,

Figure 1.
Protein haze formed in Sauvignon Blanc wine after heating at 80°C for 2 h.

the thaumatin-like protein isoforms varied in susceptibility to haze formation and interactions with phenolics and polysaccharides. These observations obtained in the model system indicated the importance of non-protein factors in affecting protein haze formation.

3.1 Phenolics

Phenolic compounds are important constituents in white wines since they contribute to many characteristics such as appearance, taste, style and quality [65]. A reduction in protein haze observed in commercial wine fined with the addition of polyvinylpolypyrrolidone (PVPP) suggested that phenolics may play a modulating role in haze formation [60]. The interaction between phenolics and proteins has been studied in relation to haze formation [59, 66]. Hydrophobic bonding was suggested as the major model of interaction in tannin-protein complexes [67]. A conceptual model for the protein-phenolics interaction is that a protein molecule has a fixed number of polyphenol binding sites and more sites were exposed when protein hydrogen bonds were broken [68]. Thus, the concentration of various phenolics in wine could have a great impact on heat stability through the interaction between phenolics and proteins.

3.2 Polysaccharides

Most polysaccharides present in musts are derived from grape cell walls, which include arabinogalactans, galacturonans, arabinans and smaller amounts of xyloglucans, cellulose and mannans. In the musts, type II arabinogalactan proteins are the main polysaccharides released from berries at the initial time of pressing; in the resulting wine, polysaccharides consist essentially of type II arabinogalactan proteins and rhamnogalacturonan-II [69]. Wine polysaccharides can affect the characteristic pattern of haze formation, increasing protein instability under moderately high temperature (40–50°C) [58]. However, mannoproteins, the polysaccharides derived from yeast, have been described as protecting wines from protein haze formation [70]. This polysaccharide is considered a promising prospect for preventing protein haze formation in white wine.

3.3 Wine pH

In a model wine system, maximum haze formed at pH 4.0–4.5 when ethanol was 12%, with less haze at lower or higher pH values [71]. However, a study using six Portuguese varietal wines indicated that wine proteins were increasingly heat stable when the pH increased from wine pH to 7.5 [58]. When the wine pH was adjusted to a typical wine pH value at 2.80, 3.00, 3.34, 3.65 and 3.85, lower bentonite dosages for stability was observed in lower pH wines, which was likely due to the improved efficiency of protein adsorption by bentonite at reduced wine pH [72]. This observation is in agreement with another study which investigated the protein haze formation in an Italian white wine as affected by pH ranging from 3.00 to 3.60, and the increased heat stability of wine was found in wines at lower pH [73].

3.4 Sulphate

It has been found that the sulphate anion in white wine was an essential factor that is required for protein haze formation [60]. In that study, the authors investigated various common wine anions such as sulphate, acetate, chloride, citrate, phosphate and tartrate and wine cations such as iron and copper. When these ions were added into artificial model wine solutions at typical white wine concentrations, only sulphate was found to be essential for protein haze formation. Furthermore, in this model wine system, the thaumatin-like protein (150 mg/L) required approximately 150 mg/L sulphate, and the chitinase (150 mg/L) required approximately 15 mg/L sulphate, for visible haze formation. The range of sulphate in Australian wines between 1994 and 1997 was from 56 to 1780 mg/L, with a mean of 385 mg/L, which exceeds the requirement of both thaumatin-like protein and chitinase for haze formation. A recent study [74] confirmed that sulphate was essential in the aggregation of grape chitinases and thaumatin-like proteins in a model system, and furthermore, the authors pointed out that the aggregation mechanisms of thaumatin-like proteins and chitinases are different and influenced by the ionic content of the model wine.

3.5 Metal Ions

There are many ions present in wine, and these ions could play a role in white wine protein haze formation. Metal ions, particularly copper and iron, have been implicated in the formation of protein hazes in white wines, but as they are also associated with hazes of non-protein origin, their role in protein haze formation is very poorly understood [75]. The copper concentration in wine decreased after protein haze removal, suggesting that copper was part of the protein precipitation [76].

3.6 Ion concentration and electrical conductivity

In a model wine system, increasing the ionic strength and electrical conductivity could increase protein haze formation after heating by reducing electrostatic repulsion of proteins [74, 77]. A study on Chilean Sauvignon Blanc wine reported that more protein haze formation in wine was observed by increasing the electrical conductivity [78]. However, a more recent study on a range of Australian white wines showed a negative correlation between protein haze formation and electrical conductivity [79]. These contradictory results could be related to the differences in other wine components of importance to protein haze formation.

3.7 Sulphur dioxide

A recent study [80] revealed the role of sulphur dioxide in the aggregation of heat unstable wine proteins. In comparison to chitinases, TLPs are more reactive to sulphur dioxide. The aggregation of TLPs could be triggered by sulphur dioxide during cooling after heating, with aggregates held by hydrophobic interactions and intermolecular disulphide bonds.

4. Protein stability tests

To avoid protein haze formation in bottled white wine, a protein stability test is usually conducted before bottling in the winery. If the wine is not protein stabilised, a range of bentonite fining trials will be carried out to determine the minimum required dosage of bentonite addition for protein stabilisation. The most common protein stability test is the heat test, which is a heating procedure to force protein haze formation. Wine samples are normally heated to 80°C for 6 h and then left to cool down to 4°C overnight. The turbidity in heated wine samples is measured by a nephelometer and expressed as nephelometer turbidity units (NTU). Turbidity measurement of less than 2.0 NTU is usually recommended. Different temperatures and durations of heating could have a great impact on the resulting haze formation [81]. A recent study suggests that the less severe condition of heating at 80°C for 2 h is more appropriate to predict bentonite requirement for wine stored in the short term to medium term [82]. The cooling temperature and time are also critical to the accuracy of heat test results. A recent study [83] investigated the influence of heating and cooling conditions on protein heat test results. In this study, white wines were heated at 80°C for a time ranging from 0.5 to 6.0 h and then cooled down for 0.5–18 h at 0, 4 or 20°C, respectively. The results indicated that heating at 80°C for 2 h and then cooled at 20°C for 3 h enabled the repeatable production of haze and bentonite requirement.

As traditional heat test is very time-consuming, near infrared (NIR) spectroscopy has been studied for its potential to predict protein stability with high efficiency [84]. Results from 111 white wines representing multiple regions and varieties in California showed that the turbidity of wine could be predicted from the short-wavelength NIR spectra, but further IR analysis on a large number of wines will be required for the application of NIR in the global wine industry, and the high cost of equipment may limit its widespread use.

There are also commercial reagents, i.e. Bentotest and Proteotest, available for winemakers to check the protein stability in white wine. In general, these commercial reagents are considered to be harsher than the heat test, and as a result, the addition of bentonite is normally more than the actual requirement for protein stabilisation. Other tests including trichloroacetic acid (TCA), tannin and ethanol tests were originally established for protein quantification by precipitation of proteins in wine, but they are rarely used in the winery for checking the protein stability as bentonite requirement can be significantly overestimated using those tests.

5. Protein removal by bentonite and potential alternatives

Removal of the proteins remaining in finished wine before bottling is critical for wine protein stabilisation. In the wine industry, bentonite, a swelling 2:1 aluminosilicate clay, is usually added to wine to remove the proteins [85]. As proteins in wine are positively charged at wine pH and bentonite carries a net negative charge, wine proteins can be absorbed onto bentonite by cationic exchange [1, 85]. Hsu and Heatherbell [86] have shown that wine proteins with higher pI (5.8–8.0) and intermediate MW (32–45 kDa) are preferentially removed by bentonite fining, but a proportion of wine proteins, which have a MW range from 60 to 65 kDa and pI range from 4.1 to 8.0, are highly resistant to removal by bentonite fining. In that study, the authors also pointed out that the removal of proteins with lower pI (4.1–5.8) and lower MW (12.6 and 20–30 kDa) was necessary for protein stabilisation. In contrast to this conclusion, another study [56] showed that the amount of protein depletion correlated linearly with the level of bentonite addition, implying no bentonite selectivity based on isoelectric point. Different conclusions from these two studies might be partly attributed to the different methods used to separate and quantify the proteins [87].

Bentonite fining is effective in removing proteins to stabilise the wine, but the use of bentonite could also remove some important aroma and flavour compounds [88], as well as result in loss of wine as lees [89]. Thus, alternative treatments for protein stabilisation that can reduce or eliminate the use of bentonite are of great interest for winemakers. Potential bentonite alternatives that have been studied are summarised in **Table 2**.

Ultrafiltration is very effective in removing proteins in wine, and up to 90% of wine proteins can be retained using a 10 kDa molecular weight cut-off membrane; as a result, bentonite requirement is greatly reduced [90]. However, in addition to the high setup and running costs, ultrafiltration could also result in the loss of beneficial aroma compounds in wine [91]. Studies on Riesling and Gewürztraminer wines using ultrafiltration with membrane nominal MW cut-off from 10 to 50 kDa showed that ultrafiltration can significantly decrease the overall aroma intensity and fruity, floral, sweet and honey/caramel aromas, but also increase vegetative aroma [92]. Furthermore, a reduction in browning colour (A420 nm) and total phenolics was also observed in filtered Riesling and Gewürztraminer wines [93].

Short-term pasteurisation (90°C for 1 min) can reduce bentonite requirement up to 70% [94], but heating of juice could impart negative sensory implications into resultant wine [95]. A recent study reported that applying heating of juice at 75°C for 1 min in the presence of a heat-tolerant protease, aspergillopepsin (AGP), derived from fungus *Aspergillus niger*, before fermentation showed a significant reduction of PR proteins without damaging wine quality [96]. In general, thaumatin-like proteins and chitinases are highly resistant to proteases, but heating unfolds these proteins and thus they can be degraded by protease. Another promising protease BcAP8 (aspartic acid protease) from *Botrytis cinerea* has been proven to

Bentonite alternatives	Advantages	Disadvantages	Reference
	Ultrafiltration		
Ultrafiltration	Efficient in protein removal and reduction of bentonite requirement	High setup and running costs and possible loss of flavour compounds	[90, 91]
	Heating		
Short-time pasteurisation	Acceleration of protein denaturation in juice and thus lower bentonite requirement in wine	Impart negative sensory implications into wine	[94, 95]
	Enzymes		
Aspergillopepsin	Very active towards wine proteins during flash pasteurisation	Heating required for better efficiency, which may alter sensory properties of wine	[96]
Aspartic acid protease	Capable of degrading chitinases without heat denaturation	Does not remove all PR proteins	[97]
	Polysaccharides		
Carrageenan	Effective in heat stabilising white wines at low addition rates without deleterious sensory impacts	Not a good settling agent	[99, 100]
Mannoproteins	Protective effect on protein haze formation in wine	Long-term prevention of protein haze formation is unknown	[101]
Chitin and chitosan	Interact and remove PR proteins and protect wine from browning	Interact with phenolics which may alter wine colour and texture	[103]
	Nanomaterials		
Nanoparticles	High surface area and efficiency in removing proteins	Removal of nanoparticles and the costs	[104, 105]

Table 2.
Summary of bentonite alternatives for white wine protein stabilisation.

be effective against grape chitinases during juice fermentation without the need for heating [97]. This protease could potentially reduce the bentonite requirement, but it is less effective on the degradation of thaumatin-like proteins.

Novel fining agents such as polysaccharides could be another potential class of bentonite alternatives [98]. Carrageenan, a food grade polysaccharide extracted from seaweeds, is effective in heat stabilising white wines at low addition rates (125–250 mg/L) without deleterious sensory impacts compared to bentonite treated wines [99, 100]. However, technical issues including frothing, slower filterability and risk of over-fining should be considered when applying carrageenan for protein stabilisation, particularly when it is used prior to or during fermentation. In addition, yeast mannoproteins, the highly glycosylated polypeptides present in yeast cell walls, have also been reported to have a protective effect on protein haze formation in wine [101]. Thus, mannoproteins extracted from purified yeast cell walls could be added into juice/wine to reduce the addition of bentonite, but further research is required to understand whether this protection against protein aggregation is suitable for long-term wine storage. Chitin [102] and chitosan [103], polysaccharides principally from *Aspergillus niger*, also have potential to remove haze-forming proteins in wine, but the wine colour and texture could be affected as chitosan interacts with phenolics and organic acids in wine.

Nanotechnology is currently a very active research topic in food science. Recently some researchers also become interested in using nanomaterials to remove proteins in wine. Magnetic steel nanoparticles coated with acrylic acid have been tested and are highly efficient in attracting and thus removing haze-forming proteins [104]. Another study using mesoporous nanomaterials to fine Muscat Ottonel and Pedro Ximenez wines also confirmed the high efficiency of nanomaterials in removing haze-forming proteins, and the loss of aroma compounds in wine due to addition of nanomaterials was even less than bentonite [105]. Nanomaterials have shown great potential as bentonite alternative to remove proteins in wine, but the cost of coating the nanoparticles and the removal of nanoparticles from wine after the treatment are the main concerns for their wide application in the wine industry.

6. Conclusions

Pathogenesis-related proteins in grapes play an important role in plant defence mechanisms. The concentration of PR proteins in grapes generally increases during ripening, but a number of factors, e.g. fungal infection and UV radiation, can influence the accumulation of PR proteins in grapes. Extraction of PR proteins from grapes into juice during grape processing, influenced by factors like harvesting (manual or mechanical), skin contact time and grape pressing, could have a great impact on PR protein extraction. The concentration of PR proteins in juice can largely predict their final concentration in wine. If PR proteins were not removed before white wine bottling, they could potentially denature and could form a haze in bottled wine, especially when wines are stored under unfavourable conditions, e.g. high temperatures. Bentonite is commonly used in white wine protein stabilisation, but bentonite is not specific to absorb proteins; it can also remove positive aroma compounds and cause the loss of some wine volume as bentonite lees. Alternatives to bentonite have been widely studied; however, they are not yet enough efficient to replace the cheap bentonite. Carrageenan and some proteolytic enzymes can be in a near future interesting tools to get white protein stabilisation, perhaps more efficient than the classic bentonite.

Author details

Bin Tian* and Roland Harrison
Faculty of Agriculture and Life Sciences, Lincoln University, New Zealand

*Address all correspondence to: bin.tian@lincoln.ac.nz

References

[1] Ferreira RB, Picarra-Pereira MA, Monteiro S, Loureiro VB, Teixeira AR. The wine proteins. Trends in Food Science & Technology. 2002;**12**(7):230-239

[2] Ferreira RB, Monteiro S, Picarra-Pereira MA, Tanganho MC, Loureiro VB, Teixeira AR. Characterization of the proteins from grapes and wines by immunological methods. American Journal of Enology and Viticulture. 2000;**51**(1):22-28

[3] Dambrouck T, Marchal R, Marchal-Delahaut L, Parmentier M, Maujean A, Jeandet P. Immunodetection of proteins from grapes and yeast in a white wine. Journal of Agricultural and Food Chemistry. 2003;**51**(9):2727-2732

[4] Kwon SW. Profiling of soluble proteins in wine by nano-high-performance liquid chromatography/tandem mass spectrometry. Journal of Agricultural and Food Chemistry. 2004;**52**(24):7258-7263

[5] Hsu JC, Heatherbell DA. Isolation and characterization of soluble proteins in grapes, grape juice, and wine. American Journal of Enology and Viticulture. 1987;**38**(1):6-10

[6] Santoro M. Fractionation and characterization of must and wine proteins. American Journal of Enology and Viticulture. 1995;**46**(2):250

[7] Pueyo E, Dizy M, Polo MC. Varietal differentiation of must and wines by means of protein fraction. American Journal of Enology and Viticulture. 1993;**44**(3):255

[8] Lamikanra O. The proteins of muscadine grapes. Journal of Food Science. 1987;**52**(2):483-484

[9] Lamikanra O, Inyang ID. Temperature influence on muscadine wine protein characteristics. American Journal of Enology and Viticulture. 1988;**39**(2):113-116

[10] Brissonnet F, Maujean A. Characterization of foaming proteins in a champagne base wine. American Journal of Enology and Viticulture. 1993;**44**(3):297-301

[11] Monteiro S, Piçarra-Pereira MA, Mesquita PR, Loureiro VB, Teixeira A, Ferreira RB. The wide diversity of structurally similar wine proteins. Journal of Agricultural and Food Chemistry. 2001;**49**(8):3999-4010

[12] Waters EJ, Wallace W, Williams PJ. Identification of heat-unstable wine proteins and their resistance to peptidases. Journal of Agricultural and Food Chemistry. 1992;**40**(9):1514-1519

[13] Waters EJ, Shirley NJ, Williams PJ. Nuisance proteins of wine are grape pathogenesis-related proteins. Journal of Agricultural and Food Chemistry. 1996;**44**(1):3-5

[14] Hung WF, Harrison R, Morton JD, Trought MCT, Frost A. Protein concentration and bentonite requirement in Marlborough Sauvignon Blanc wines. Australian Journal of Grape and Wine Research. 2014;**20**(1):41-50

[15] Datta SK, Muthukrishnan S. Pathogenesis-Related Proteins in Plants. Boca Raton: CRC Press; 1999. DOI: 10.1201/9781420049299

[16] Van Loon LC, Van Kammen A. Polyacrylamide disc electrophoresis of the soluble leaf proteins from *Nicotiana tabacum* var. 'Samsun' and 'Samsun NN': II. Changes in protein constitution after infection with tobacco mosaic virus. Virology. 1970;**40**(2):199-211

[17] Van Loon LC, Pierpoint WS, Boller TH, Conejero V. Recommendations

for naming plant pathogenesis-related proteins. Plant Molecular Biology Reporter. 1994;**12**(3):245-264

[18] Robinson SP, Davies C. Molecular biology of grape berry ripening. Australian Journal of Grape and Wine Research. 2000;**6**(2):175-188

[19] Robinson SP, Jacobs AK, Dry IB. A class IV chitinase is highly expressed in grape berries during ripening. Plant Physiology. 1997;**114**(3):771-778

[20] Tattersall DB, Van Heeswijck R, Hoj PB. Identification and characterization of a fruit-specific, thaumatin-like protein that accumulates at very high levels in conjunction with the onset of sugar accumulation and berry softening in grapes. Plant Physiology. 1997;**114**(3):759-769

[21] Peng Z, Pocock KF, Waters EJ, Francis IL, Williams PJ. Taste properties of grape (*Vitis vinifera*) pathogenesis-related proteins isolated from wine. Journal of Agricultural and Food Chemistry. 1997;**45**(12):4639-4643

[22] Pocock KF, Hayasaka Y, McCarthy MG, Waters EJ. Thaumatin-like proteins and chitinases, the haze-forming proteins of wine, accumulate during ripening of grape (*Vitis vinifera*) berries and drought stress does not affect the final levels per berry at maturity. Journal of Agricultural and Food Chemistry. 2000;**48**(5):1637-1643

[23] Jaillon O, Aury J-M, Noel B, Policriti A, Clepet C, Casagrande A, et al. The grapevine genome sequence suggests ancestral hexaploidization in major angiosperm phyla. Nature. 2007;**449**(7161):463-467

[24] Velasco R, Zharkikh A, Troggio M, Cartwright DA, Cestaro A, Pruss D, et al. A high quality draft consensus sequence of the genome of a heterozygous grapevine variety. PLoS One. 2007;**2**(12):e1326

[25] Marzia G, Gabriella GM. Heard it through the grapevine: Proteomic perspective on grape and wine. Journal of Proteomics. 2010;**73**(9):1647-1655

[26] Deytieux C, Geny L, Lapaillerie D, Claverol S, Bonneu M, Doneche B. Proteome analysis of grape skins during ripening. Journal of Experimental Botany. 2007;**58**(7):1851-1862

[27] Jellouli N, Ben Jouira H, Skouri H, Ghorbel A, Gourgouri A, Mliki A. Proteomic analysis of Tunisian grapevine cultivar Razegui under salt stress. Journal of Plant Physiology. 2008;**165**(5):471-481

[28] Okuda T, Fukui M, Takayanagi T, Yokotsuka K. Characterization of major stable proteins in Chardonnay wine. Food Science and Technology Research. 2006;**12**(2):131-136

[29] Tian B, Harrison R, Morton J, Deb-Choudhury S. Proteomic analysis of Sauvignon Blanc grape skin, pulp and seed and relative quantification of pathogenesis-related proteins. PLoS One. 2015;**10**(6):e0130132 [p. 15]

[30] López-Miranda S, Hernández-Sánchez P, Serrano-Martínez A, Hellín P, Fenoll J, Núñez-Delicado E. Effect of ripening on protein content and enzymatic activity of crimson seedless table grape. Food Chemistry. 2011;**127**(2):481-486

[31] Murphey JM, Spayd SE, Powers JR. Effect of grape maturation on soluble protein characteristics of Gewürztraminer and White Riesling juice and wine. American Journal of Enology and Viticulture. 1989;**40**(3):199-207

[32] Serrano-Megías M, Núñez-Delicado E, Pérez-López AJ,

López-Nicolás JM. Study of the effect of ripening stages and climatic conditions on the physicochemical and sensorial parameters of two varieties of *Vitis vinifera* L. by principal component analysis: Influence on enzymatic browning. Journal of the Science of Food and Agriculture. 2006;**86**:592-599

[33] Tian B, Harrison R, Morton J, Jaspers M. Changes in pathogenesis-related proteins and phenolics in *Vitis vinifera* L. cv. 'Sauvignon Blanc' grape skin and pulp during ripening. Scientia Horticulturae. 2019;**243**:78-83

[34] Tattersall DB, Pocock KF, Hayasaka Y, Adams K, Van Heeswijck R, Waters EJ, et al. Pathogenesis related proteins—Their accumulation in grapes during berry growth and their involvement in white wine heat instability. Current knowledge and future perspectives in relation to winemaking practices. Molecular Biology and Biotechnology of the Grapevine: Kluwer Academic. 2001:183-202

[35] Jacobs D. Induction of different pathogenesis-related cDNAs in grapevine infected with powdery mildew and treated with ethephon. Plant Pathology. 1999;**48**(3):325-336

[36] Monteiro S, Piçarra-Pereira MA, Loureiro VB, Teixeira AR, Ferreira RB. The diversity of pathogenesis-related proteins decreases during grape maturation. Phytochemistry. 2007;**68**(4):416-425

[37] Salazar FN, López F, Chiffelle I, López R, Peña-Neira A. Evaluation of pathogenesis-related protein content and protein instability of seven white grape (*Vitis vinifera* L.) clones from Casablanca Valley, Chile. European Food Research and Technology. 2012;**234**:509-515

[38] Marsh E, Alvarez S, Hicks LM, Barbazuk WB, Qiu WP, Kovacs L, et al. Changes in protein abundance during powdery mildew infection of leaf tissues of Cabernet Sauvignon grapevine (*Vitis vinifera* L.). Proteomics. 2010;**10**(10):2057-2064

[39] Giribaldi M, Perugini I, Sauvage FX, Schubert A. Analysis of protein changes during grape berry ripening by 2-DE and MALDI-TOF. Proteomics. 2007;**7**(17):3154-3170

[40] Falconer RJ, Marangon M, Van Sluyter SC, Neilson KA, Chan C, Waters EJ. Thermal stability of Thaumatin-like protein, Chitinase, and Invertase isolated from Sauvignon Blanc and Semillon juice and their role in haze formation in wine. Journal of Agricultural and Food Chemistry. 2010;**58**(2):975-980

[41] Marchal R, Tabary I, Valade M, Moncomble D, Viaux L, Robillard B, et al. Effects of *Botrytis cinerea* infection on champagne wine foaming properties. Journal of the Science of Food and Agriculture. 2001;**81**(14):1371-1378

[42] Crippen DD Jr, Morrison JC. The effects of sun exposure on the compositional development of Cabernet Sauvignon berries. American Journal of Enology and Viticulture. 1986;**37**(4):235-242

[43] Crippen DD Jr, Morrison JC. The effects of sun exposure on the phenolic content of Cabernet Sauvignon berries during development. American Journal of Enology and Viticulture. 1986;**37**(4):243-247

[44] Spayd SE, Tarara JM, Mee DL, Ferguson JC. Separation of sunlight and temperature effects on the composition of *Vitis vinifera* cv. Merlot berries. American Journal of Enology and Viticulture. 2002;**53**(3):171-182

[45] Bergqvist J, Dokoozlian N, Ebisuda N. Sunlight exposure and

temperature effects on berry growth and composition of Cabernet Sauvignon and Grenache in the Central San Joaquin Valley of California. American Journal of Enology and Viticulture. 2001;**52**(1):1-7

[46] Ristic R, Downey MO, Iland PG, Bindon K, Francis IL, Herderich M, et al. Exclusion of sunlight from Shiraz grapes alters wine colour, tannin and sensory properties. Australian Journal of Grape and Wine Research. 2007;**13**(2):53-65

[47] Schultz HR, Löhnertz O, Bettner W, Bálo B, Linsenmeier A, Jähnisch A, et al. Is grape composition affected by current levels of UV-B radiation? Vitis. 1998;**37**(4):191-192

[48] Tian B, Harrison R, Jaspers M, Morton J. Influence of ultraviolet exclusion and of powdery mildew infection on Sauvignon Blanc grape composition and on extraction of pathogenesis-related proteins into juice. Australian Journal of Grape and Wine Research. 2015;**21**(3):417-424

[49] Marchal R, Berthier L, Legendre L, Marchal-Delahaut L, Jeandet P, Maujean A. Effects of *Botrytis cinerea* infection on the must protein electrophoretic characteristics. Journal of Agricultural and Food Chemistry. 1998;**46**(12):4945-4949

[50] Girbau T, Stummer BE, Pocock KF, Baldock GA, Scott ES, Waters EJ. The effect of *Uncinula necator* (powdery mildew) and *Botrytis cinerea* infection of grapes on the levels of haze-forming pathogenesis-related proteins in grape juice and wine. Australian Journal of Grape and Wine Research. 2004;**10**(2):125-133

[51] Monteiro S, Barakat M, Piçarra-Pereira MA, Teixeira AR, Ferreira RB. Osmotin and thaumatin from grape: A putative general defense mechanism against pathogenic fungi. Phytopathology. 2003;**93**(12):1505-1512

[52] Pocock KF, Waters EJ. The effect of mechanical harvesting and transport of grapes, and juice oxidation, on the protein stability of wines. Australian Journal of Grape and Wine Research. 1998;**4**(3):136-139

[53] Pocock KF, Hayasaka Y, Peng Z, Williams PJ, Waters EJ. The effect of mechanical harvesting and long-distance transport on the concentration of haze-forming proteins in grape juice. Australian Journal of Grape and Wine Research. 1998;**4**(1):23-29

[54] Tian B, Harrison R, Morton J, Jaspers M, Hodge S, Grose C, et al. Extraction of pathogenesis-related proteins and phenolics in Sauvignon Blanc as affected by grape harvesting and processing conditions. Molecules. 2017;**22**(7):1164

[55] Tian B, Harrison R, Morton J, Jaspers M. Influence of skin contact and different extractants on extraction of proteins and phenolic substances in Sauvignon Blanc grape skin. Australian Journal of Grape and Wine Research. 2020;**26**(2):180-186

[56] Dawes H, Boyes S, Keene J, Heatherbell D. Protein instability of wines: Influence of protein isolelectric point. American Journal of Enology and Viticulture. 1994;**45**(3):319-326

[57] Sarmento MR, Oliveira JC, Slatner M, Boulton RB. Influence of intrinsic factors on conventional wine protein stability tests. Food Control. 2000;**11**(6):423-432

[58] Mesquita PR, Picarra-Pereira MA, Monteiro S, Loureiro VB, Teixeira AR, Ferreira RB. Effect of wine composition on protein stability. American Journal of Enology and Viticulture. 2001;**52**(4):324-330

[59] Siebert KJ. Effects of protein-polyphenol interactions on beverage haze, stabilization, and analysis. Journal of Agricultural and Food Chemistry. 1999;**47**(2):353-362

[60] Pocock KF, Alexander GM, Hayasaka Y, Jones PR, Waters EJ. Sulfate—A candidate for the missing essential factor that is required for the formation of protein haze in white wine. Journal of Agricultural and Food Chemistry. 2007;**55**(5):1799-1807

[61] Batista L, Monteiro S, Loureiro VB, Teixeira AR, Ferreira RB. The complexity of protein haze formation in wines. Food Chemistry. 2009;**112**(1):169-177

[62] Marangon M, Van Sluyter SC, Neilson KA, Chan C, Haynes PA, Waters EJ, et al. Roles of grape thaumatin-like protein and chitinase in white wine haze formation. Journal of Agricultural and Food Chemistry. 2011;**59**:733-740

[63] Marangon M, Van Sluyter S, Waters EJ, Herderich MJ, Pretorius IS. Recent advances help us understand protein haze more clearly. Australian & New Zealand Wine Industry Journal. 2010;**25**(2):24-27

[64] Gazzola D, van Sluyter S, Curioni A, Waters EJ, Marangon M. Roles of proteins, polysaccharides and phenolics on haze formation in white wine via reconstitution experiments. Journal of Agricultural and Food Chemistry. 2012;**60**(42):10666-10673

[65] Parker M, Mercurio M, Jeffery D, Holt H, Herderich MJ, Smith PA. An overview of the phenolic chemistry of white juice and wine production. Australian and New Zealand Grapegrower & Winemaker. 2007;**521a**:74-80

[66] Siebert KJ, Carrasco A, Lynn PY. Formation of protein-polyphenol haze in beverages. Journal of Agricultural and Food Chemistry. 1996;**44**(8):1997-2005

[67] Oh HI, Hoff JE, Armstrong GS, Haff LA. Hydrophobic interaction in tannin-protein complexes. Journal of Agricultural and Food Chemistry. 1980;**28**(2):394-398

[68] Siebert KJ, Troukhanova NV, Lynn PY. Nature of polyphenol-protein interactions. Journal of Agricultural and Food Chemistry. 1996;**44**(1):80-85

[69] Vidal S, Doco T, Moutounet M, Pellerin P. Soluble polysaccharide content at initial time of experimental must preparation. American Journal of Enology and Viticulture. 2000;**51**(2):115-121

[70] Waters EJ, Pellerin P, Brillouet JM. A Saccharomyces mannoprotein that protects wine from protein haze. Carbohydrate Polymers. 1994;**23**(3):185-191

[71] Siebert KJ, Lynn PY. Effects of alcohol and pH on protein-polyphenol haze intensity and particle size. Journal of the American Society of Brewing Chemists. 2003;**61**(2):88-98

[72] Hung WF. Protein stabilisation of New Zealand Sauvignon Blanc [Doctoral thesis]. New Zealand: Lincoln University; 2010

[73] Lambri M, Dordoni R, Giribaldi M, Riva Violetta M, Giuffrida MG. Effect of pH on the protein profile and heat stability of an Italian white wine. Food Research International. 2013;**54**(2):1781-1786

[74] Marangon M, Sauvage FX, Waters EJ, Vernhet A. Effects of ionic strength and sulfate upon thermal aggregation of grape Chitinases and Thaumatin-like proteins in a model system. Journal of Agricultural and Food Chemistry. 2011;**59**:2652-2662

[75] Waters EJ, Alexander G, Muhlack R, Pocock KF, Colby C, O'Neill BK, et al. Preventing protein haze in bottled white wine. Australian Journal of Grape and Wine Research. 2005;**11**(2):215-225

[76] Besse C. Investigation of the role of total and free copper in protein haze formation. Australian Grapegrower & Winemaker. 2000;**437**:19-20

[77] Dufrechou M, Cl P-L, Sauvage F-X, Vernhet A. Stability of white wine proteins: Combined effect of pH, ionic strength, and temperature on their aggregation. Journal of Agricultural and Food Chemistry. 2012;**60**(5):1308-1319

[78] de Bruijn J, Loyola C, Arumí JL, Martínez J. Effect of non-protein factors on heat stability of Chilean Sauvignon Blanc wines. Chilean Journal of Agricultural Research. 2014;**74**(4):490-496

[79] McRae J, Schulkin A, Dambergs R, Smith PA. Effect of white wine composition on protein haze potential. Australian Journal of Grape and Wine Research. 2018;**24**(4):498-503

[80] Chagas R, Laia CA, Ferreira RB, Ferreira LM. Sulfur dioxide induced aggregation of wine thaumatin-like proteins: Role of disulfide bonds. Food Chemistry. 2018;**259**:166-174

[81] Pocock KF, Rankine BC. Heat test for detecting protein instability in wine. Australian Wine Brewing and Spirit Review. 1973;**91**(5):42-43

[82] Pocock KF, Waters EJ. Protein haze in bottled white wines: How well do stability tests and bentonite fining trials predict haze formation during storage and transport? Australian Journal of Grape and Wine Research. 2006;**12**(3):212-220

[83] McRae J, Barricklow V, Pocock K, Smith PA. Predicting protein haze formation in white wines. Australian Journal of Grape and Wine Research. 2018;**24**(4):504-511

[84] Versari A, Laghi L, Thorngate JH, Boulton RB. Prediction of colloidal stability in white wines using infrared spectroscopy. Journal of Food Engineering. 2011;**104**(2):239-245

[85] Blade WH, Boulton R. Adsorption of protein by bentonite in a model wine solution. American Journal of Enology and Viticulture. 1988;**39**(3):193-199

[86] Cilindre C, Jegou S, Hovasse A, Schaeffer C, Castro AJ, Clement C, et al. Proteomic approach to identify champagne wine proteins as modified by *Botrytis cinerea* infection. Journal of Proteome Research. 2008;**7**(3):1199-1208

[87] Waters EJ, Colby CB. Proteins. In: Wine Chemistry and Biochemistry. New York: Springer; 2009. pp. 213-230

[88] Vincenzi S, Panighel A, Gazzola D, Flamini R, Curioni A. Study of combined effect of proteins and bentonite fining on the wine aroma loss. Journal of Agricultural and Food Chemistry. 2015;**63**(8):2314-2320

[89] Høj P, Tattersall D, Adams K, Pocook K, Hayasaka Y, Van Heeswijck R, et al. The 'haze proteins' of wine–A summary of properties, factors affecting their accumulation in grapes, and the amount of bentonite required for their removal from wine. In: Proceedings of the 50th American Society for Enology and Viticulture Anniversary Annual Meeting. Seattle, Washington. 2001. p. 149-154

[90] Hsu JC, Heatherbell DA, Flores JH, Watson BT. Heat-unstable proteins in grape juice and wine. II. Characterization and removal by ultrafiltration. American Journal of Enology and Viticulture. 1987;**38**(1):17-22

[91] Miller G, Amon J, Gibson R, Simpson R. Loss of wine aroma attributable to protein stabilization with bentonite or ultrafiltration. Auststralian Grapegrower Winemaker. 1985;**256**:46-50

[92] Flores J, Heatherbell D, Henderson L, McDaniel M. Ultrafiltration of wine: Effect of ultrafiltration on the aroma and flavor characteristics of white Riesling and Gewürztraminer wines. American Journal of Enology and Viticulture. 1991;**42**(2):91-96

[93] Flores J, Heatherbell D, McDaniel M. Ultrafiltration of wine: Effect of ultrafiltration on white Riesling and Gewürztraminer wine composition and stability. American Journal of Enology and Viticulture. 1990;**41**(3):207-214

[94] Pocock KF, Hoj PB, Adams KS, Kwiatkowski MJ, Waters EJ. Combined heat and proteolytic enzyme treatment of white wines reduces haze forming protein content without detrimental effect. Australian Journal of Grape and Wine Research. 2003;**9**(1):56-63

[95] Lloyd F, Colby C, O'Neill B, Waters E. Combined heat/proteolytic enzyme treatment for the removal of protein haze in wine. In: Proceedings of the 33rd Australasian Chemical Engineering Conference. Brisbane, Queensland. 2005; p. 436-441

[96] Marangon M, Van Sluyter SC, Robinson E, Muhlack R, Holt H, Haynes PA, et al. Degradation of white wine haze proteins by Aspergillopepsin I and II during juice flash pasteurization. Food Chemistry. 2012;**135**(3):1157-1165

[97] Van Sluyter SC, Warnock NI, Schmidt S, Anderson P, van Kan JA, Bacic A, et al. Aspartic acid protease from *Botrytis cinerea* removes haze-forming proteins during white winemaking. Journal of Agricultural and Food Chemistry. 2013;**61**(40):9705-9711

[98] Cabello-Pasini A, Victoria-Cota N, Macias-CarranzaV,Hernandez-GaribayE, Muñiz-Salazar R. Clarification of wines using polysaccharides extracted from seaweeds. American Journal of Enology and Viticulture. 2005;**56**(1):52-59

[99] Marangon M, Stockdale VJ, Munro P, Trethewey T, Schulkin A, Holt HE, et al. Addition of carrageenan at different stages of winemaking for white wine protein stabilization. Journal of Agricultural and Food Chemistry. 2013;**61**(26):6516-6524

[100] Ratnayake S, Stockdale V, Grafton S, Munro P, Robinson A, Pearson W, et al. Carrageenans as heat stabilisers of white wine. Australian Journal of Grape and Wine Research. 2019;**25**(4):439-450

[101] Jaeckels N, Meier M, Dietrich H, Will F, Decker H, Fronk P. Influence of polysaccharides on wine protein aggregation. Food chemistry. 2016;**200**:38-45

[102] Ndlovu T, Buica A, Bauer FF. Chitinases and thaumatin-like proteins in Sauvignon Blanc and Chardonnay musts during alcoholic fermentation. Food Microbiology. 2019;**78**:201-210

[103] Colangelo D, Torchio F, De Faveri DM, Lambri M. The use of chitosan as alternative to bentonite for wine fining: Effects on heat-stability, proteins, organic acids, colour, and volatile compounds in an aromatic white wine. Food Chemistry. 2018;**264**:301-309

[104] Mierczynska-Vasilev A, Boyer P, Vasilev K, Smith PA. A novel technology for the rapid, selective, magnetic removal of pathogenesis-related proteins from wines. Food Chemistry. 2017;**232**:508-514

[105] Dumitriu G-D, de Lerma NL, Luchian CE, Cotea VV, Peinado RA. Study of the potential use of mesoporous nanomaterials as fining agent to prevent protein haze in white wines and its impact in major volatile aroma compounds and polyols. Food Chemistry. 2018;**240**:751-758

Chapter 10

Cork and Cork Stoppers: Quality and Performance

Vanda Oliveira and Helena Pereira

Abstract

Cork is a world-renowned material used for sealing wine bottles. Cork is a cellular material with chemical inertia and a set of physical and mechanical properties that provide an outstanding performance for in-bottle wine aging, by combining minute oxygen transfer with sealing, durability, and chemical stability, for example, inertness toward the liquid content and along storage, preventing sensory deterioration. Cork is a natural material, one of the most important nontimber forest products, that is associated with sustainability and ecosystem preservation. The cork industry has steadily improved production processes and adopted innovative technologies, quality control measures, and certification, regarding prevention of potential wine taints and off-flavors. This chapter makes a review of cork stoppers, their properties, and quality and role for wine bottle aging, tackling their importance for wine aroma and off-flavors while presenting the latest advances in cork research.

Keywords: cork quality, chemical composition, permeability, oxygen ingress, off-flavors, phenolic compounds

1. Introduction

The cork oak (*Quercus suber* L.) forest is a unique production system developed by men that has a substantial ecological role, for example, against desertification and in maintaining animal and plant biodiversity in their restricted area of occurrence in the Western Mediterranean countries, covering a total of 2.1 million hectares. Portugal has 34% of the world's area, which corresponds to about 720,000 hectares and 22% of the national forest [1]. Cork is obtained from the bark of the cork oak, and the production is made in a sustainable process carried out during the tree's lifetime. It is one of the world's important nontimber forest products, supporting a dedicated industry of cork products directed toward global markets.

World cork production reached 201,000 tons annually. Portugal is the leader, concentrating 49.6% of the cork production, acting also as an importer of cork raw material, therefore increasing its share on the world production of cork products. In 2017, Portugal exported 197,000 tons of cork products representing 986.3 million euros [2]. Cork stoppers lead the exports, accounting for 72% of the total value (710.7 million euros). Within the cork stopper segment, natural corks come first with 60% of the total (428.6 million euros), followed by other types of stoppers with 21%, including technical stoppers, microstoppers, and others (149.8 million euros), and champagne stoppers with 19% [2].

Cork is a world-renowned material used for sealing wine bottles. Cork is a cellular material with chemical inertia and a set of specific physical and mechanical properties that provide an outstanding performance when in-bottle wine aging is wanted, by combining the required minute oxygen transfer with mechanical sealing of the bottle, durability, and chemical stability [3, 4]. Cork is the closure material preferred by wine consumers, and a bottle corked with a natural cork stopper is indicative of high-quality or very high-quality wine, as shown by recent surveys (Wine Opinions, CTR, Iniciativa Cork, Gfk, Opinion Way).

2. The sustainable management of cork production

Cork oak forests are often a part of a multifunctional agroforestry-pastoral system called "montado" that is considered a high nature value farming system, according to the European classification proposed by the European Environmental Agency [5]. These ecosystems are also recognized as habitats of conservation value listed in the Habitats Directive [6].

In this nonwood forest production system, cork harvesting is the major economic activity, and cork is the most valuable product. The entire cork chain from the forest to the consumer relies on the regular and sustainable production of cork. To maintain cork production capacity and provide the mentioned environmental services, it is necessary that cork oak forests are adequately managed, being the sustainability a matter of general concern [3].

The exploitation of the cork oak tree for cork production needs its periodical removal from the stem and branches in a degree that is considered well suited with the maintenance of the tree in good physiological conditions [3]. Cork production yields depend on the tree and cork growth, as well as on management variables such as the intensity of cork extraction and the interval between strippings that are regulated with strict rules by the Portuguese legislation (Decreto-Lei nr. 155/2004).

The production of cork relies on specific forest management and silvicultural model, often called subericulture that is based on the biological characteristics of the cork oak bark development [7]. The extraction of cork, or cork stripping, is done manually by cutting large rectangular planks and pulling them out of the tree when the cork oak is physiologically active in late spring and early summer. By law, young trees can only be stripped when they reach at least 70 cm of the perimeter at 1.3 m of height, corresponding to about 25 years of age. Moreover, cork cannot be stripped above a stem height equal to twice the perimeter of the stem in the first stripping, or not more than three times, for a mature tree in full production, the so-called stripping coefficient. The trees are then debarked every 9 years (the legal minimum allowed in Portugal and Spain) or more, called the production cycle [8]. The decision of longer production cycles is often related to an adjustment to cork growth and productivity in order to achieve the cork plank thickness required by the industry.

This means that cork growth is the main criteria to consider in cork management planning since it determines the thickness of the cork plank that is available for industrial processing, which is primarily oriented toward the production of wine stoppers (requiring a minimum thickness of 27 mm after the cork postharvest boiling operation).

The annual cork growth along the production cycle varies with the number of years in the cork growth sequence (growth is usually higher in the first years of the cycle) and is influenced by environmental and tree conditions [9, 10]. A recent study encompassing a large time span (24 years) and sampling (1584 cork samples), reported an average annual cork ring width ranging from 1.2 to 7.3 mm with an average value of 3.3 mm [11].

Cork oaks show high resiliency to interannual precipitation variability, with rapid and complete recovery from extreme dry years or from rainfall exclusion, but have a high sensitivity to the amount and timing of late spring precipitation [12]. In Mediterranean conditions, access to water resources and the relationship to soil-site conditions are key factors for cork oak development [13]. Soils with low depth and high compactness have a negative influence on the development of the cork oak deep root system, thereby diminishing the access to direct groundwater resources, namely, during summer drought [14–16].

It is foreseen that in the Iberian Peninsula, spring precipitation will be reduced and more severe droughts will occur as predicted by the Intergovernmental Panel on Climate Changes (IPCC). In such a scenario, cork growth will be slower resulting in narrower annual rings that will decrease the overall thickness of the cork layer at the end of the production cycle. Given the importance of the cork plank thickness for the industrial processing into cork stoppers, it is foreseeable that the silvicultural cork management will require an adjustment to mitigate the effects of drought by postponing the cork striping (i.e., increasing the duration of the cork production cycle) [17].

3. The cork formation

Cork is a protective tissue located in the outer bark of the cork oaks as part of the periderm. The formation of cork in the periderm results from the activity of a secondary meristem, the phellogen: each phellogen mother cell originates by cellular division cork cells that grow unidirectionally outward in the tree's radial direction and phelloderm cells to the inside [18]. In the cork oak, the first phellogen maintains its activity year after year, producing successive layers of cork. The phellogen may be functional for many years, probably during the tree's life, although the intensity of its activity decreases with age [3].

If the cork layer of the initial periderm (virgin cork) formed in the young cork oaks is removed (by an operation called cork stripping), a new phellogen is formed inside the phloem and rebuilds a traumatic periderm and its subsequent cork layer (second cork). At this time of life in the young cork oaks, the radial growth of the stem is still important, and the second cork external regions are subject to large tangential stress that may result into deep fractures of the cork [19]. If the second cork is removed, the process is repeated with the formation of a new phellogen and the production of a new cork layer (reproduction cork) that will endure few fractures due to the low tangential stress caused by the radial growth of the mature tree. Upon removal of this reproduction cork, the process is repeated, therefore allowing exploitation during the tree's lifetime by successive removals of the reproduction cork. The second and reproduction corks are covered at the outside by a thin lignocellulosic layer of phloem, corresponding to the part of the phloem that remained to the outside when the traumatic phellogen was regenerated inside the phloem (**Figure 1**).

The cork oak periderm has lenticels that originate from the activity of particular regions of the phellogen, called lenticular phellogen, and differ from the surrounding cork tissue. The activity of lenticular phellogen is maintained year after year, and therefore, the lenticels prolong radially from the phellogen to the external surface of the periderm forming approximate cylinders named lenticular channels [3]. The lenticular channels are loosely filled with a lenticular filling tissue of rigid unsuberified cells with thick walls and show ruptures and intercellular voids to a great extent [3]. The region bordering the lenticular channels has often higher density than the surrounding material due to the presence of lignified and thick-walled cells at their borders.

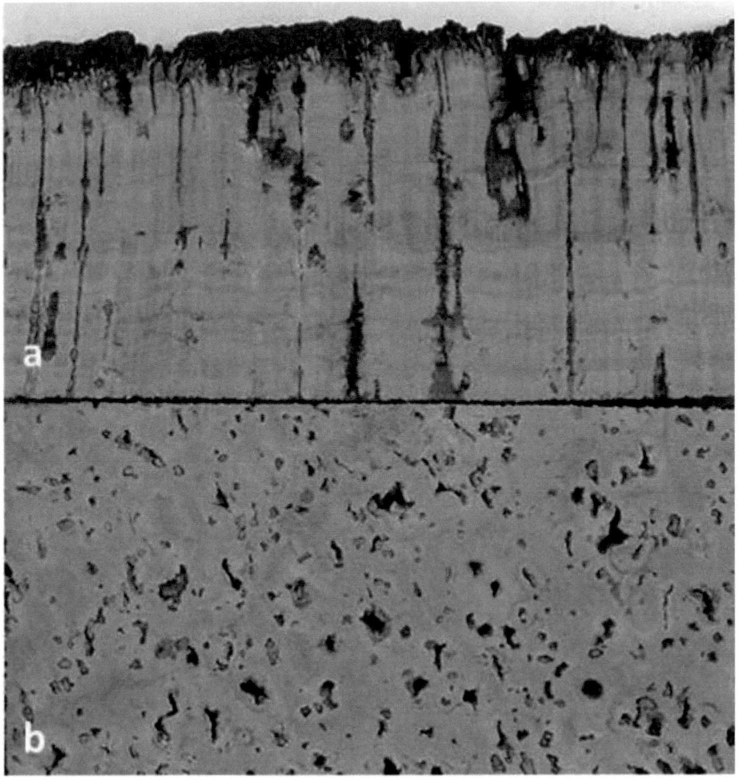

Figure 1.
Lenticular channels crossing the cork layer: (a) in cross section and (b) in the tangential section of the belly (inner part of the cork layer). The thin layer of phloemic lignocellulosic tissue can be observed in the outer side of the cross section (adapted from Ref. [3]).

The lenticular channels (**Figure 1**) are the most important and characteristic features of cork visual heterogeneity, variable in number and dimension between trees, and related directly to the quality and value of the cork material accounting for the so-called cork porosity [3, 8, 10].

4. What matters for corking

The most valuable products in the cork industry are the natural cork stoppers, and therefore, the value of the cork planks as raw material is related to their suitability to produce cork stoppers. Only the cork obtained after the second stripping is used, but the plank thickness that is related to the cork annual growth is a major parameter to establish the technological quality required to produce natural cork stoppers [9, 10]. Given its importance, the cork plank thickness is normalized by caliper classes (NP 298:1993 and ISO 1219:1998). Having the suitable thickness, the plank quality is given mostly by the cork porosity [20, 21] that also determines the suitability and quality yields of the products [22–24]. Porosity quantification by surface image analysis is the basis of cork stoppers classification into quality grades [24, 25].

4.1 Structure

Cork is a cellular material with a compact three-dimensional structure of closed prismatic, on average hexagonal, cells that are assembled base to base creating rows that are aligned in the radial direction in the tree and disposed of in parallel, forming a honeycomb-type structure [19, 26]. The structure of cork observed by scanning electron microscopy in the three principal sections is shown in **Figure 2**. In adjacent rows, the prism bases of neighbor cells most often lay in staggered positions. **Table 1** summarizes the main structural features of cork. The individual cell volume is on average 1.7×10^{-5} mm^3, and the solid cell wall content is 10% of the total volume. The cork cell walls, especially those that constitute the lateral prism faces, show ab initio some bending and undulations of varying intensity that can attain strong corrugation derived from constraints during cork growth in the tree [19, 26].

The formation of cork rings (**Figure 1a**) that give cork a layered structure is the consequence of the biological annual growth rhythm: in the main growth period from April to July, the cells formed are bigger (i.e., with a larger prism height) and have thinner walls than those formed in high summer and autumn, at the end of the growth period, which are smaller and thicker walled (earlycork and latecork cells, respectively) [26].

Cork cell walls are composed of a suberinic secondary wall and are flexible enough to undulate or corrugate with variable intensity under compression without fracture. The latecork cells are more rigid and less compressible than the earlycork cells due to their thick walls and small size (**Figure 3**). When the meristematic activity starts in spring, usually in early April, the initial cork cells formed are compressed against the existing cork layers causing the undulation of the cell walls [3].

As referred previously, the cork tissue is not completely homogeneous, and the cellular structure contains discontinuities that influence several properties of the material and the in-use performance of cork products and are thereby closely associated with the commercial value of raw cork and of cork products [3].

The occurrence of lenticular channels crossing radially the cork tissue is one of the most important features of cork heterogeneity: they cross the cork layers

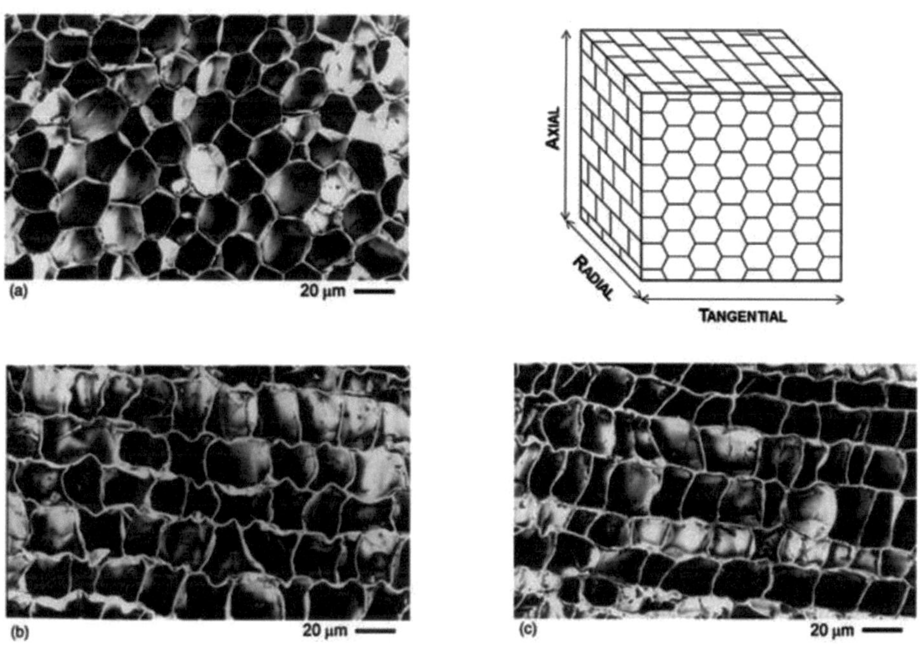

Figure 2.
Schematic representation of the cellular structure of cork with scanning electron micrographs of sections of reproduction cork: (a) tangential, (b) radial, and (c) transverse sections (adapted from Ref. [3]).

Material	Natural suberized lignocellulosic composite
Density	120–170 kg m^{-3}
Type of cells	Closed
Mean edges/face	$n = 6$
Mean faces/cell	$f = 14$
Individual cell shape	Hexagonal prism
Symmetry of structure	Axisymmetric
Cell thickness	1–1.5 μm
Fraction of solid material	10%
Largest principal cell dimension	40 μm
Smallest principal cell dimension	20 μm
Intermediate principal cell dimension	30 μm
Shape anisotropy ratios	R_{13} = 1.5–1.7, R_{12} = 1–1.1
Other specific features	Growth rings, lenticular channels

Table 1.
Main characteristics of cork structure [3].

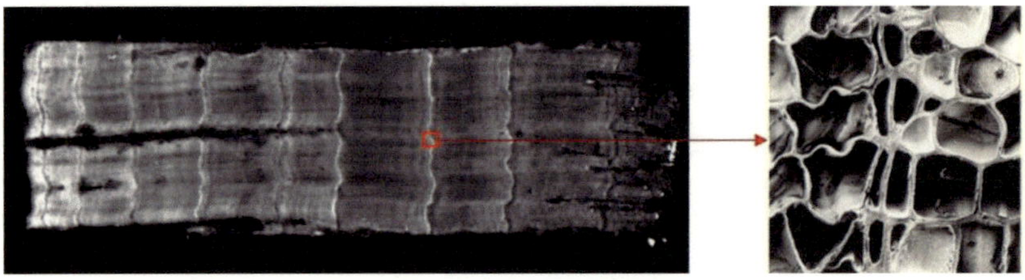

Figure 3.
Fluorescence image of the transverse section of a cork piece showing the different cork rings and a scanning electron micrograph detailing the transition between latecork and earlycork.

from the outside to the inner tissue and are loosely filled with a dark brown colored, unsuberified material, usually conspicuous to visual observation [20]. The lenticular channels appear differently shaped in the three sections of cork (**Figure 1**): (a) in the transverse and radial sections, they are thin elongated rectangular channels oriented radially, and (b) in the tangential section, they present circular to elliptical form. This feature contributes to increase the structural anisotropy of cork.

4.2 Chemical composition

Cork performance depends on structure and chemistry [27], although the impacts of variation are far from being well established, for example, it is believed that the cell wall chemical variation is related to contents in suberin (23.1–54.2%) and lignin (17.1–36.4%) and the suberin-to-lignin ratio plays a determining role in properties, namely, in compression [28–30].

Table 2 summarizes the chemical composition of cork and the range of its natural variability [29]. Cork is mainly composed of suberin, representing on average 43% of the total dry mass. Suberin is a polyester macromolecule obtained by polymerization of linear long-chain fatty acids and hydroxy acids and glycerol, which are assembled and develop spatially as flexible ribbon-like structures. Suberin is, therefore, the component responsible for the elastic properties of cork and allows the bending and compression of cell walls [3, 27].

Chemical parameter (% o.d. cork)	Mean	Std.
Extractives, total	16.2	3.9
Dichloromethane	5.8	0.8
Ethanol	5.9	3.0
Water	4.5	1.6
Suberin, total	42.8	6.2
Long-chain lipids	41.0	5.2
Glycerol	3.8	0.6
Lignin, total	22.0	3.3
Klason lignin	21.1	3.3
Acid soluble lignin	0.9	0.2
Monosaccharide composition, % of neutral sugars		
Glucose	46.1	3.6
Xylose	25.1	3.7
Arabinose	18.0	3.0
Mannose	3.0	2.8
Galactose	7.3	1.2
Rhamnose	0.5	0.5

Table 2.
Chemical composition of cork, in % of oven-dried initial material (mean of 58 samples and standard deviation) [29].

Lignin is an aromatic cross-linked polymer that represents on average 22% of the cork dry mass and is responsible for the structural rigidity of the cells and their resistance to compression [29]. Cork lignin is mostly constituted of G units (guaiacyl units) with an important proportion of H units (hydroxyphenyl units) and minor contents of S units (syringyl units) (H:G:S 1:2.5:0.3, S/G 0.12) [31].

The cellulose and hemicelluloses are less representative and amount up to 16.2% of the cell wall structural components [31]. Hemicelluloses are mainly composed of arabinoxylans with a significant proportion of galactose including uronic acids. The suberin-to-lignin ratio is 2.0, and the cellulose-to-hemicellulose ratio, determined by the ratio of glucose to other sugars, is 1:1.2 [27, 29, 31].

Cork contains a substantial proportion of nonstructural compounds that may be removed by solubilization with suitable solvents without impairing the core properties of cork. These so-called extractives represent on average 16.2% of cork: 5.8% are nonpolar compounds (e.g., lipids and terpenes), and 10.4% are polar compounds of phenolic and polyphenolic nature [29].

4.3 Looking at the surface and inside of a cork stopper

The stoppers are punched out from cork strips so that their cylindrical axis is parallel to the axial direction of cork. Therefore, the surface of the cork stoppers is not homogeneous relative to the section of cork: (a) the circular tops correspond to transverse sections with the lenticular channels crossing the surface as thin rectangular channels perpendicular to the growth rings and (b) the lateral surface of the body ranges from regions corresponding to tangential and radial sections of cork (**Figure 4**). The lenticular channels appear differently shaped in these two sections: in the radial section, they look like elongated rectangular channels, and in the tangential section, they have an approximately circular to elliptical form [3].

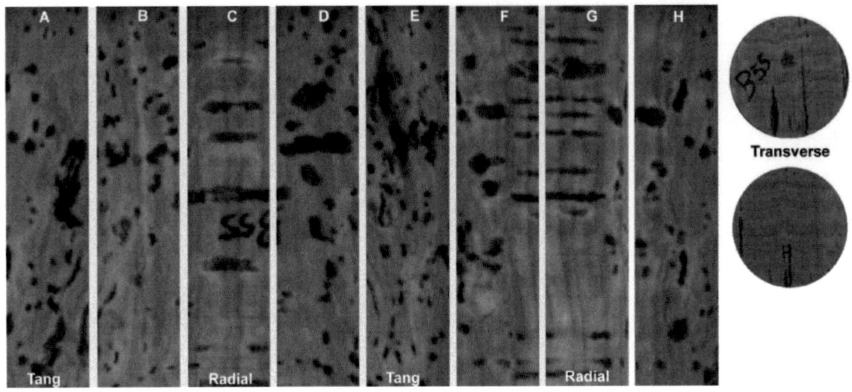

Figure 4.
Image analysis frames (eight frames in the lateral surface and two tops) and the corresponding cork sections ranging from tangential (A and E) to radial (C and G) sections, with in-between sections (B, D, F, and H).

The lenticular channels, woody inclusions, and other defects that give the cork surface its typical visual heterogeneity are together considered as the porosity of cork [20, 32]. Nowadays the evaluation of cork quality is made by visual analysis of the outer surface (lateral body surface and tops) using automated image-based inspection systems with high throughput rates based on line-scan cameras and a computer embedded in an industrial sorting machine capable of acquiring and processing in real time the surface image of the stoppers [33].

The comparison of porosity between the two tops of a cork stopper confirms the existence of axial variation in the tree, for example, one top may have significantly lower porosity than the other top. This fact can be used in practical terms in the production of capsulated natural cork stoppers for spirits by selecting the top with the lowest porosity as the visible one [21].

X-ray tomography was used as a nondestructive technique to acquire knowledge on the internal structure of natural cork stoppers and quantify the lenticular channels present in different classes of cork stoppers [34, 35]. Due to the relationship between X-ray absorption and material density, this technique allowed the visualization and identification of some defects within the cork stopper.

The image resolution with a voxel size of 50 μm achieved by Oliveira et al. [34] allowed the observation of lenticular channel development and geometry (**Figure 5**). The channels are loosely filled with a tissue of rigid unsuberified cells with thick walls, showing ruptures and intercellular voids to a great extent [3]. The region bordering the lenticular channels showed a higher density than the surrounding material due to the presence of lignified and thick-walled cells at their borders.

4.4 Cork properties as closures

Cork has an unusual combination of properties: low density, very low permeability to liquids and gases, low conductivity, chemical stability and durability, and high compressibility with dimensional recovery.

These are properties that are fundamental for the application of cork as a sealing closure in bottles. To a cork stopper, it is requested that (a) it does not allow any leakage from the liquid content either through the stopper itself or at the interface between stopper and bottle; (b) it does not negatively alter the liquid chemical and sensory features; (c) it is durable and preserves its physical and chemical characteristics during storage; and (d) it can be removed from the bottle for consumption easily. To satisfy these requirements, namely, the first requirement of liquid sealing,

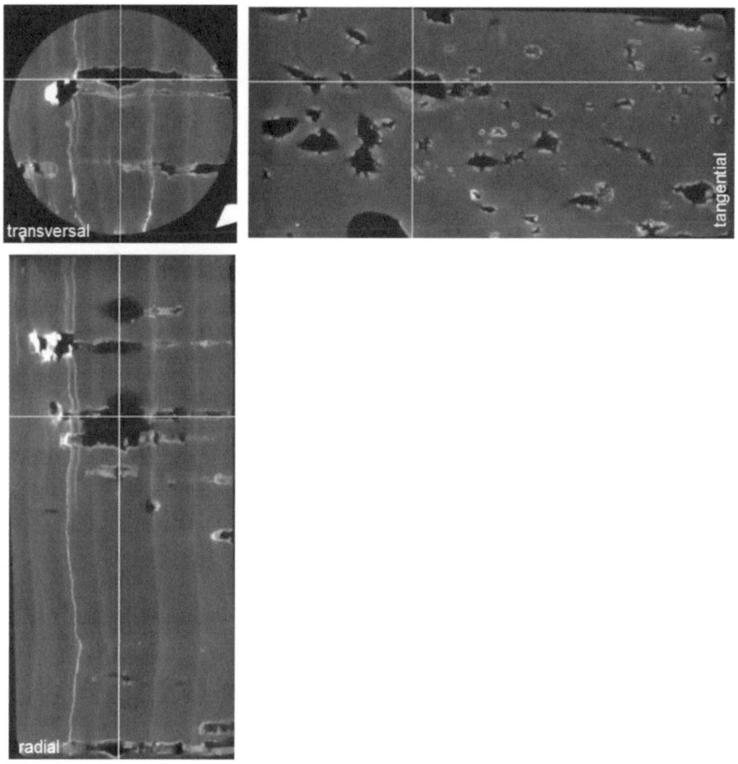

Figure 5.
Typical 2D "slices" from a 3D grayscale image of a cork stopper scanned at 50 µm resolution. Different sections through the cork stopper in three orthogonal planes: Transversal, radial, and tangential. Crossing lines represent the same point in the three images.

it is necessary to have an appropriate compression against the bottleneck and close contact between the stopper and the bottle surface to avoid liquid percolation as well as material's impermeability to avoid diffusion through the stopper.

Cork is a viscoelastic material which allows large deformations under compression without fracture, with substantial dimensional recovery when stress is relieved [36]. The stress-strain compression curve of cork is characteristic of a cellular material: it shows an elastic region up to 5% strain, followed by a wide plateau where strong dimensional reductions occur for small stress increases due to the undulation of cells and their collapse, until a densification phase with a strong stress increase but without cell or cell wall fractures.

The compressive behavior of cork is anisotropic although the differences between directions are not very large: the strength in the radial direction is only slightly higher than in axial and tangential directions, which are more similar (**Figure 6**) [30]. This anisotropy in the stress distribution is also noticed in the compression of a stopper with the radial direction corresponding to the maximal stress and the tangential direction to the minimum value [3].

The cork stoppers have a diameter well above the inside diameter of the bottleneck, and when inserted in the bottle, they will be compressed exerting pressure against the bottle glass. The strains in bottled cork average 30%, a value that is in the plateau region of the stress-strain curve corresponding to the physical phenomenon of the buckling of the cells and to stress values in the range of 1 MPa (**Figure 6**). Industry technical guides refer that cork stopper compression must never be more than 33% of their diameter, as there is a risk that this could compromise its elasticity, with loss of part of the recovery and consequent difficulty in the correct sealing of the wine in the bottle.

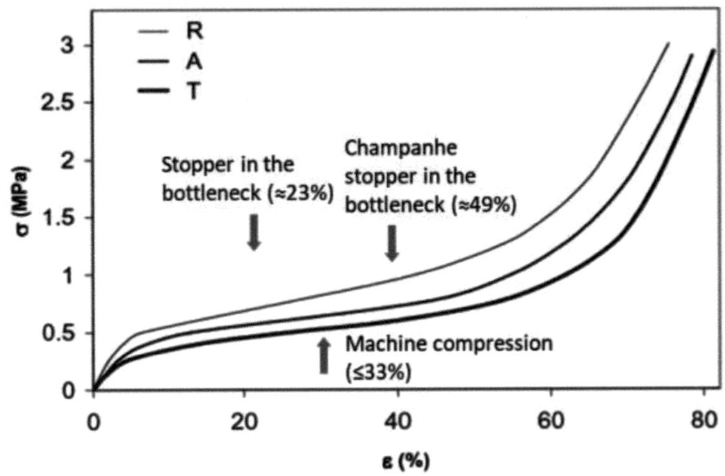

Figure 6.
Average stress-strain curves for the compression of cork in the radial (R) and nonradial (A, T) directions (adapted from [3]). The arrows indicate the compression region of a stopper in different conditions: in the bottling machine, in the bottleneck of a still wine, and a champagne stopper inserted in the bottle neck.

After insertion in the bottle, there will be a quick stress relaxation [3]. Following compression to 30%, recovery is almost total after approximately 20 days; however, the recovery rate decreases appreciably with time and increases with the degree of deformation previously imposed [37]. The standard (ISO 9727-4:2007) specifies a test method for determining the percentage of diameter recovery of cylindrical cork stoppers, after compression, and specifies that this recovery after 5 minutes shall be greater than 90%.

The extraction force applied in the longitudinal direction of the bottleneck after fixing the cork to a pulling device (usually a corkscrew) depends on the compressive stress against the bottle and on the sliding friction between cork and glass. The extraction force increases with the increase of dimensions of the stopper: a longer stopper will increase the contact surface while a larger diameter stopper will increase the compressive stress [3]. The surface treatments (silicon and paraffin coatings) can reduce the extraction force. It is considered that the extraction force of natural cork stoppers should be 20–40 decanewtons (daN). The test method for the determination of maximum extraction force is specified in ISO 9727-5:2007.

The sealing performance of a cork stopper involves two possibilities for liquid passages: leakage between the stopper and the bottleneck and diffusion through the cork material. In the first possibility, the sealing capacity is evaluated by the penetration of the liquid in the interface, and it is usually measured by applying liquid pressures over the atmospheric pressure and observing the depth of penetration of the liquid [3]. The standard (ISO 9727-6:2007) specifies a test method for determining the liquid tightness of a cylindrical cork stopper: the liquid seal capability is expressed as the maximum internal pressure that the stopper can support in a bottle (at 1.2 bar internal pressure).

On the other hand, a cork stopper absorbs water/wine, and this liquid penetration through the cork is governed by the diffusion of the liquid in contact with the surface of the stopper [3]. González-Adrados et al. [38] evaluated the magnitude and evolution in time of absorption phenomena under conditions as close to reality as possible and described the transport of liquid as a combination of liquid progression by the cork-glass interface and diffusion through cell walls.

5. Oxygen transmission rate (OTR) properties of cork stoppers during bottled wine aging

The oxygen transmission rate (OTR) into the closed bottle is one important parameter for the wine cellars given its relation to the quality development of the wines [4, 39–43]. Therefore, the OTR properties of cork stoppers will define their ability as a quality sealant, also in comparison with other types of wine closures [41, 44, 45]. The OTR behavior depends on the type of wine closure. As shown by Lopes et al. [41] and presented in **Figure 7**, technical stoppers allow the lowest value of OTR into the closed bottle (1.0–1.2 mg of oxygen over 36 months), while synthetic closures present the highest value with an average oxygen ingress of 1.6 mg of oxygen in the first month. The path of oxygen ingress into the bottle was also experimentally studied by Lopes et al. [46]: the oxygen coming into the bottle and the wine during the storage period originates from the cork stopper itself, that is, from its macroscopic and cellular structure and not from an interface flow. In fact, the closed cells of cork contain air-filled lumens while lenticular channels or other tissue voids may provide additional air-filled pockets [26]. The cork itself has very low permeability to oxygen [47, 48], and correspondingly, the cork stoppers are essentially impermeable to atmospheric oxygen [46].

The kinetics of oxygen ingress into the bottle could be adjusted to logarithmic models, with an initial high ingress rate, followed by a decreasing ingress rates during the 1st month and further on, until stabilizing a low and rather constant ingress rate from the 3rd to the 12th month and thereafter (**Figure 7**) [49].

As primarily suggested by Ribéreau-Gayon [50], oxygen ingress into bottles occurs mainly out of the cork structure due to the high internal pressure in the cork cells created when the cork stoppers are compressed into the bottleneck. Natural cork stoppers with 24 mm diameter and 45 mm length have a volume of 20.4 mL of which 80–85% is air contained in the cell lumen, implying the existence of 4.9–5.2 mg of oxygen within their structure [3]. Oliveira et al. [49] showed that, in average, 1.88 and 2.35 mg of oxygen diffuses from the natural cork stoppers that

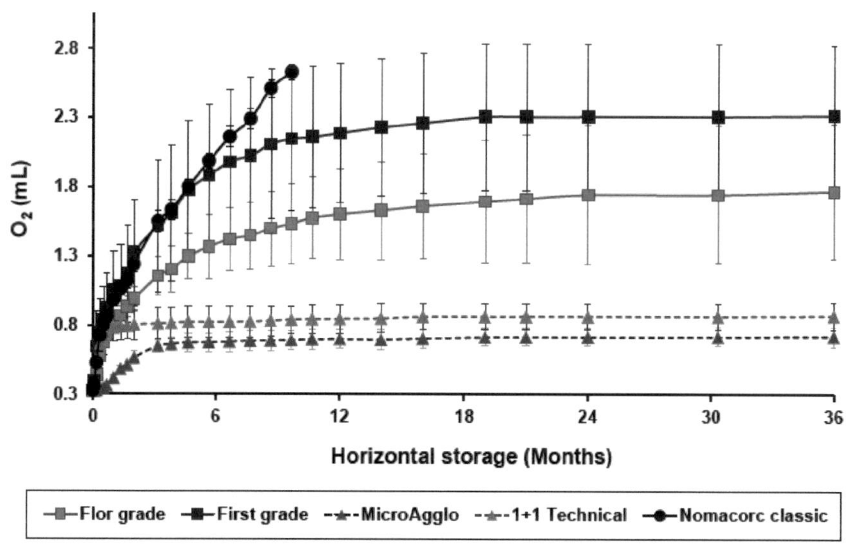

Figure 7.
Kinetics of oxygen ingress through different closures (natural cork stoppers of "flor" and extra quality grade, microagglomerated cork stoppers, 1 + 1 technical stoppers, and Nomacorc synthetic stoppers) into commercial bottles stored horizontally over 36 months. Error bars represent the standard deviation of four replicates (adapted from [41]).

were bored out of cork planks with 27–32 and 45–54 mm calipers, respectively, representing 36–38 and 47–50% of the theoretically total oxygen present in the cell structure. Moreover, Oliveira et al. [34] suggested that the high oxygen ingress rates immediately after bottling are due to the transfer of the air trapped in the voids in the bottom part of the cork stopper. After this first period, gas transport through the cork cells occurs with very low diffusion rates through small channels (i.e., through the plasmodesmata) present in the cork cell walls [47].

6. Off-flavors and beneficial compounds

One of the most important taints in wine is a moldy taste and aroma due to the presence of haloanisoles, namely, of 2,4,6-trichloroanisole (TCA), traditionally named as corkiness or cork taint. The olfactory detection threshold (ODT) of TCA is very low (1–4 ng/L) although variable depending on the consumer, the type of wine, and the occasion at which it is consumed, among other factors.

TCA is a nonpolar compound that has a large affinity for lipids such as those found in cork (monomeric extractives and polymeric suberin) that therefore can absorb TCA as well as other chloroanisoles [3]. The accumulation of chloroanisoles and chlorophenols in cork could be related to the use of polychlorinated phenolic biocides in forest, the chlorine bleaching of cork, the hypochlorite washing of wine barrels, the use of chlorinated phenolic biocides in packaging material, and the environmental contamination of wine cellars [3]. Therefore, the contamination sources have been eliminated in the cork industry in the last years. The International Code of Cork Stopper Manufacturing Practices prohibits nowadays the use of chlorine and other materials containing this compound at any stage of stopper production in order to reduce the possibility of chlorophenol development. Stoppers are currently washed with hydrogen peroxide as means of disinfecting and whitening.

Once the TCA can have origin in microbial sources (such as yellow stain), cork industry implemented new and changed old procedures aiming at the elimination of microbial contamination: the cork planks are screened before processing and during the process line for the elimination of those that contain signs of microbial activity; the industrial yards are nowadays cemented to avoid the contact of cork planks with soil in the preprocessing storage; and the time was decreased and the conditions controlled for the storage between the boiling of cork planks and their processing into stoppers.

In addition to the recognized standards in the International Code of Cork Stopper Manufacturing Practices, other processes to eradicate TCA have been implemented by companies in the sector: new boiling systems, controlled steam distillation, volatilization by dragging a controlled temperature and humidity, volatilization by dragging in the gaseous phase of adjusted polarity, under controlled temperature and humidity, and supercritical extraction with CO_2 [51].

Quality control of cork stoppers also increased, with procedures given by standard protocols, for example, ISO 20752:2014 specifies a test method to determine releasable TCA from cork stoppers, and ISO/PRF 22308 specifies a method for detecting and quantifying by sensory analysis several aromas including mold. The Cork Quality Council in the USA developed a research project using SPME-GC/MS analysis which allows technologically complex and very sensitive equipment to be used in the quantification of TCA in cork lots. Recently, Amorim launched NDtech, a natural cork that they guarantee with a nondetectable TCA (releasable TCA content at or below the 0.5 ng/L quantification limit; analysis performed in accordance to ISO 20752), based on the deployment of fast and unitary cork stopper chromatography.

In addition, research studies demonstrated that TCA does not permeate through cork showing that cork stoppers can act as an effective barrier against environmental TCA contamination [52].

Wine is, from a chemical point of view, a very complex fluid composed of a mixture of water, alcohols, organic acids, phenolic compounds, sugars, amino acids, and various minerals [53]. Several phenolic compounds present in wine, especially red wine, have gathered scientific interest in medical applications, including nonflavonoid phenolic acids (coumaric, cinnamic, caffeic, ferulic, and vanillic acids), trihydroxy stilbenes (resveratrol and polydatin), and flavonoids (catechin, epicatechin, and quercetin) [54].

Extrinsic sources of phenolic compounds found in wine include the barrels used in oak aging and cork stoppers, where small molecules are susceptible to migration from oak wood or cork into the wine.

Cork polyphenols were reported to migrate into wine after bottling, which may benefit the aging of wine through the release of phenolics and volatiles from the structure of the closure. In recent years, various studies have been carried out to analyze the intrinsic chemical features of cork regarding its advantages to health. The phenolic composition of cork has been studied, and several polyphenols were identified in cork: low molecular weight polyphenols such as acids, aldehydes, flavanones, and coumarins; hydrolysable tannins; and highly antioxidant molecules, such as gallotannins and ellagitannins and condensed tannins [55, 56]. Hydrolysable tannins are not naturally present in grapes, and their presence in wine stems from enological practices such as oak aging and natural wood extraction or the addition of extrinsic tannins in commercial products [57].

Polyphenols are easily extracted by a hydroalcoholic solution [55] and may react with the main components of wine [58], impacting on color and sensory properties of a bottled wine such as odor, flavor, and astringency. Azevedo et al. [59] identified and quantified the phenolic compounds from cork that were able to migrate from different cork stoppers to wine model solutions during 27 months. The phenolic compounds that solubilized in higher amounts were gallic acid, protocatechuic acid and aldehyde, and vanillin, while in lesser amounts, more complex structures have migrated such as valoneic acid, ellagic acid pentose, and castalagin/vescalagin. Mislata et al. [60] characterized the aromatic composition of different wine cork stoppers and granulates, reporting that vanillins are by far the most important aromatic family in cork extracts (**Table 3**).

Aromatic compound	**Aromatic descriptor**	**Content (μg/g)**
Vanillins		
Vanillin	Vanilla	9–170
Acetovanillone	Vanilla	0.6–14
Volatile phenols		
Guaiacol	Wood, smoked, sweet, medicine	0.03–5.0
4-Vinylguaiacol	Wood, spice cloves, curry	0.5–23
Eugenol	Spice cloves, honey	0.01–0.3
Isoeugenol	Carmination	0.06–2.4
Cerulignol	Spicy	0.04–2.2
Aldehydes		
Benzaldehyde	Almonds, sweet, caramel	0.02–0.21
Nonenal	Wax, citrus	0.03–0.47
Phenylacetaldehyde	Green, grass, honey	0.05–4.5

Aromatic compound	Aromatic descriptor	Content (μg/g)
Alcohols		
Phenylethyl alcohol	Flowers, honey, pollen	0.01–2.26
Benzyl alcohol	Roses, almond	0.05–0.13
Terpenols		
Camphor	Mint	0–0.23
Borneol	Pine tree	0–0.2
4-Terpineol	Spices, wood, soil	0.02–0.14
α-Terpineol	Flowers, lilac, sweet	0–0.2
Lactones		
γ-Nonalactone	Coconut, peach	0.03–0.11
Fatty acids		
Nonanoic acid	Wax, dry, fatty	0.12–0.67
Vanillic acid	Vanilla	0.07–0.86
Octanoic acid	Coconut, lactic, rancid, Cheese, sweet	0.14–3.38
Dodecanoic acid	Coconut, fatty, metallic	0–6.3
Benzeneacetic acid	Honey, fruity, sour	0–3.0
Furans		
Furfural	Caramel, candy	0–0.19

Table 3.
Aromatic compounds, families, descriptors, and minimum and maximum content found in the studied granulates and cork macerates [60].

7. Conclusion

The natural cork stoppers are world known, to many years, as closures for high-quality wines. The cork is a natural and sustainable material that provides an outstanding performance for in-bottle wine during aging, by combining minute oxygen transfer with sealing, durability, and chemical stability, for example, inertness toward the liquid content and along storage, preventing its sensory deterioration. Both the cellular structure features and the chemical composition of cork are at the base of a set of physical and mechanical properties that are important for bottle sealing, namely, the very low permeability to liquids and gases and the behavior under compression with an outstanding recovery upon stress relief, which is absolutely essential for the sealing in the bottle neck preventing liquid leakage.

Acknowledgements

This work was supported by the Fundação para a Ciência e a Tecnologia, Portugal (FCT), by financing the Forest Research Center (UID/AGR/00239/2019). Vanda Oliveira was supported by FCT through a postdoctoral grant (SFRH/BPD/118037/2016).

***Figures 1, 2,** and **6** and **Table 1** were adapted and reprinted from* Cork: Biology, Production and Uses. 1st ed., Pereira H, *Copyright 2007, with permission from Elsevier.*

Figure 7 *was adapted with permission from Lopes P, Saucier C, Teissedre P-L, and Glories Y. Impact of storage position on oxygen ingress through different closures into wine bottles. Journal of Agricultural and Food Chemistry. Copyright 2006 American Chemical Society.*

Conflict of interest

The authors declare no conflict of interest.

Author details

Vanda Oliveira* and Helena Pereira
Forest Research Center, School of Agriculture, University of Lisbon, Lisbon, Portugal

*Address all correspondence to: vandaoliveira@isa.ulisboa.pt

References

[1] 6° Inventário Florestal Nacional. 2015. Available from: http://www2.icnf.pt/portal/florestas/ifn/resource/doc/ifn/ifn6/IFN6_Relatorio_completo-2019-11-28.pdf [Accessed: 16 December 2019]

[2] APCOR's Cork Yearbook 18/19. 2018. Available from: http://www.apcor.pt/wp-content/uploads/2018/12/Anuario_APCOR_2018.pdf [Accessed: 16 December 2019]

[3] Pereira H. Cork: Biology, Production and Uses. 1st ed. Amsterdam: Elsevier Science; 2007. p. 346

[4] Lopes P, Saucier C, Glories Y. Nondestructive colorimetric method to determine the oxygen diffusion rate through closures used in winemaking. Journal of Agricultural and Food Chemistry. 2005;**53**:6967-6973. DOI: 10.1021/jf0404849

[5] Pinto-Correia T, Ribeiro N, Sá-Sousa P. Introducing the montado, the cork and holm oak agroforestry system of southern Portugal. Agroforestry Systems. 2011;**82**:99-104. DOI: 10.1007/s10457-011-9388-1

[6] Catry FX, Moreira F, Cardillo E, Pausas JG. Post-fire Management of Cork oak Forests. In: Moreira F, Arianoutsou M, Corona P, de las Heras J, editors. Post-Fire Management and Restoration of Southern European Forests. Dordrecht: Springer Netherlands; 2012. pp. 195-222. DOI: 10.1007/978-94-007-2208-8_9

[7] Pereira H, Tomé M. Cork oak. In: Evans J, Youngquist JA, editors. Encyclopedia of Forest Sciences. Oxford: Elsevier; 2004. pp. 613-620

[8] Lauw A, Oliveira V, Lopes F, Pereira H. Variation of cork quality for wine stoppers across the production regions in Portugal. European Journal of Wood and Wood Products. 2018;**76**:123-132. DOI: 10.1007/s00107-017-1196-5

[9] Costa A, Pereira H, Oliveira A. Influence of climate on the seasonality of radial growth of cork oak during a cork production cycle. Annals of Forest Science. 2002;**59**:429-437. DOI: 10.1051/forest:2002017

[10] Ferreira A, Lopes F, Pereira H. Caractérisation de la croissance et de la qualité du liège dans une région de production. Annals of Forest Science. 2000;**57**:187-193. DOI: 10.1051/forest:2000169

[11] Oliveira V, Lauw A, Pereira H. Sensitivity of cork growth to drought events: Insights from a 24-year chronology. Climatic Change. 2016;**137**:261. DOI: 10.1007/s10584-016-1680-7

[12] Besson CK, Lobo-do-Vale R, Rodrigues ML, Almeida P, Herd A, Grant OM, et al. Cork oak physiological responses to manipulated water availability in a Mediterranean woodland. Agricultural and Forest Meteorology. 2014;**184**:230-242. DOI: 10.1016/j.agrformet.2013.10.004

[13] Costa A, Pereira H, Madeira M. Landscape dynamics in endangered cork oak woodlands in southwestern Portugal (1958-2005). Agroforestry Systems. 2009;**77**:83-96. DOI: 10.1007/s10457-009-9212-3

[14] Costa A, Madeira M, Oliveira AC. The relationship between cork oak growth patterns and soil, slope and drainage in cork oak woodland in southern Portugal. Forest Ecology and Management. 2008;**255**:1525-1535. DOI: 10.1016/j.foreco.2007.11.008

[15] David TS, Henriques MO, Kurz-Besson C, Nunes J, Valente F, Vaz M, et al. Water-use strategies in two

co-occuring Mediterranean evergreen oaks: Surviving the summer drought. Tree Physiology. 2007;**27**:793-803

[16] David TS, Pinto CA, Nadezhdina N, Kurz-Besson C, Henriques MO, Quilhó T, et al. Root functioning, tree water use and hydraulic redistribution in *Quercus suber* trees: A modeling approach based on root sap flow. Forest Ecology and Management. 2013;**307**: 136-146. DOI: 10.1016/j.foreco.2013.07.012

[17] Leite C, Oliveira V, Lauw A, Pereira H. Cork rings suggest how to manage *Quercus suber* to mitigate the effects of climate changes. Agricultural and Forest Meteorology. 2019;**266-267**:12-19. DOI: 10.1016/j.agrformet.2018.11.032

[18] Graça J, Pereira H. The periderm development in *Quercus suber*. IAWA Bulletin. 2004;**25**:325-336. DOI: 10.1163/22941932-90000369

[19] Fortes MA, Rosa ME. Growth stresses and strains in cork. Wood Science and Technology. 1992;**26**: 241-258. DOI: 10.1007/BF00200160

[20] Pereira H, Lopes F, Graça J. The evaluation of the quality of cork planks by image analysis. Holzforschung. 1996;**50**:111-115. DOI: 10.1515/hfsg.1996.50.2.111

[21] Oliveira V, Knapic S, Pereira H. Natural variability of surface porosity of wine cork stoppers of different commercial classes. Journal International des Sciences de la Vigne et du Vin. 2012;**46**(4):331-340

[22] Lopes F, Pereira H. Definition of quality classes for champagne cork stoppers in the high quality range. Wood Science and Technology. 2000;**34**:3-10. DOI: 10.1007/s002260050002

[23] Pereira H, Melo B, Pinto R. Yield and quality in the production of cork stoppers. Holz als Roh- und Werkstoff. 1994;**52**:211-214

[24] Oliveira V, Knapic S, Pereira H. Classification modelling based on surface porosity for the grading of natural cork stoppers for quality wines. Food and Bioproducts Processing. 2015;**93**:69-76. DOI: 10.1016/j.fbp.2013.11.004

[25] Costa A, Pereira H. Decision rules for computer-vision quality classification of wine natural cork stoppers. American Journal of Enology and Viticulture. 2006;**57**:210-219

[26] Pereira H, Rosa ME, Fortes MA. The cellular structure of cork from *Quercus suber* L. IAWA Bulletin. 1987;**8**:213-218. DOI: 10.1163/22941932-90001048

[27] Pereira H. The rationale behind Cork properties: A review of structure and chemistry. BioResources. 2015;**10**:6207-6229. DOI: 10.15376/biores.10.3.Pereira

[28] Pereira H. Chemical composition and variability of cork from *Quercus suber* L. Wood Science and Technology. 1988;**22**:211-218. DOI: 10.1007/BF00386015

[29] Pereira H. Variability of the chemical composition of cork. BioResources. 2013;**8**(2):2246-2256. DOI: 10.15376/biores.8.2.2246-2256

[30] Oliveira V, Rosa ME, Pereira H. Variability of the compression properties of cork. Wood Science and Technology. 2014;**48**(5):937-948. DOI: 10.1007/s00226-014-0651-2

[31] Costa R, Lourenço A, Oliveira V, Pereira H. Chemical characterization of cork, phloem and wood from different *Quercus suber* provenances and trees. Heliyon. 2019;**5**(12):e02910. DOI: 10.1016/j.heliyon.2019.e02910

[32] Gonzalez-Adrados J, Pereira H. Classification of defects in cork planks using image analysis. Wood Science and Technology. 1996;**30**:207-215. DOI: 10.1007/BF00231634

[33] Lima J, Costa P. Real-time cork classification method: A colour image processing approach. International Journal of Factory Automation, Robotics and Soft Computing. 2006;**2**:1-10

[34] Oliveira V, Van den Bulcke J, Van Acker J, de Schryver T, Pereira H. Cork structural discontinuities studied with X-ray microtomography. Holzforschung. 2015;**70**(1):87-94. DOI: 10.1515/hf-2014-0245

[35] Lagorce-Tachon A, Karbowiak T, Loupiac C, Gaudry A, Ott F, Alba-Simionesco C, et al. The cork viewed from the inside. Journal of Food Engineering. 2015;**149**:214-221. DOI: 10.1016/j.jfoodeng.2014.10.023

[36] Mano JF. The viscoelastic properties of cork. Journal of Materials Science. 2002;**37**(2):257-263. DOI: 10.1023/A:1013635809035

[37] Rosa ME, Fortes MA. Rate effects on the compression and recovery of dimensions of cork. Journal of Materials Science. 1988;**23**:879-885. DOI: 10.1007/BF01153983

[38] González-Adrados J, González-Hernández F, García de Ceca J, Cáceres-Esteban M, García-Vallejo M. Wine absorption by cork stoppers. Spanish Journal of Agricultural Research. 2008;**6**(4):645-649. DOI: 10.5424/sjar/2008064-356

[39] Caillé S, Samson A, Wirth J, Diéval J-B, Vidal S, Cheynier V. Sensory characteristics changes of red Grenache wines submitted to different oxygen exposures pre and post bottling. Analytica Chimica Acta. 2010;**660**:35-42. DOI: 10.1016/j.aca.2009.11.049

[40] Mas A, Puig J, Lladó N, Zamora F. Sealing and storage position effects on wine evolution. Journal of Food Science. 2002;**67**:1374-1378. DOI: 10.1111/j.1365-2621.2002.tb10292.x

[41] Lopes P, Saucier C, Teissedre P-L, Glories Y. Impact of storage position on oxygen ingress through different closures into wine bottles. Journal of Agricultural and Food Chemistry. 2006;**54**:6741-6746. DOI: 10.1021/jf0614239

[42] Kontoudakis K, Biosca P, Canals R, Fort F, Canals JM, Zamora F. Impact of stopper type on oxygen ingress during wine bottling when using an inert gas cover. Australian Journal of Grape and Wine Research. 2008;**14**:116-122. DOI: 10.1111/j.1755-0238.2008.00013.x

[43] Silva MA, Julien M, Jourdes M, Teissedre P-L. Impact of closures on wine post-bottling development: A review. European Food Research and Technology. 2011;**233**:905-914. DOI: 10.1007/s00217-011-1603-9

[44] Escudero A, Asensio E, Cacho J, Ferreira V. Sensory and chemical changes of young white wines stored under oxygen. An assessment of the role played by aldehydes and some other important odorants. Food Chemistry. 2002;**77**:325-331. DOI: 10.1016/S0308-8146(01)00355-7

[45] Karbowiak T, Gougeon RD, Alinc J-B, Brachais L, Debeaufort F, Voilley A, et al. Wine oxidation and the role of cork. CRC Critical Reviews in Food Science and Nutrition. 2010;**50**:20-52. DOI: 10.1080/10408398.2010.526854

[46] Lopes P, Saucier C, Teissedre P-L, Glories Y. Main routes of oxygen ingress through different closures into wine bottles. Journal of Agricultural and Food Chemistry. 2007;**55**:5167-5170. DOI: 10.1021/jf0706023

[47] Faria DP, Fonseca AL, Pereira H, Teodoro OMND. Permeability of cork to gases. Journal of Agricultural and Food Chemistry. 2011;**59**:3590-3597. DOI: 10.1021/jf200491t

[48] Lequin S, Chassagne D, Karbowiak T, Simon J-M, Paulin C, Bellat J-P. Diffusion of oxygen in cork. Journal of Agricultural and Food Chemistry. 2012;**60**:3348-3356. DOI: 10.1021/jf204655c

[49] Oliveira V, Lopes P, Cabral M, Pereira H. Kinetics of oxygen ingress into wine bottles closed with natural cork stoppers of different qualities. American Journal of Enology and Viticulture. 2013;**64**(3):395-399. DOI: 10.5344/ajev.2013.13009

[50] Ribéreau-Gayon J. Dissolution d'oxygène dans les vins. In: Delmas, editor. Contribution à l'étude des oxidations et réductions dans les vins. In: Application à l'étude de vieillissement et des cases. Bordeaux, France: Delmas; 1933

[51] APCOR's Technical Guide, Cork Stoppers. 2016. Available from: https://www.apcor.pt/wp-content/uploads/2016/01/ManualRolhas_EN_VF.pdf [Accessed: 16 December 2019]

[52] Capone DL, Skouroumounis GK, Sefton MA. Permeation of 2,4,6-trichloroanisole through cork closures in wine bottles. Australian Journal of Grape and Wine Research. 2002;**8**:196-199. DOI: 10.1111/j.1755-0238.2002.tb00256.x

[53] Waterhouse AL. Wine phenolics. Annals of the New York Academy of Sciences. 2002;**957**:21-36. DOI: 10.1111/j.1749-6632.2002.tb02903.x

[54] Soleas GJ, Diamandis EP, Goldberg DM. Wine as a biological fluid: History, production, and role in disease prevention. Journal of Clinical Laboratory Analysis. 1997;**11**:287-313. DOI: 10.1002/(SICI)1098-2825(1997)11:5<287::AID-JCLA6>3.0.CO;2-4

[55] Fernandes A, Sousa A, Mateus N, Cabral M, de Freitas V. Analysis of phenolic compounds in cork from *Quercus Suber L.* by HPLC–DAD/ESI–MS. Food Chemistry. 2011;**125**(4):1398-1405. DOI: 10.1016/j.foodchem.2010.10.016

[56] Santos SAO, Pinto PCRO, Silvestre AJD, Neto CP. Chemical composition and antioxidant activity of phenolic extracts of cork from *Quercus Suber L.* Industrial Crops and Products. 2010;**31**(3):521-526. DOI: 10.1016/j.indcrop.2010.02.001

[57] Moutounet M, Souquet J-M, Meudec E, Leaute B, Delbos C, Doco T, et al. Analyse de la composition de tanins oenologiques. Revue Francaise d'Oenologie. 2004;**208**:22-27

[58] Azevedo J, Fernandes A, Oliveira J, Brás NF, Reis S, Lopes P, et al. Reactivity of cork extracts with (+)-catechin and malvidin-3-O-glucoside in wine model solutions: Identification of a new family of ellagitannin-derived compounds (Corklins). Journal of Agricultural and Food Chemistry. 2017;**65**(39):8714-8726. DOI: 10.1021/acs.jafc.7b02845

[59] Azevedo J, Fernandes I, Lopes P, Roseira I, Cabral M, Mateus N, et al. Migration of phenolic compounds from different cork stoppers to wine model solutions: Antioxidant and biological relevance. European Food Research and Technology. 2014;**239**:951-960. DOI: 10.1007/s00217-014-2292-y

[60] Mislata AM, Puxeu M, Ferrer-Gallego R. Aromatic potential and bioactivity of cork stoppers and cork by-products. Food. 2020;**9**(2):133. DOI: 10.3390/foods9020133